THE MODIFIED NUCLEOSIDES IN NUCLEIC ACIDS

The Modified Nucleosides in Nucleic Acids

ROSS H. HALL

COLUMBIA UNIVERSITY PRESS

NEW YORK AND LONDON

1971

Ross H. Hall is Professor and Chairman
of the Department of Biochemistry
at McMaster University in Hamilton,
Ontario, Canada.

Copyright © 1971 Columbia University Press
Library of Congress Catalog Card Number: 73-122745
ISBN: 0-231-03018-5
Printed in the United States of America

PREFACE

Before 1950 each of the known classes of nucleic acid was thought to consist of only four basic nucleoside monomers. This concept has proved to be about as naive as the earlier tetranucleotide theory of RNA structure, since numerous additional components have now been discovered in both RNA and DNA. Thirty-five of these additional components (here termed the modified nucleosides) have been discovered in RNA and five in DNA. It is not possible to estimate how many nucleic acid components actually exist; in addition, it is becoming apparent that some of these components are peculiar to the nucleic acids of certain organisms. For the most part, investigations have been conducted with only a limited range of organisms; consequently, as the nucleic acids of other organisms are studied, many unique chemical structures are likely to be uncovered.

The modified nucleosides have been found in all major classes of nucleic acid except mRNA and 5S-ribosomal RNA. The greatest number and structural variety per molecule occur in tRNA. Since the announcement of the structure of tRNA^{Ala} by Holley and his coworkers in 1965, a great deal of attention has been focused on the structure and function of this class of RNA. Many data about the modified nucleosides have been obtained from studies on tRNA; as a result, this book may seem to be concerned primarily with tRNA and might well have been entitled "The Biochemistry of tRNA Components." This bias was not intentional on my part but is merely a reflection of the present state of knowledge in the field.

The importance of the modified nucleosides to the structure and function of nucleic acid molecules is now evident, although their exact significance is not clear. Before this significance can be fully understood, basic facts about the structure, chemical reactivity, and location of these components in specific

v

nucleic acid molecules must be ascertained. A considerable body of data has already accumulated concerning the chemistry and biochemistry of the modified nucleosides. The main purpose of this book is to gather the pertinent facts related to their properties and to correlate these data with the known and postulated concepts of nucleic acid structure and function. With this intention, I hope that the book will serve as a useful source of information for investigators in this area of research.

I am indebted to many colleagues who have assisted in a variety of ways in the preparation of this book. In particular, I wish to thank Drs. David B. Dunn, Hara P. Ghosh, and Byron G. Lane, each of whom read portions of the manuscript and made many valuable suggestions. I also acknowledge the assistance of Mrs. B. Bell, Miss L. Csonka, Mr. J. Mozejko, and Mr. R. Thedford in preparing samples and obtaining some of the data used in the text. I am particularly fortunate in having had the competent editorial assistance of Mrs. H. Santa-Barbara and the skilled secretarial help of Mrs. L. Flynn and Mrs. L. Marsh in the preparation of the manuscript.

Much of my own work cited in this book was carried out at the Department of Experimental Therapeutics, Roswell Park Memorial Institute, Buffalo, New York. Work there was partially supported by grants from the National Cancer Institute of the United States Public Health Service, CA-04640 and CA-05697. At McMaster University my work is supported by grants from the Medical Research Council of Canada, MT-2738, and the National Cancer Institute of Canada.

Hamilton, Ontario, Canada   February, 1969   Ross H. Hall

CONTENTS

ABBREVIATIONS

THE RECOMMENDATIONS of the IUPAC-IUB Commission on Biochemical Nomenclature have been followed. See the articles in *J. Biol. Chem.* **241,** 527 (1966) or *Biochemistry* **5,** 1445 (1966).

Some of the commonly used conventions are:

N	= nucleoside
Np	= nucleoside 3′-phosphate
pN	= nucleoside 5′-phosphate
NpN, i.e., UpU	= Uridylyl-(3′-5″)-uridine
tRNA^{Ala}	= Alanine tRNA
tRNA$^{Ser}_1$	= Serine tRNA, isoaccepting species number 1
poly A	= Polyadenylic acid

The symbols for the modified nucleosides have been suggested by Dr. W. E. Cohn and have appeared in the *Handbook of Biochemistry,* edited by H. A. Sober, The Chemical Rubber Co., Cleveland, Ohio, 1968, p. G-3. These symbols are listed in Chapter 2, p. 23. With respect to the methylated nucleosides, the convention is as follows:

Nm = 2′-*O*-methylribonucleoside

mN = base methylated nucleoside with a superscript number to indicate ring position methylated, and a subscript 2 to indicate dimethyl groups, i.e., m$^2_2$G = N^2,N^2-dimethylguanosine

A_{260} units	one ml of a solution with an optical density at 260 mμ of one
ATP	Adenosine triphosphate
α	alpha
Å	angstroms
β	beta
BD Cellulose	Benzoylated cellulose (Gillam et al., 1967)
bp	boiling point
cpm	counts per minute
CTP	Cytidine triphosphate
DEAE	diethylaminoethyl-
dec.	decomposition
diam.	diameter
DMF	dimethyl formamide
DMSO	dimethyl sulfoxide
DNA	Deoxyribonucleic acid
E(p)	extinction coefficient based on phosphorus content
γ	microgram
grade BAP-C	Bacterial alkaline phosphatase, chromatographed (Worthington Biochemical Corp.)
Kieselguhr column	diatomaceous earth
λ	wavelength
M	molar
mg	milligram
μg	microgram
$\mu\mu$m	micro-micro-mole
min	minute
mls/cm^2	flow rate of column as a function of cross section of the column
M.P., dec.	melts with decomposition
mRNA	messenger RNA
mμ	millimicron
M.W.	molecular weight
nmole	nanamole
NMR	nuclear magnetic resonance
O.D.	optical density
π	pi

rRNA	ribosomal ribonucleic acid
S (as in 5S)	Svedberg units
Tm	temperature at which double-stranded nucleic acid molecule denatures
TMV	tobacco mosaic virus
TRIS	2-amino-2-(hydroxymethyl)-3-propanediol
tRNA	transfer ribonucleic acid
UDPG	uridine diphosphoglucose
UTP	uridine triphosphate
g	gravity

THE MODIFIED NUCLEOSIDES IN NUCLEIC ACIDS

Chapter 1

GENERAL INTRODUCTION: THE NATURE OF STRUCTURAL MODIFICATION OF NUCLEIC ACIDS

1. INTRODUCTION

THE NUCLEIC ACID MOLECULE can be visualized as a polymer constructed from four major nucleoside monomers: adenosine, guanosine, cytidine, and uridine for RNA, and deoxyadenosine, deoxyguanosine, deoxycytidine, and thymidine for DNA. The basic polymeric structure may be modified in a specific manner at specific locations by the addition or substitution of a variety of groups. Such structural modifications give rise to a class of nucleic acid components referred to variously as the minor, odd, or rare nucleosides. These terms are not properly descriptive for the following reasons: the compounds so designated may well play a major role in nucleic acid function, they are certainly not odd (they are natural compounds), and many of them are by no means rare, but actually occur rather commonly. In this book they will be called the modified nucleosides since all known members of this group are structural modifications of one of the eight major nucleosides.

The presence of modified nucleosides in nucleic acids has been realized only since Hotchkiss (1948) detected the first known modified component of a nucleic acid, 5-methylcytosine, in a sample of calf thymus DNA. Since that time five modified nucleosides in DNA and thirty-five modified nucleosides in RNA have been identified. A major question that now concerns us is the significance of these components to the function of the nucleic acids; even after much study, it is difficult to attribute any specific biological function to the modified nucleosides, although a considerable body of indirect evidence suggests that these components are significant to the structure and essential to the biological function of nucleic acid molecules. The purpose of this chapter is to examine the data in the light of the known function of each of the different classes of nucleic acids and to attempt to relate the specific occurrence of modified nucleosides to specific functions of the nucleic acids.

The principal inference one can draw is that the number and types of modified nucleosides within a given molecule of nucleic acid reflect the degree of complexity of the chemical interactions which the molecule undergoes. The modifying groups undoubtedly affect the physical, chemical, and biological properties of the nucleic acid molecule. Studies with model compounds have shown that the chemical modifications produce subtle changes in the secondary and tertiary structures of the nucleic acid molecule, and consequently these changes could influence the overall properties of the molecule. Subtle changes in the physical structure, for example, could affect the substrate specificity of the nucleic acid molecule with respect to enzymic reactions. Modification of a nucleic acid molecule by the addition of a chemically reactive group could provide a biochemical handle unique to a specific location in the molecule.

The degree of structural modification of nucleic acids is by no means uniform among various organisms, a fact that is most apparent in tRNA. Mammalian tRNA, for example, is much richer in methyl groups than bacterial tRNA. These concepts and others will be developed in some detail, but first it will be helpful to consider the general pattern of occurrence of the modified nucleosides in each of the principal classes of nucleic acids.

2. MODIFIED NUCLEOSIDES IN DNA

In the DNA of plant and mammalian tissue, only one modified component, 5-methyldeoxycytidine, has ever been detected. The amount of this component varies between 2 and 8 % of the deoxycytidine content of the DNA of mammalian cells (Chapter 5). The 5-methyl substitution does not affect normal base pairing but could perturb the secondary structure of the DNA. The relatively low content of modifying groups in the DNA molecule contrasts with the large number and structural variety of the modifying groups found in the tRNA. Although the individual DNA molecule interacts with other molecules in the course of replication and transcription, the basic chemical reactions are essentially repetitive. The required chemical specificity in DNA molecular interaction, apart from complementary base pairing, may be provided by associated molecules. In biochemical terms, therefore, DNA seems to have little need for the type of structural variety imparted by modifying groups.

The methylated nucleosides found in the DNA of animals, plants, and bacteria represent subtle modifications of the DNA structure at relatively few specific locations and hence resemble analogous modifications in the RNA. But, in addition, in the DNA of several bacterial viruses one of the major nucleosides is completely replaced by a modified nucleoside. The normal base pairing is retained, however, and the DNA apparently functions normally. No RNA has been isolated in which one of the four major nucleosides is completely substituted.

3. MODIFIED NUCLEOSIDES IN mRNA

The mRNA does not appear to contain any modified nucleosides, although accurate data on this point remain somewhat sparse. Because of technical difficulties, chemical analyses of mRNA samples have not been carried out; therefore we must rely on indirect evidence. Moore (1966), for example, attempted to determine whether mRNA contains any methylated nucleosides by growing *E. coli* in the presence of $[^{14}C\text{-methyl}]$-methionine. No methyl groups are incorporated into the mRNA fraction, whereas extensive labeling of the tRNA and rRNA takes place.

Several reasons may be given *a priori* why extensive modification of mRNA is not likely to occur. Perhaps the most important reason is that, after the mRNA molecule is sequentially synthesized on the corresponding DNA molecule, any structural modification could affect the efficiency and/or the fidelity of transmission of the message. Indirect experimental evidence obtained by using synthetic messenger molecules lends support to this idea. When guanosine residues of synthetic oligonucleotides are replaced with 7-methyl-guanosine, loss of normal coding properties results (Ludlum et al., 1964). Methylation of only 13% of the bases of poly A reduces the messenger activity of this polymer to less than 5% of normal (Wilhelm and Ludlum, 1966). Replacement of uridine residues with dihydrouridine in oligonucleotide triplets or in poly U results in a loss of template activity (Rottman and Cerutti, 1966; Smrt et al., 1966). In all probability, therefore, perturbation of the secondary structure of natural messengers cannot be tolerated, at least if the message is to be translated.

From the viewpoint of the biosynthetic mechanism of mRNA and its subsequent participation in protein synthesis, there might seem to be little opportunity for extensive modification of the molecule. In *E. coli*, for example, mRNA apparently becomes attached to the ribosome while it is still being synthesized (Byrne et al., 1964; Das et al., 1967; Revel et al., 1968). This argument may not be relevant, however, since rRNA appears to be methylated during or shortly after it is synthesized (Greenberg and Penman, 1966; Zimmerman and Holler, 1967), and DNA appears to be methylated close to or at the replication point (Billen, 1968). Another factor that may or may not have a bearing on the function of mRNA concerns the observation that in many cases internucleotide bonds involving modified nucleosides are hydrolyzed more slowly by nucleases (Bayev et al., 1963; Gray and Lane, 1967). If modified components were present in the mRNA, they might stabilize this molecular species of RNA.

4. MODIFIED NUCLEOSIDES IN rRNA

The role of rRNA in the structure and function of the ribosome is not well understood. Relatively few chemical studies have been made on the structure of rRNA, but the limited data suggest that in order for this species to function specific modifications of the basic polymeric structure are necessary. A study by Fellner and Sanger (1968) of partial sequences of rRNA of *E. coli* provides the most definitive data obtained on the nature of structural modifications. Their data show first that the 16S and 23S RNAs are homogeneous with respect to methylated sequences and second that methylation is a highly specific process. The 16S and 23S RNAs of *E. coli* contain about 17 and 11 methyl groups, respectively, per 1000 nucleotides [16S and 23S molecules consist of about 1640 and 3090 nucleotides each (Midgley, 1965)], and these groups are concentrated in a few loci. Major oligonucleotide fragments containing methylated components occur twice in the 23S molecule, and this fact led Fellner and Sanger to suggest that 23S RNA consists of two sections, perhaps identical to each other.

The ribosomal RNA of plant and mammalian tissue contains about the same proportion of methyl groups as that of *E. coli*. For example, the 16S and 28S RNAs of HeLa cells contain 19 and 13 methyl groups, respectively, per 1000 nucleotides (Vaughan et al., 1967). A major difference between the rRNA of *E. coli* and the tissues of higher organisms lies in the pattern of methylation. Methylation in *E. coli* rRNA occurs principally on the heterocyclic bases; the

2'-O-methylribonucleosides account only for about 1 in 1000 nucleotides (Dubin and Gunalp, 1967; Hall, 1964b; Nichols and Lane, 1966b), In contrast, wheat germ 18S + 28S rRNA contains 17 2'-O-methylribonucleosides per 1000 nucleotides (Singh and Lane, 1964b; Lane, 1965). Also, of the methyl groups in the 16S and 28S rRNAs in HeLa cells, approximately 80% are attached to the 2' position of the ribose moiety (Brown and Attardi, 1965; Hall, 1964b; Wagner et al., 1967).

The results of Vaughan et al. (1967) provide a glimpse of the relevance of methyl groups to the function of rRNA. When HeLa cells are grown under conditions of methionine deprivation, the 45S ribosomal precursor RNA becomes undermethylated. Although the methyl-deficient 45S RNA can yield 32S RNA, functional ribosomes are not formed under these conditions. These data do not indicate whether the lack of ribose methyl groups or the lack of base methyl groups is responsible for the lack of formation of functional ribosomes. A reason for the observed effect may be that the methylated nucleosides are necessary for the rRNA to adopt the proper configuration with the ribosomal protein.

Ribosomes contain a small molecule of RNA termed 5S-ribosomal RNA, which Rosset et al. (1964) isolated and characterized in *E. coli*. The complete sequence of this RNA molecular species (Brownlee et al., 1968) and that of the 5S-ribosomal RNA of KB cells (human epidermoid carcinoma line) (Forget and Weissman, 1967, 1968) have been determined. No modified nucleosides were detected in either of these molecules.

5. MODIFIED NUCLEOSIDES IN tRNA

Much of the current research on nucleic acid structure has centered on tRNA. This attention is understandable, since the tRNA molecule is relatively small and individual molecular species can now be readily isolated. The primary sequences of several tRNA molecules have already been determined (Chapter 4), and a great deal has been learned about structural details. More than thirty modified nucleosides have been detected in tRNA, and it is becoming apparent that they make a significant contribution to the structure and function of the molecules. In fact, most of the available information about the modified nucleosides and their possible role in the structure and function of nucleic acids has been derived from studies on tRNA. Therefore, the rest of this chapter will be concerned with the modified nucleosides of tRNA and their properties.

General Functions of tRNA. Transfer RNA is perhaps the most versatile of the known classes of nucleic acids, in terms of the variety and complexity of the chemical reactions in which it participates. Each tRNA molecule carries a specificity for a particular amino acid, recognizes the corresponding aminoacyl synthetase, and maintains its reading fidelity for the codon. In a purely mechanical sense, the basic tRNA structure would seem to be sufficient to carry out all these functions. The individual tRNA molecule, however, must perform these functions in a complex molecular environment. The genetic code, for example, contains more than one triplet for each amino acid, and although each tRNA molecule is capable of reading more than one codon *in vitro*, there could be at least one corresponding tRNA molecule for each triplet. In some cases there seems to be more than one molecular species of tRNA for a given triplet (Bergquist et al., 1968; Söll et al., 1966, 1967). Consequently, dozens, perhaps hundreds, of molecular species of tRNA may exist in a cell.

How does each tRNA molecule achieve an individual identity that provides for unique recognition among its companion tRNA molecules? The tRNA molecule is composed of about eighty nucleotides, permitting a multitude of sequential combinations. Variation in sequence alone might be considered sufficient to provide for functional individuality. For example, the sequences of the related pair, yeast $tRNA_I^{Ser}$ and $tRNA_{II}^{Ser}$, differ by only three nucleosides (Chapter 4). A recognition mechanism, supplemental to sequential variation, could be provided through subtle modification of the tRNA structure. This type of recognition mechanism also offers the cell a convenient method for modifying the molecular structure of tRNA to meet changing functional requirements. Such structural modification undoubtedly has functions other than just creating molecular fingerprints, and some of these possibilities will be developed in the light of relevant concepts of metabolic processes in the cell, particularly those related to differential protein synthesis.

tRNA Involved in Regulatory Processes. The growth and development of an organism depend on its ability to synthesize the particular cell components required at a given time in the life cycle, and these requirements are constantly changing. The basic mechanism underlying the control of selective or differential synthesis of enzymes, therefore, represents one of the most critical processes taking place during the life of a cell or organism.

Two models for the mechanism controlling differential protein synthesis have been advanced. One model, that of Jacob and Monod (1961), states that control is exerted at the transcription level by selective messenger synthesis. Thus protein synthesis is dictated by the supply of the corresponding messenger molecules, implying that the other components of the protein-synthesizing system are always available in adequate amounts and do not exert any controlling effect. The alternative model, suggested by several workers and summarized in an article by Strehler et al. (1967) (see also review by Novelli, 1967), postulates that selective translation of messages is a guiding force in differential protein synthesis.

The transcription model was advanced to explain regulation of the activity of the genetic material in *E. coli*, and Jacob and Monod suggest that the mechanism also operates in higher organisms. The more complex activities of the cells of higher organisms, however, may require supplemental control mechanisms operating at the translational level. A translation mechanism, superimposed on the transcription mechanism, may also exist to some extent in bacteria, but in the more sophisticated cells of higher organisms the translational mechanism may play a more significant role. A pertinent consideration embodied in these concepts is that tRNA would occupy a pivotal spot in the regulatory mechanism.

tRNA and Synthesis of the α-Chain of Hemoglobin. What is the evidence that tRNA is involved in control of differential protein synthesis, and what are the probable molecular mechanisms? The available data obtained from experimental models do not as yet provide any clear answers to these questions but do supply some strong indications. Evidence obtained in one model system, an *in vitro* protein-synthesizing system, suggests that certain molecular species of tRNA function in a specific manner. In the experimental protocol rabbit hemoglobin is synthesized in a standard *in vitro* polypeptide-synthesizing system utilizing the tRNA fraction isolated from *E. coli* or yeast. This system is particularly useful, since the amino acid sequence of the hemoglobin is known and techniques for its degradation are standardized. Gonano (1967) found that the tRNA molecule responsible for inserting the serine residue into position 3 of the α peptide chain was not the same molecule that inserted serine into either position 49 or 52 of the chain. Similarly, Weisblum et al. (1967) found that $tRNA_I^{Arg}$ and $tRNA_{II}^{Arg}$, purified from yeast tRNA, inserted arginine residues into different positions of the α chain of the hemoglobin. Weisblum et al. (1965) demonstrated that two species of *E. coli* $tRNA^{Leu}$ distributed leucine differently into various peptides of the α chain of rabbit hemoglobin.

These results, although obtained in a heterogeneous system, indicate that a relatively rare molecular species of tRNA, responsible for inserting amino acid X into one location of a given protein molecule, could become the rate-limiting factor in the synthesis of this molecule. This form of control would occur even though an excess of other tRNA molecules capable of transferring amino acid X was available. A corollary of this reasoning is that for some molecular species of proteins specific tRNA molecules may exist.

tRNA and Viral-Host Relationships. A study of the initial events of viral infection provides additional support for the idea that some degree of protein specificity exists for certain molecular species of tRNA. These studies indicate that, together with the synthesis of viral-specific protein in the host, unique species of tRNA are also synthesized. Foft et al. (1968) detected the presence of new tRNA molecules specific for the infecting virus in *E. coli* cultures infected with T4 bacteriophage. Subak-Sharpe et al. (1966), working with an L-cell system, noted that on infection with herpes virus a new species of tRNA^{Arg} appeared. Kano-Sueoka et al. (1968), Sueoka et al. (1966), and Waters and Novelli (1967) have observed the presence of a new species of tRNA^{Leu} in *E. coli* after bacteriophage infection. Whether the observed changes in tRNA^{Leu} result from *de novo* synthesis of a new species of tRNA^{Leu} or from alteration of host tRNA^{Leu} is not clear. If the latter alternative is correct, modification of the host tRNA^{Leu} might result in cessation of host protein synthesis.

Additional information concerning this question comes from the results of Hsu et al. (1967). They examined the tRNA of *E. coli* B infected with T2 or T4 bacteriophage for structural alterations. As a probe they made use of the fact that *E. coli* tRNA contains thionucleosides (4-thiouridine, the principal thio-nucleoside, accounts for about one mole %). tRNA samples of bacteria grown in the presence of [^{35}S] and infected for varying lengths of time was fractioned on methylated albumin-Kieselguhr columns and the radioactive profiles were compared. Hsu et al. noted that a significant change in the elution profile occurs early after infection. The alteration is blocked by chloramphenicol, indicating that protein synthesis is required for the structural alteration. They conclude that the thiolation of tRNA molecules after infection represents *de novo* addition of sulfur. Futhermore, the phage-induced species of tRNA appears to hybridize preferentially with the phage DNA (Weiss et al., 1968).

tRNA and Tissue Differentiation. Cell differentiation represents another biological process involving differential protein synthesis, and investigators have applied an analogous approach to ascertain whether specific tRNA molecules relate to specific protein molecules. In this experimental approach it is proposed that differentiation should be accompanied by the appearance or disappearance of certain molecular species of tRNA. Techniques devised for the resolution of multiple forms of each amino acyl tRNA have now become sufficiently sensitive to make this approach feasible. The experimental design consists of isolating samples of tRNA at various stages of growth in an organism undergoing differentiation, and resolving each tRNA sample into aminoacyl-isoaccepting fractions on sensitive ion-exchange or partition columns.

Lee and Ingram (1967, 1968) applied this method to a study of the erythrocytes of developing chicks. They separated two forms of tRNA^{Met} from the erythrocytes and found that the ratio of the two species changed during the chicks' development. Vold and Sypherd (1968) observed a change in the elution pattern of three different tRNA^{Lys} fractions of germinating wheat seeds from embryos to 48-hour-old seedlings. Yang and Comb (1968) found that the patterns of distribution of multiple forms of tRNA^{Lys} and tRNA^{Ser} changed during the early embryogenesis of the sea urchin, *Lytechinus variegatus*. They did not observe similar changes, however, in the distributions of seven other aminoacyl tRNAs. Analysis of the tRNA patterns of sporulating *Bacillus subtilis* shows that during the process of morphogenesis a shift occurs in the ratio of two molecular species of tRNA^{Val} (Doi et al., 1968). Goehler et al. (1966) compared on a column of methylated albumin-Kieselguhr the distribution patterns of 16 aminoacyl tRNAs in *B. subtilis* grown in both a normal medium and a nutritionally poor medium (sporulation medium). The elution profiles for 15 of the aminoacyl tRNAs were identical. With respect to serine acceptance, the profile of the tRNA obtained from cells grown in the normal medium had three peaks and the profile of the tRNA from cells grown in the nutritionally poor medium had two peaks.

The foregoing comparisons in animal and plant systems were made at different stages of growth. It may be asked whether comparable differences exist between isoaccepting profiles from different tissues of the same organism. Holland et al. (1967) and Taylor et al. (1968) have attempted to answer this question by comparing the elution patterns of specific aminoacyl tRNAs from different organs of the same animal (rabbit, mouse, cow, chicken, and hamster) on methylated albumin-Kieselguhr columns. The elution profiles of such chromatographs are usually similar, both between different organs in the same animal and across widely divergent mammalian species. Species differences are found in the elution profiles of the avian tRNA^{Leu} and tRNA^{Tyr} molecular species, but not among other tRNA molecular species. In contrast, the elution profiles of *Ehrlich ascites* tumor-cell phenylalanine, serine, glycine, and tyrosine tRNA differ from those of mouse organ tRNA. An extra peak of tRNA^{Tyr} is present in HeLa cells and adeno-7 virus-transformed cells but not in all tumor cells. The elution pattern of tRNA^{Tyr} of cultured mammalian and chick fibroblasts is markedly different from that of tRNA^{Tyr} of cultured epithelial cells and cells of normal organs.

A more complete evaluation of these data will have to wait until the isoaccepting fractions are correlated with differences in tRNA structure. The fractionation procedures, however, appear to be sensitive to subtle changes in structure; i.e., the presence or absence of a methyl group can shift the position of a tRNA molecule in the profile. The question arises whether observed differences in amino acid accepting profiles represent original structural differences *in vivo*.

In view of the presence of multiple species of tRNA for each amino acid in a cell, the question arises whether corresponding multiple aminoacyl synthetase systems are also present for each of the aminoacyl tRNAs. Some evidence bearing on this question has been obtained from studies on *Neurospora crassa*. The mitochondria contain species of aspartic acid, phenylalanine, leucine, serine, and methionine tRNA which are distinct from the corresponding species found in the cytoplasm (Barnett and Brown, 1967; Brown and Novelli, 1968). Two aminoacyl synthetases have been detected for both tRNA^{Asp} and tRNA^{Phe}, which catalyze the aminoacylation of the mitochondrial and cytoplasmic tRNA species, respectively. Neidhardt and Earhart (1966) detected the presence of a new tRNA^{Val} synthetase activity in *E. coli* after infection with T4 phage. This enzyme activity results not from *de novo* synthesis but rather from a modification of an enzyme existing at the time of infection, according to Chrispeels et al. (1968). These authors speculate that the conversion is important with respect to the differential regulation of early versus late protein synthesis. Ceccarini et al. (1967) have presented evidence that the aminoacyl synthetase activities change quantitatively during the growth cycle of the sea urchin, *Paracentrotus lividus*. These authors suggest that such changes may control both the amount and the types of proteins being synthesized.

Regulation at the translational level of protein synthesis, therefore, could involve both the tRNA and the aminoacyl synthetase, inferring that highly specific combining sites for both these molecules are required. This factor would tend to place a great deal of importance on structural subtleties in the regions of the tRNA molecule which bind with the synthetase molecule.

6. GENERAL STRUCTURAL REQUIREMENTS OF tRNA TO MATCH ITS BIOLOGICAL RESPONSIBILITIES

Having examined some of the experimental observations and current thinking with respect to the functions of tRNA, I will now consider the structural characteristics of the molecule and attempt to relate them to its known and postulated biological activities. If tRNA is to assume all the biological responsibilities suggested above, its molecular structure must fulfill certain stringent requirements. In view of the cursory state of knowledge with respect to the actual mechanisms of tRNA interactions with other molecules, it is difficult to define these structural requirements in detail. Some interesting generalizations, however, can be deduced.

The elements of the mechanism of protein synthesis are common to all organisms. A prime requirement, therefore, is for a basic structure that enables the tRNA molecule to recognize and transfer its specific amino acid in response to the corresponding codon. This basic requirement can probably be fulfilled by a tRNA molecule constructed from only the four major nucleosides. Even within the limitations imposed by a common secondary structure the sequential permutations would permit the existence of a very large number of molecular species. The question can be raised, however, whether a structure based on only four monomers would suffice to enable the tRNA to carry out its function in a complex molecular environment, particularly if some kind of a regulatory function is superimposed on the amino acid transfer function.

Two points concerning this concept need to be made. First, the basic tRNA structure does not allow for subtle variations in the secondary structure of the molecule. A polynucleotide consisting of the four major ribonucleosides, for example, presents a relatively smooth exterior. In distinct contrast, the structural variety of the individual amino acids provides polypeptides with a highly contoured exterior. In an analogous manner the modification of the primary rRNA chain by the attachment of additional groups represents a mechanism for conferring subtle variations in the secondary and tertiary structures. Second, the basic structure of tRNA probably does not provide for a sufficient variation in chemically active sites. All the nucleotide components possess chemically reactive groups such as phosphate, sugar hydroxyl, and amino groups, but these are common to all tRNA components. Sequential variation and macromolecular conformation would bring certain combinations of the four major bases into proximity to form active centers; however, the range and subtlety of the combinations are extended enormously by including the modified nucleosides in the possible combinations. All the active centers on tRNA molecules need not involve a group of nucleotides, and some active sites might consist of only one nucleotide. A polynucleotide made up of only four monomers would not permit much specificity in this regard, whereas attachment of a unique chemically reactive group to a major nucleoside at a specific location in the tRNA chain would create a highly specific active center.

Extension of the latter point to a consideration of a possible role of tRNA in regulatory activities suggests a mechanism for quantitative and possibly qualitative control of the activity of tRNA. Active centers specific to a given tRNA molecule could undergo a reaction that would change the activity of the whole molecule. In this manner, a single enzymically catalyzed reaction could quantitatively activate or inactivate the molecule and/or qualitatively modify its coding specificity. Thus, one can visualize how the activity of tRNA molecules might be modulated by relatively simple but nonetheless specific enzyme-catalyzed changes in structure.

7. SPECIFIC TYPES OF MODIFICATION FOUND IN tRNA AS RELATED TO THE STRUCTURAL REQUIREMENTS

Relatively Simple Modifications. I will now turn to a consideration of the known modified nucleosides and discuss their biochemical properties in relation to the general structural requirements of tRNA discussed above. What characteristics, for example, do the methyl nucleosides confer on tRNA structure? They could have a pronounced influence on secondary structure; apart from a bulk effect, the addition of methyl groups to the polynucleotide changes the hydrogen-bonding capability and modifies the base-stacking pattern. Methylation is the most common form of tRNA modification, and more than twenty different methylated nucleosides (including both heterocyclic and ribose methylation) have been identified in tRNA. Yeast tRNA, for example, contains an average of six methylated nucleosides per molecule. The variety of methyl nucleosides suggests that the secondary and tertiary structures of the tRNA molecule can be individually tailored by adding methyl groups to specific locations in the primary structure.

Other types of modification which also are relatively simple chemically can contribute equally to subtle changes in secondary and tertiary structure. Some examples of these modifications are the following: reduction of the 5,6 double bond of a pyrimidine (5,6-dihydrouridine), replacement of a hydroxyl group with a sulfur atom (4-thiouridine), replacement of an amino group with a hydroxyl group (inosine), and rotation of a uracil residue on the sugar stem (5-ribosyluracil). The properties of the modified nucleoside in some cases differ substantially from those of the parent nucleoside. For example, the shortened π-electron system of dihydrouridine modifies the neighbor-neighbor stacking interaction normally present in uridine-containing oligonucleotides. An acidic hydrogen in 5-ribosyluracil would provide a specific reactive site in the tRNA polymer not provided by uridine. These modifications would thereby cause a considerable change in the chemical properties of the tRNA at their point of location.

Hypermodifications. Some modifications of tRNA structure are relatively more elaborate and result from the attachment of a more complex side chain. The occurrence of these "hyper modified" nucleosides has only recently been recognized, and three have now been identified in the primary sequences: N^6-(Δ^2-isopentenyl)adenosine, in tRNA^{Ser}, tRNA^{Tyr}; N^6-(Δ^2-isopentenyl)-2-methylthioadenosine, in Su$_{III}^+$ tRNA^{Tyr}; and N-(nebularin-6-ylcarbamoyl)-threonine, in tRNA^{Ile}. These particular modified components occur in the anticodon loop adjacent to the 3′ end of the presumed anticodon of the respective tRNA molecules (see Chapter 4). The chemistry of the hypermodified nucleosides and their significance to tRNA structure are discussed in Chapter 7.

Compounds such as N^6-(Δ^2-isopentenyl)adenosine and N-(nebularin-6-ylcarbamoyl)threonine represent components of tRNA in which a relatively prominent side chain protrudes from the basic polynucleotide backbone. These side chains, moreover, carry a functional group (organic chemistry definition) which must confer some unique properties on the surrounding oligonucleotide region. The allylic double-bond, hydroxyl, or carboxyl groups of these compounds, for example, could readily bond covalently with other molecules. Because of their critical location in the sequence and their unique structure, these groups would constitute reactive centers of high specificity. One can visualize the role of such specific groups in controlling a function of the tRNA molecule. For example, esterification/de-esterification of the carboxyl group of N-(nebularin-6-ylcarbamoyl)threonine might represent a facile inactivation/activation process for the transfer function of tRNA^{Ile} or perhaps a mechanism for changing the reading fidelity of the anticodon. Regardless of the exact mechanism by which these groups interact, evidence (see Chapter 7, p. 328) shows that for tRNA molecules containing N^6-(Δ^2-isopentenyl)adenosine, the

N^6-(Δ^2-isopentenyl)adenosine N^6-(Δ^2-isopentenyl)-2-methylthioadenosine

N-(nebularin-6- ylcarbamoyl)threonine

Δ^2-isopentenyl group is essential for binding the tRNA molecule with ribosomes.

This type of reversible conversion need not be limited to the larger modified components. As Cerutti and Miller (1967) have suggested, a reversible conversion of uridine residue into 5,6-dihydrouridine could represent a means for regulating the activity of the particular tRNA molecule. Although supporting evidence has yet to be obtained, in theory any one of the modifications of tRNA could be reversible and could therefore constitute a process for reversibly changing either quantitatively or qualitatively the activity of the tRNA molecule.

The mechanisms of biosynthesis of the modified components of tRNA support the general concepts outlined above. In the known cases, the basic mechanism adheres to a single principle: modification of tRNA, giving rise to a modified component, occurs after the tRNA polymer has been formed. Thus methylation and attachment of the Δ^2-isopentenyl group are known to occur at the polymeric level. (These synthetic pathways are described in Chapter 6.)

8. DOES A RELATIONSHIP EXIST BETWEEN THE COMPLEXITY OF tRNA STRUCTURE AND THE COMPLEXITY OF CELLULAR ACTIVITY?

As more analytical data become available, an interesting and logical correlation between the complexity of an organism and the complexity of its tRNA structure as reflected in the number of modified nucleosides per tRNA molecules is becoming evident. Mycoplasma represents the most primitive organism known to be capable of self-replication on artificial medium. The tRNA and rRNA of certain strains contain very few modified nucleosides (Hall et al., 1967b; Hayashi et al., 1969). Protein synthesis, therefore, appears to proceed in mycoplasma without the benefit of extensive modification in the structure of the tRNA or rRNA; this observation indicates that modification of the tRNA structure such as that found in the higher organisms is probably not essential to the fundamental mechanism of amino acid transfer. It follows, therefore, that the structural modifications of tRNA in higher organisms relate to additional functions such as regulatory mechanisms and/or mechanisms that enable the basic transfer function to be carried out in a more complex molecular environment.

The data presented in Chapter 2 show clearly a striking increase in the number of modified nucleosides per tRNA molecule as one proceeds up the evolutionary scale from *E. coli* to yeast and mammalian tissue. For example, *E. coli* tRNA[Phe] (Uziel and Gassen, 1968) contains only two methylated nucleosides, whereas yeast tRNA[Phe] (RajBhandary et al., 1967) has nine. In the highly modified tRNA molecules of mammalian cells almost 25% of the nucleoside components are modified. Not only are the individual molecules structurally more complex in the tRNA of mammalian tissue than in the tRNA of organisms of lower phyla, but also there appears to be a greater degree of heterogeneity of the tRNA (Apgar and Holley, 1962; Glebov et al., 1965).

9. SUMMARY

In summary, the data quoted and the correlations drawn in this chapter suggest that subtle modification of nucleic acid structure is critical to the fundamental mechanism of protein synthesis and in particular to the effective functioning of the nucleic acid molecules in a complex molecular environment. The more sophisticated is the nature of cellular activity, the greater is the degree of modification. The available evidence has pointed to the fact that the fine structure of nucleic acid molecules, as reflected in the modified nucleoside content, principally rRNA and tRNA, is complex. Much remains to be learned, but the technology in this field is improving rapidly, and we can optimistically expect a great deal more information concerning the structure of nucleic acids in the next few years. This knowledge should contribute to a better understanding of the mechanisms of key cellular processes, for it is obvious that a clear comprehension of nucleic acid function must be based on knowledge of the details of chemical reactivity and structure.

Chapter 2

MODIFIED NUCLEOSIDES: PHYSICAL PROPERTIES, CHEMICAL SYNTHESES, AND NATURAL OCCURRENCE

Introduction

THIS CHAPTER lists the natural occurrence, basic properties, and chemical syntheses of all the modified nucleosides isolated from RNA and DNA. The modified components of nucleic acid included in this list have been detected and identified over the past fifteen years. In the earlier years of this period, the variety and number of modified nucleosides to be found were certainly not anticipated, and consequently the approaches used in various laboratories for the detection of "new" nucleic acid components were somewhat adventitious. In general, a sample of the nucleic acid was prepared and characterized by means of the techniques available at the time and then hydrolyzed enzymically or chemically to yield the constituent mononucleotides, nucleosides, or bases. The hydrolyzate was then fractionated by using one of a variety of chromatographic techniques, and the modified nucleoside was isolated and identified. In practical terms this type of approach has merit, since relatively large samples of nucleic acid can be worked up and the newly detected modified components can be furnished in amounts adequate for unequivocal characterization.

Most of the modified nucleosides reported were originally identified as the free bases, nucleosides, or mononucleotides in nucleic acid hydrolyzates. The subsequent identification of the majority of the known modified components in a single molecular species of tRNA confirms the original work. Some judgment is required to decide whether nucleosides so far found only in nucleic acid hydrolyzates are genuine constituents of the nucleic acid. The method of detection, together with information about possible sources of error that could arise during isolation, is mentioned in the description for each nucleic acid component.

With respect to the accuracy of the reported data, two important questions need to be considered, apart from the type of procedure used to isolate a given nucleic acid component. The first question concerns the identity of the sample as finally isolated. The general techniques of organic chemistry have been applied to the identification of the components, and the adequacy of identification can be judged on the evidence presented. In recent years, physical techniques such as mass spectroscopy and NMR spectroscopy have played important roles and will undoubtedly have even greater significance in the identification of still unknown nucleic acid components. Most of the reported compounds have been synthesized chemically, and the properties of the synthetic sample have been compared with those of the isolated material.

The second question, sometimes difficult to answer, should be posed for many of the components. The possibility exists that changes occur in the structure of a compound during the isolation procedure; this problem becomes particularly acute with respect to components that have elaborate structures. For example, N-[9-(β-D-ribofuranosyl)purin-6-ylcarbamoyl]-L-threonine has been well characterized chemically and has been found in a specific sequence of yeast tRNA^{Ile}. The structure as it exists in the native tRNA, however, appears to be more complex—that is, part of the molecule may have been lost during the isolation procedure (see Chapter 7). Other unidentified components may be first recognized in the form of a partial structure or perhaps in a form altered with respect to the native state.

In most cases, a decision as to whether a nucleoside listed in this chapter is a genuine nucleic acid component has been made on the basis of what is considered to be firm evidence. In a few cases, the evidence is insufficient to decide one way or the other and this fact is mentioned. Three compounds originally reported as components of RNA are not included in the list, since more recent evidence suggests definitely that either they may be artifacts of the isolation procedure or their structural assignment may be incorrect. Neoguanosine, isolated by Hemmens (1963) and identified as N^2-ribosylguanine by Shapiro and Gordon (1964), appears to result from an acid-catalyzed interaction of ribose and guanine (Hemmens, 1964). α-Cytidylic acid, identified by Gassen and Witzel

(1965), may result from alkali-catalyzed anomerization during the hydrolysis procedure (Dekker, 1965). The properties of isolated "1,5-diribosyluracil" (Lis and Lis, 1962) do not correspond to a chemically synthesized sample (Dlugajczyk and Eiler, 1966a).

A word of caution must be interjected concerning the isolated amounts in unfractionated nucleic acid of each compound listed in the tables. The values reported for many of the analyses are the actual amounts of the compound obtained in an isolation procedure and should be considered only as minimum amounts.

The methods of reporting also vary among investigators. Some report the amount of a modified component relevant to the amount of major component isolated simultaneously. In order to list the values in the table in a comparable manner, such values have been converted to a total nucleotide basis by using a factor derived from the known major nucleotide composition. The values are listed to two significant figures; but when all factors are taken into account— preparation of the nucleic acid sample, precision of the method, etc.—it is doubtful whether many of the values can be justified to the second figure.

Errors inherent in the isolation procedure could affect the reported values, and one source of error that may not be adequately controlled concerns the release of the modified nucleoside from the nucleic acid. Enzymic hydrolysis of a nucleic acid sample may appear to be quantitative when, in reality, a small proportion of oligonucleotide fragments refractory to hydrolysis may remain undetected during the subsequent workup. Gray and Lane (1967), for example, noted that internucleotide bonds adjacent to the $2'$-O-methyl nucleoside residues are slowly hydrolyzed by snake venom diesterase. Fittler et al. (1968a) reported that the internucleotide bond adjacent to N^6-$(\Delta^2$-isopentenyl)- adenosine is more slowly hydrolyzed by whole snake venom than other inter- nucleotide bonds. The technique of acid hydrolysis of nucleic acid samples, either with dilute hydrochloric acid, which releases the purine bases, or with perchloric acid, which releases all the bases, reduces errors due to inconsistent rates of hydrolysis, but the risk of partial or complete decomposition of labile components is greatly increased. The technique of alkaline hydrolysis has been used for many analyses, and most of the known nucleic acid components are stable under the conditions of this procedure. Nevertheless, some components are alkali labile, and moreover certain combinations of nucleotides may be resistant to alkaline hydrolysis. For example, the dinucleotide $m_2^6Apm_2^6Ap$, which occurs in 16S RNA, is very resistant to hydrolysis (Nichols and Lane,

1966b). (After hydrolysis in 1.0 N sodium hydroxide for 90 hours at 37°, 38% of this dinucleotide remains; see also discussion, p. 213.) It is clear that no one technique of hydrolysis is completely satisfactory for the release and subsequent detection of all the components of a nucleic acid sample.

An additional factor to be considered concerns the preparation of the sample. The values reported in the tables were determined on the major classes of nucleic acids: tRNA, rRNA, and DNA. The method of preparation of each nucleic acid sample varied from laboratory to laboratory, and it is difficult to judge whether some inadvertent fractionation may have occurred that eliminated a fraction of the nucleic acid richer in certain of the modified components. The designation of rRNA, in many cases, probably signifies only high-molecular-weight RNA. Only in a few cases has there been any fractionation of rRNA into 18S (16S) and 28S (23S) components, and some modified components may be located in 18S (16S) and not in 28S (23S) species, and vice versa. For example, m^4Cm and m_2^6A occur only in 16S RNA of *E. coli* (Fellner and Sanger, 1968; Nichols and Lane 1966a, 1966b, 1967, 1968a), and m^2A, m^1G, and possibly T occur only in the 23S molecular species (Fellner and Sanger, 1968). Therefore, the total rRNA mentioned in many papers should be considered as only a high-molecular-weight fraction. It is reasonably certain that most of the rRNA preparations used in these studies were free of tRNA; many of the constituents known to be present in tRNA were not detected in the rRNA fractions.

FORMAT OF CHAPTER

The material in this chapter is divided into three sections. All the known components of RNA and DNA are first listed as the free bases. The components are then grouped first as the ribonucleosides and then as the deoxyribonucleosides. The natural occurrence of each RNA component is reported in the nucleoside sections. Some of these data are reassembled in Table 1 according to the source, and the nucleoside composition of the tRNA and rRNA of selected organisms and tissues is listed. The natural occurrence of the modified nucleosides in DNA is reviewed separately in Chapter 3. With respect to the values reported for microorganisms it should be kept in mind that the modified nucleoside content varies with strains. Thus, unless the strains are rigorously characterized, the reported values may not be entirely reproducible.

Abbreviations were suggested by W. E. Cohn in accord with the principles set out by the IUPAC-IUB Commission on Biochemical Nomenclature in

Section 5 of "Abbreviations and Symbols for Chemical Names of Special Interest in Biological Chemistry" [1965 revision, published in *J. Biol Chem.* **241,** 527 (1966), *Biochemistry* **5,** 1445 (1966), and elsewhere]. The "three-letter" abbreviations are proposed for use in tables, figures, and equations involving the monomeric units themselves; the "one-letter" abbreviations, for polymer sequences. For deoxyribonucleosides in sequences, "d" may precede the sequence and thus be eliminated from each residue.

The properties of the compounds have been recorded, where possible, on chemically synthesized samples. For the optical rotation, the number in parentheses refers to the concentration (c). References have been given for the known chemical syntheses of each compound, and the more relevant methods are outlined schematically. The percentage yield listed on the schemes refers to the immediate step. If the yield is based on another precursor, the information is given.

Spectral Data

Unless indicated on the chart the ultraviolet absorption spectra were obtained on a Cary Model 14 Recording Spectrophotometer in my laboratory. Data for which there is no chart were taken from a compilation of properties of nucleosides and bases by Dunn and Hall (1968). The mass spectra unless indicated on the chart were run by Dr. L. Baczynskyj of the Department of Biochemistry, McMaster University, on a CEC Model 21–110B Spectrometer, direct inlet, 70 ev.

The samples used were obtained from commercial sources or had been synthesized or isolated by ourselves or by investigators in other laboratories. I wish to thank Drs. J. A. Carbon, M. Fleysher, M. Honjo, N. J. Leonard, R. K. Robins, and C. B. Reese who kindly sent me samples as well as data of compounds they have prepared.

The class of RNA is listed as (t)-transfer or (r)-ribosomal. If no designation is given, the sample consisted of a mixture of all classes of RNA from the organism or was not specified in the original article.

(a) The presence of 1-methyladenosine was confirmed by isolation of either 1-methyladenine (Dunn, 1961a) or 1-methyladenylic acid (Dunn, 1963).

(b) The presence of N^6-methyladenine was established as a component by acid hydrolysis (Dunn et al., 1960b).

(c) The value was originally reported relative to uridine. The uridylic acid contents of yeast and mammalian liver are 20 mole % and 18.2 mole %, respectively (Monier et al., 1960; Osawa, 1960). The reported values are, therefore, divided by 5 except those for E. coli, which has a uridylic acid content of 15% (Dunn et al., 1960a).

Many of the data reported in this section have been obtained by Drs. D. B. Dunn, B. G. Lane, and their colleagues. The analytical methods used in these two laboratories, as well as those used in my own, are described in Chapter 3.

Index of Compounds Listed in Order of Appearance

MODIFIED NUCLEOSIDES

Compound Number		Abbreviations	
		Three Letter	*One Letter*

Ribonucleosides

35.	Adenosine	Ado	A
36.	1-Methyladenosine	1-meAdo	m^1A
37.	2-Methyladenosine	2-meAdo	m^2A
38.	N^6-Methyladenosine	6-meAdo	m^6A
39.	N^6,N^6-Dimethyladenosine	6-Me$_2$Ado	m_2^6A
40.	2'-O-Methyladenosine	OMeAdo	Am
41.	N^6-(Δ^2-Isopentenyl)adenosine	6-ipAdo	i^6A
42.	N^6-(*cis*-4-Hydroxy-3-methylbut-2-enyl)-adenosine		
43.	N^6-(Δ^2-Isopentenyl)-2-methylthioadenosine		m^2SiA
44.	N-[9-(β-D-Ribofuranosyl-9H-purin-6-yl)-carbamoyl]-L-threonine (N-(nebularin-6-ylcarbamoyl)-L-threonine)	6-Thr(co)Ado	(ThrCO)^{6}A
45.	2'(3')-O-Ribosyladenosine	ORibAdo	
46.	Cytidine	Cyd	C
47.	3-Methylcytidine	3-MeCyd	m^3C
48.	N^4,O^2-Dimethylcytidine	O,4-Me$_2$Cyd	m^4Cm
49.	5-Methylcytidine	5-MeCyd	m^5C
50.	2'-O-Methylcytidine	OMeCyd	Cm
51.	N^4-Acetylcytidine	4-AcCyd	ac^4C
52.	2-Thiocytidine	2-SCyd	s^4C
53.	Guanosine	Guo	G
54.	1-Methylguanosine	1-MeGuo	m^1G
55.	N^2-Methylguanosine	2-MeGuo	m^2G
56.	N^2,N^2-Dimethylguanosine	2-Me$_2$Guo	m_2^2G
57.	7-Methylguanosine	7-MeGuo	m^7G
58.	2'-O-Methylguanosine	OMeGuo	Gm
59.	Inosine	Ino	I
60.	1-Methylinosine	1-MeIno	m^1I
61.	Uridine	Urd	U
62.	3-Methyluridine	3-MeUrd	m^3U
63.	5-Methyluridine	5-MeUrd	m^5U or T
64.	2'-O-Methyluridine	OMeUrd	Um
65.	5-(β-D-Ribofuranosyl)uracil (Pseudouridine)	ψrd	ψ
66.	5-(2'-O-Methylribosyl)uracil (2'-O-Methylpseudouridine)	OMeψrd	ψm
67.	2-Thio-5-carboxymethyluridine methyl ester		mcm^{5}s^{2}U

Compound Number		Abbreviations	
		Three Letter	*One Letter*
68.	4-Thiouridine	4-Srd	s^4U
69.	5-Hydroxyuridine	5-OH-Urd	ho^5U
70.	2-Thio-5-(*N*-methylaminomethyl)uridine		mnm^5s^2U
71.	5-Carboxymethyluridine		cm^5U
72.	5,6-Dihydrouridine	h_2Urd	hU
73.	5,6-Dihydro-5-methyluridine		hT

Deoxyribonucleosides

74.	Deoxyadenosine	dAdo	dA
75.	N^6-Methyldeoxyadenosine	6-MedAdo	m^6dA
76.	Deoxycytidine	dCyd	dC
77.	5-Methyldeoxycytidine	5-MedCyd	m^5dC
78.	5-Hydroxymethyldeoxycytidine	5-HmdCyd	hm^5dC
79.	Deoxyguanosine	dGuo	dG
80.	Deoxyuridine	dUrd	dU
81.	5-Methyldeoxyuridine (thymidine)	dThd	dT
82.	5-Hydroxymethyldeoxyuridine	5-HmUrd	hm^5dU

Free Bases

1. ADENINE   ADE

$C_5H_5N_5$

M.W. 135.13

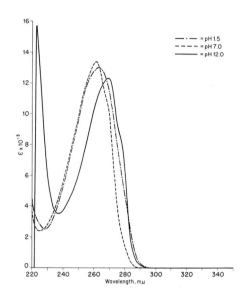

Physical Properties

M.P. 360°   Baddiley et al. (1943b); Richter et al. (1960)
pK (basic) 4.15   Levene and Bass (1913); Harkins and Freiser (1958)
(acidic) 9.8   Levene and Bass (1931); Harkins and Freiser (1958)

Additional References for Synthesis

Baddiley et al. (1943b); Levene and Bass (1931)

Synthesis

1. Ichikawa et al. (1965)

2. Wakamatsu et al. (1966)

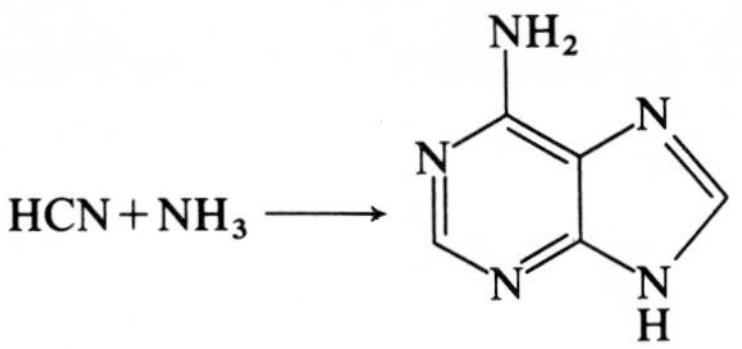

$$HCN + NH_3 \longrightarrow$$

3. Richter et al. (1960)

$$\xrightarrow[\text{DMF}]{HC(OC_2H_5)_3}$$

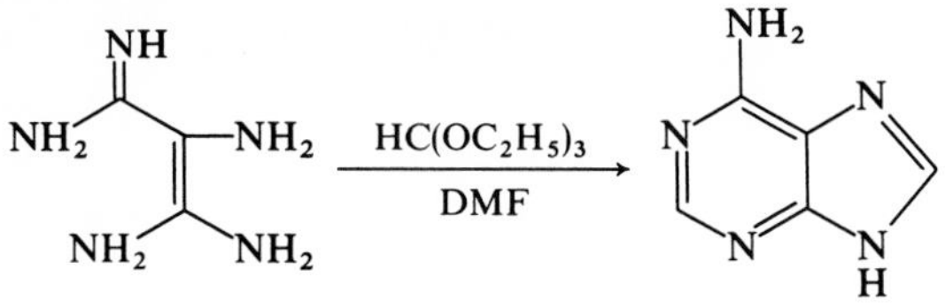

(72% yield)

4. Robins et al. (1953)

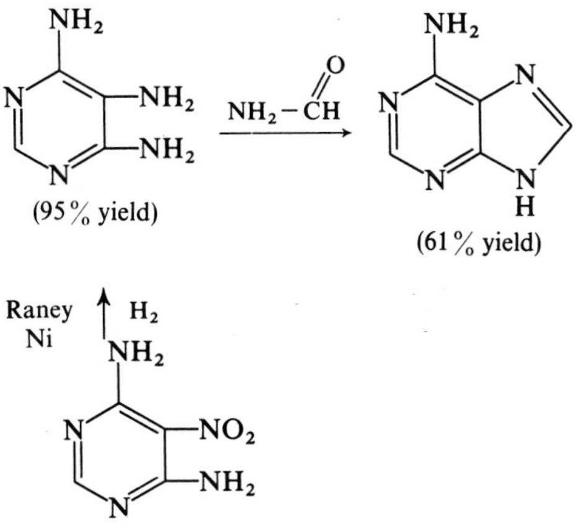

$$\xrightarrow{NH_2 - \overset{\overset{\displaystyle O}{\|}}{CH}}$$

(95% yield)

(61% yield)

Raney Ni $\uparrow$ H_2

5. Shaw (1950)

$$CH_2 \begin{smallmatrix} CN \\ CN \end{smallmatrix} \xrightarrow[NH_3]{C_2H_5OH} \quad (60\%) \xrightarrow[\text{aniline}]{\text{diazotized}} \quad (100\%)$$

$$\xrightarrow{HCOOH} \quad (45\% \text{ yield}) \xrightarrow[HCOOH]{\text{acetic anhydride}} \quad (80\% \text{ yield})$$

6. Cavalieri et al. (1949)

$$\xrightarrow{NH_3} (95\% \text{ yield}) + \xrightarrow[n-C_4H_9OH]{NaOC_4H_9}$$

$$\xrightarrow{Zn, H_2O} (90\% \text{ yield}) \xrightarrow[\underset{O}{HC-NH_2}]{HCOOH} (87\% \text{ yield})$$

7. Ochiai et al. (1968)

(43.5% yield)

8. Ferris and Orgel (1966)

(36% yield) (68% yield)

2. 1-METHYLADENINE   1-MeAde

$C_6H_7N_5$

M.W. 149.16

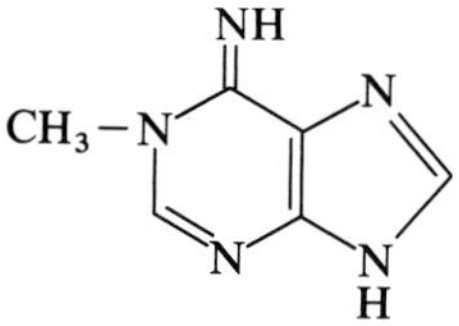

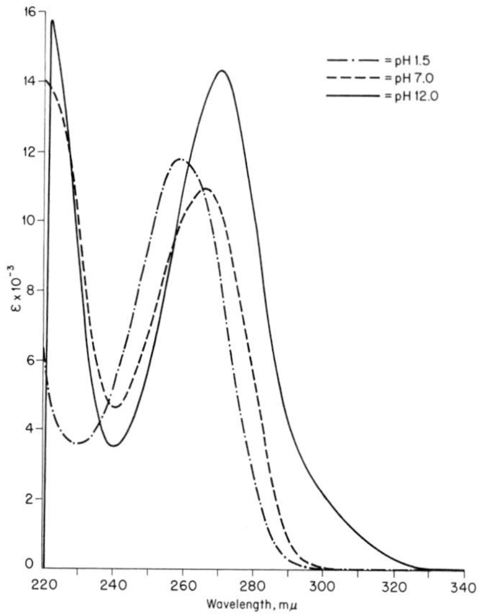

— · — = pH 1.5
— — — = pH 7.0
———— = pH 12.0

Physical Properties

M.P. 296–299° (dec.)	Brookes and Lawley (1960)
pK (basic) 7.2	Brookes and Lawley (1960)
(acidic) 11.0	Brookes and Lawley (1960)

Synthesis

(1) Jones and Robins (1963); (2) Brookes and Lawley (1960)

(1) CH$_3$I in *N,N*-dimethylacetamide

(2) (CH$_3$)$_2$SO$_4$ in DMF + K$_2$CO$_3$

HCl

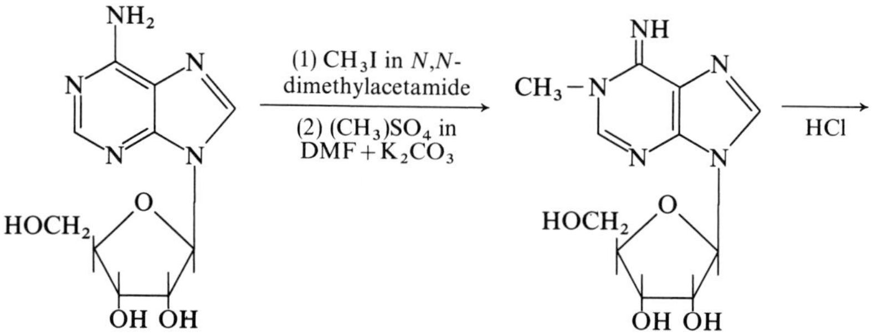

[(1) 81 % yield]

[(1) yield not reported;
(2) 31 % yield overall]

Note. Method 2 gives other methylated products which must be separated.

3. 2-METHYLADENINE   2-MeAde

$C_6H_7N_5$

M.W. 149.16

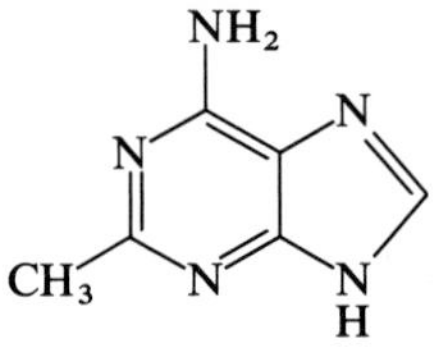

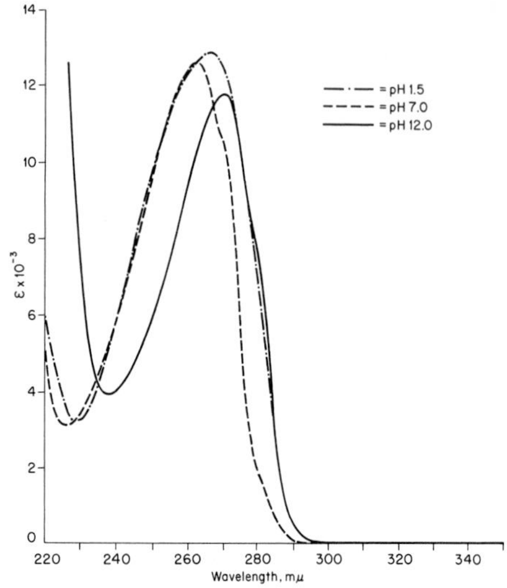

= pH 1.5
= pH 7.0
= pH 12.0

Physical Properties

M.P. $> 350°$ Robins et al. (1953); Taylor et al. (1959b)

pK (basic) ~ 5.1 Lynch et al. (1958)

Synthesis

1. Taylor et al. (1959b)

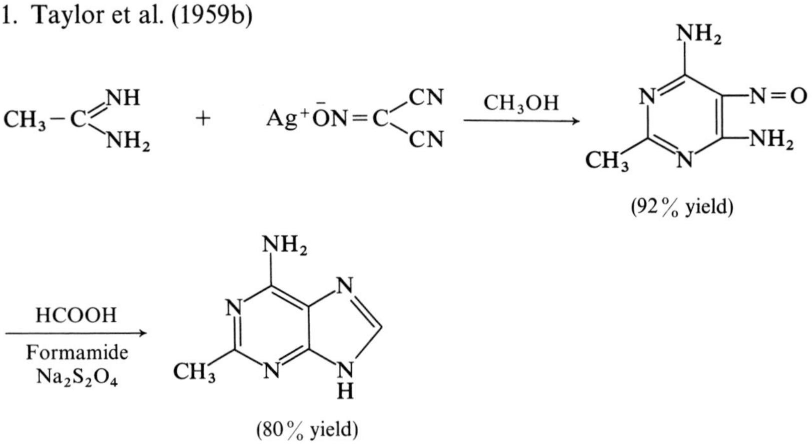

(92% yield)

(80% yield)

2. Robins et al. (1953)

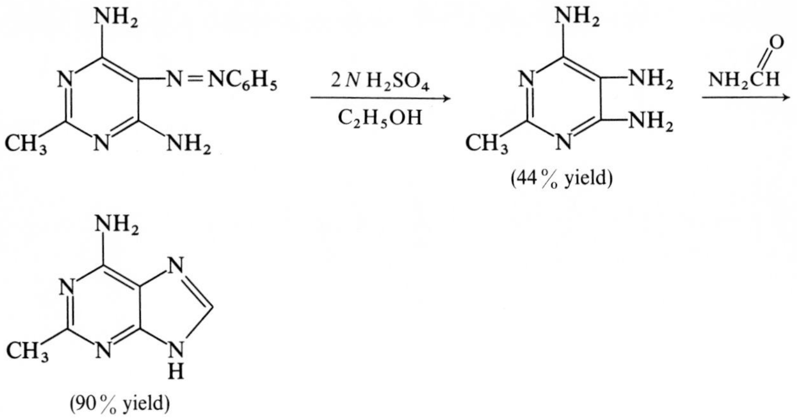

(44% yield)

(90% yield)

3. Davoll and Lowy (1952)

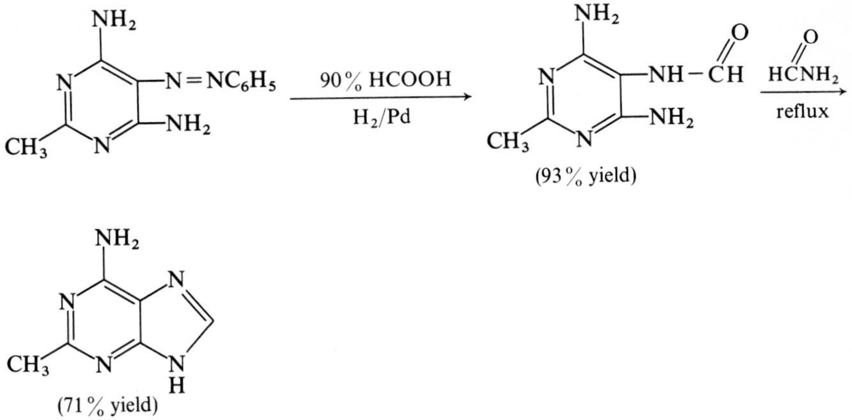

(93% yield)

(71% yield)

4. Baddiley et al. (1943a)

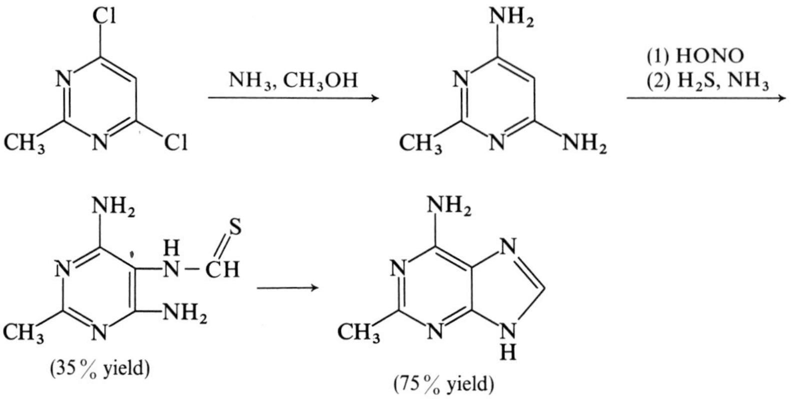

CH₃ (35% yield)

(75% yield)

4. N⁶-METHYLADENINE   6-MeAde

$C_6H_7N_5$

M.W. 149.16

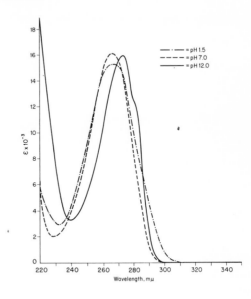

Physical Properties

M.P. 319–320°	Broom et al. (1964)	
pK (basic) <1, 4.2	Albert and Brown (1954)	
(acidic) 10.0	Albert and Brown (1954)	

Synthesis

1. Broom et al. (1964)

CH₃NH₂

(77% yield)

2. Elion et al. (1952)

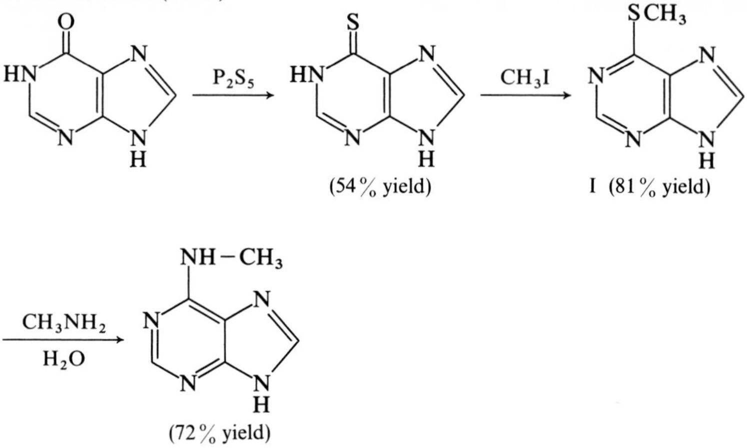

(54% yield)   I (81% yield)

(72% yield)

5. *N⁶,N⁶*-DIMETHYLADENINE   6-Me₂Ade

$C_7H_9N_5$

M.W. 163.18

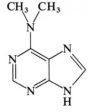

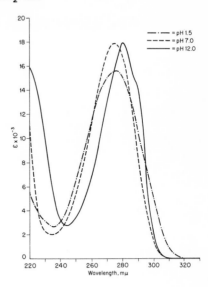

Physical Properties

M.P. 257–258°   Baker et al. (1954)
pK (basic) <1, 3.9 Albert and Brown (1954)
   (acidic) 10.5   Albert and Brown (1954)

Additional Reference for Synthesis

Baker et al. (1954)

Synthesis

1. Breshears et al. (1959)

CH₃—NH₂
CH₃
C₂H₅OH, H₂O,
NaOAc

(67% yield)

(71% yield)

2. Elion et al. (1952)

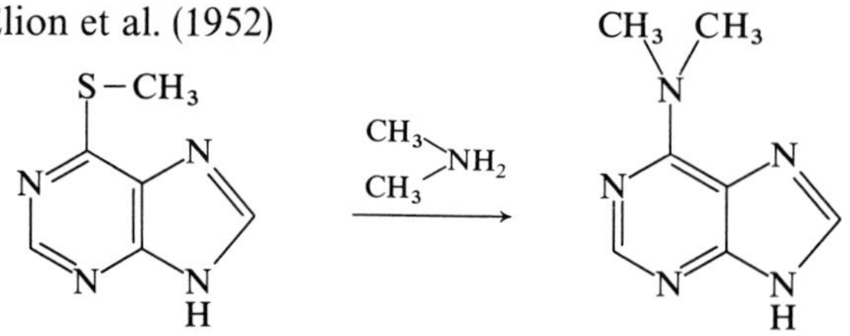

Intermediate I
in synthesis of *N*⁶-methyladenine

(79% yield)

3. Albert and Brown (1954)

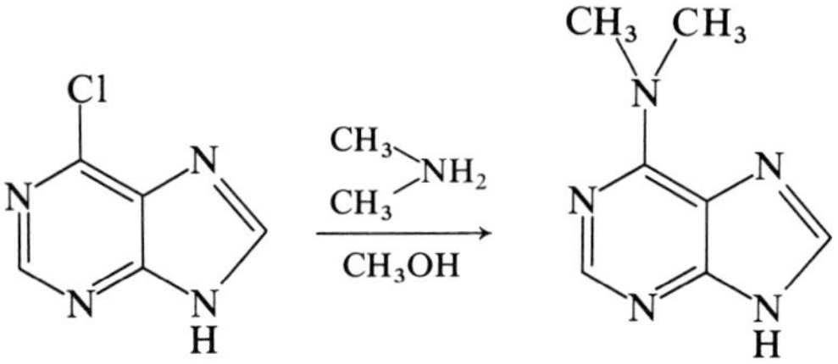

(42% yield)

6. *N⁶-(Δ²-ISOPENTENYL)ADENINE*   6-ipAde

$C_{10}H_{13}N_5$

M.W. 203.24

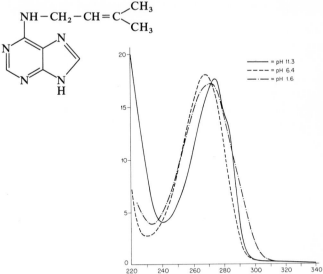

= pH 11.3
= pH 6.4
= pH 1.6

Physical Properties

M.P. 212–214°   Hall et al. (1966)
pK (basic)   3.4   Leonard and Fujii (1964)
   (acidic) 10.4   Leonard and Fujii (1964)

Additional References for Synthesis

Hall et al. (1966); Hecht et al. (1968)

Synthesis

1. Hall and Robins (1968a)

(70% yield)

2. Leonard and Fujii (1964)

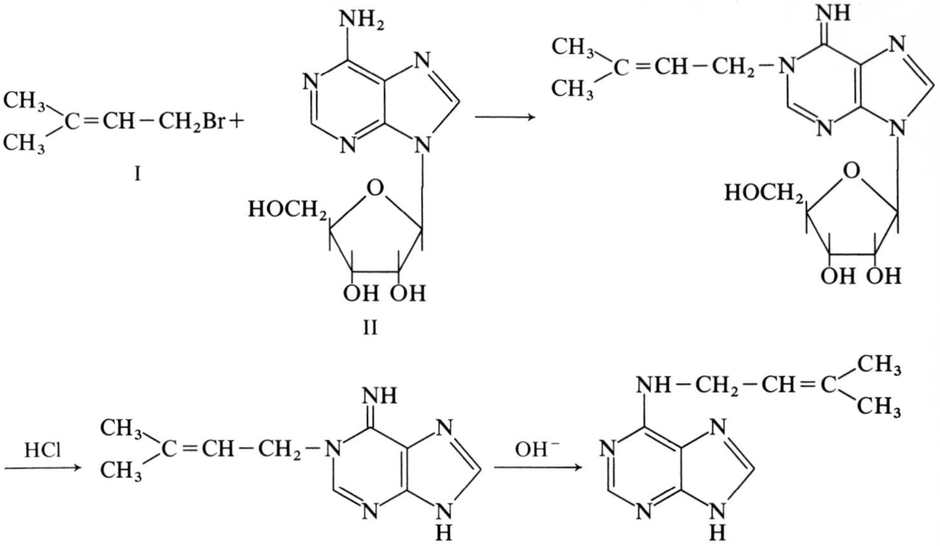

(51% yield from II)

γ,γ-Dimethylallylbromide, I, can be synthesized according to Hall and Fleysher (1968).

7. *N*⁶-(*cis*-4-HYDROXY-3-METHYLBUT-2-ENYL)ADENINE

$C_{10}H_{13}N_5O$

M.W. 219.24

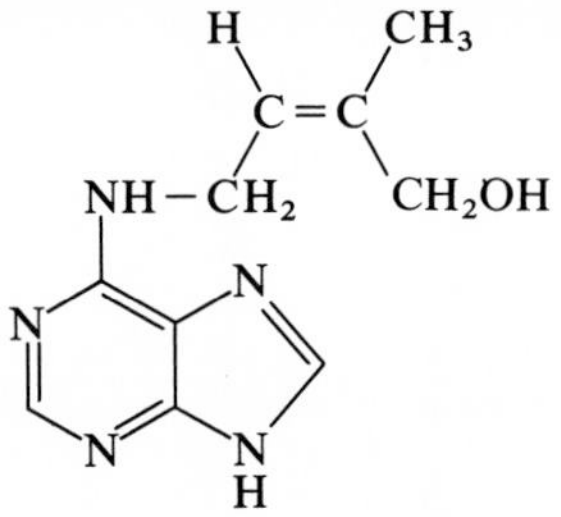

Ultraviolet absorption spectra same as those of **6**.

8. N^6-(Δ^2-ISOPENTENYL)-2-METHYLTHIOADENINE

$C_{11}H_{15}N_5S$

M.W. 249.32

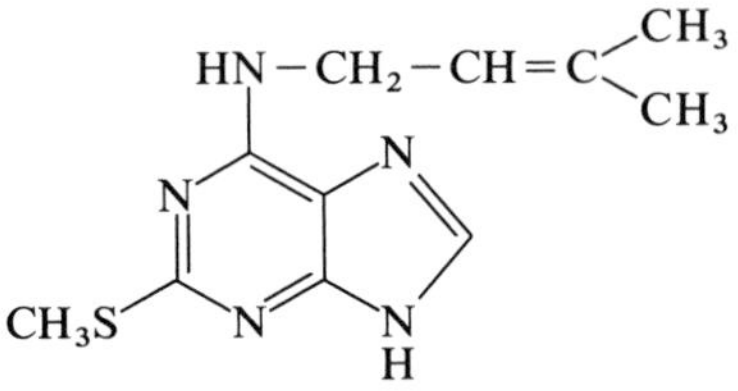

pH	λ_{max}	ε_{max} $(\times 10^{-3})$	λ_{min}
1	292		275
	252		218
7	278		257
	241		219
12	286		255
	230		220

N. J. Leonard, personal communication

Physical Properties

M.P. 259–260°   Burrows et al. (1968)

Synthesis

1. Burrows et al. (1968)

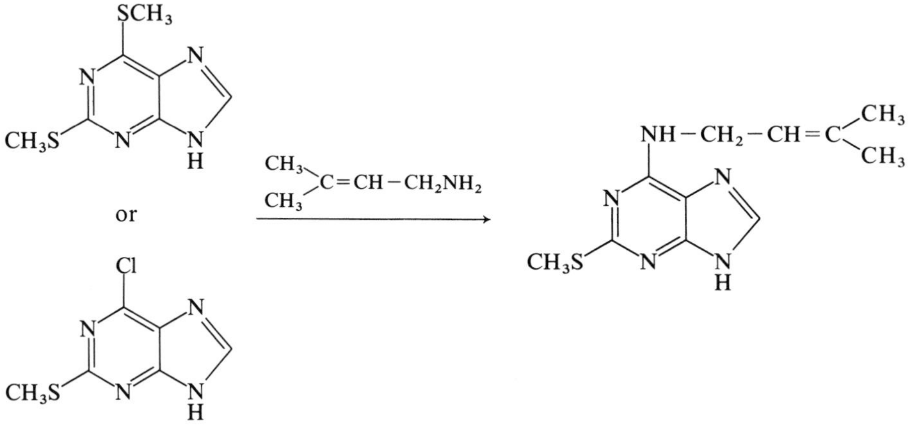

9. *N*-(PURIN-6-YLCARBAMOYL)-L-THREONINE

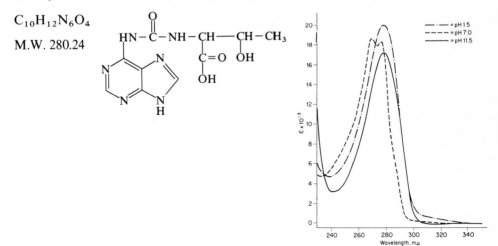

$C_{10}H_{12}N_6O_4$

M.W. 280.24

Synthesis

1. Chheda (1969)

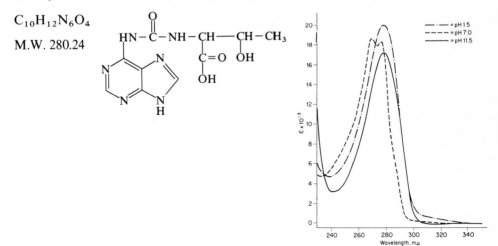

Compound I can be prepared according to Giner-Sorolla and Bendich (1958).

10. CYTOSINE   Cyt

$C_4H_5N_3O$

M.W. 111.10

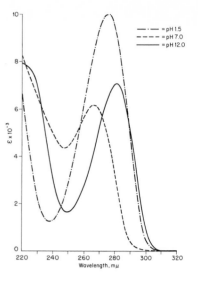

Physical Properties

M.P. 308° (dec.)   Hilbert et al. (1935)
pK (basic)   4.60   Szer and Shugar (1966)
   (acidic) 12.4   Shugar and Fox (1952)

Additional References for Synthesis

Hilbert and Johnson (1930); synthesis of [2–^{14}C]-cytosine by Codington et al. (1958)

Synthesis

1. Hilbert et al. (1935)

2. Hitchings et al. (1949)

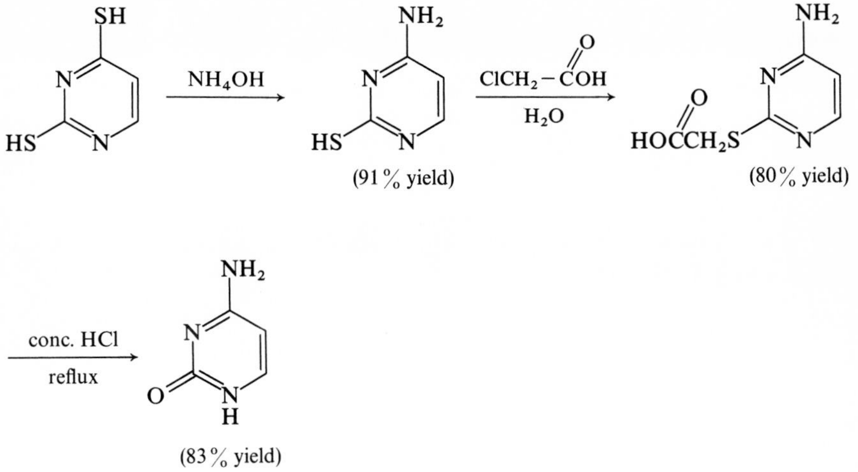

(91% yield) → (80% yield)

(83% yield)

3. Wheeler and Johnson (1903); Wheeler and Merriam (1903)

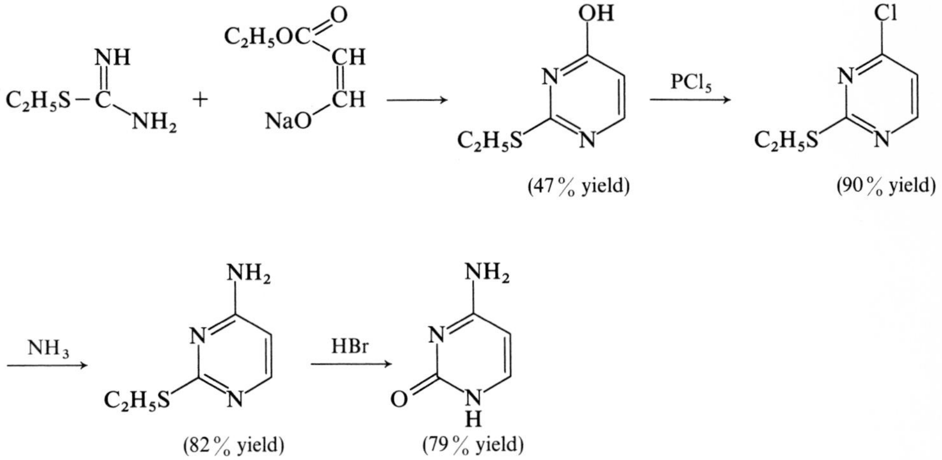

(47% yield) → (90% yield)

(82% yield) → (79% yield)

4. Wempen et al. (1965)

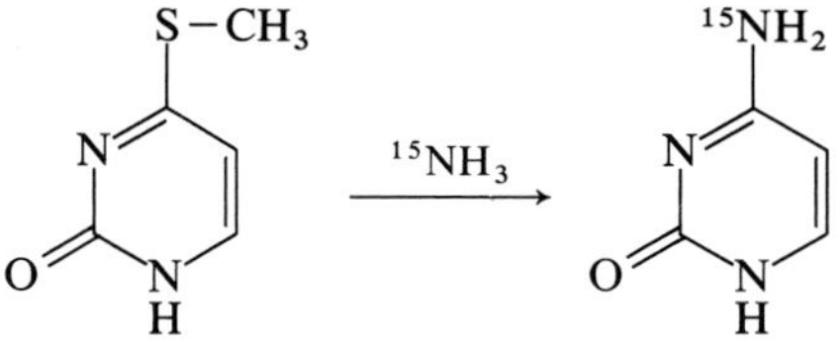

(89% yield)

5. Arantz and Brown (1968)

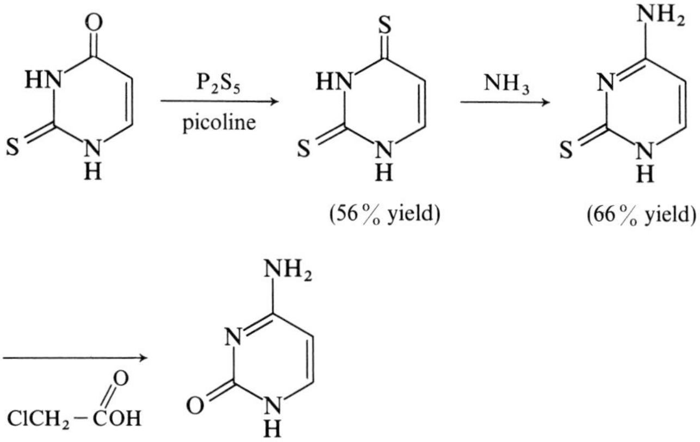

(56% yield)   (66% yield)

(95% yield)

11. 3-METHYLCYTOSINE   3-MeCyt

$C_5H_7N_3O$

M.W. 125.13

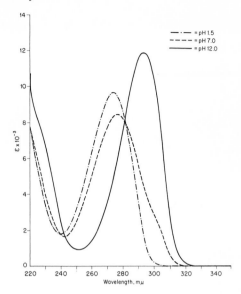

Physical Properties

M.P. 242–245° (HCl salt)   Brookes and Lawley (1962)
pK (basic) 7.4   Ueda and Fox (1963)
   (acidic) > 13   Ueda and Fox (1963)

Additional Reference for Synthesis

Bredereck et al. (1948)

Synthesis

1. Brookes and Lawley (1962)

(86% yield)

12. *N*⁴-METHYLCYTOSINE   4-MeCyt

$C_5H_7N_3O$

M.W. 125.13

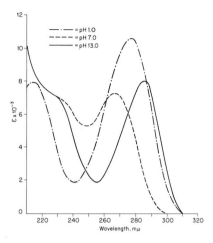

Szer and Shugar (1966)

Physical Properties

M.P. 277–280° (dec.)   Ueda and Fox (1964)
pK (basic)   4.40   Szer and Shugar (1968a)
   (acidic) 12.7   Szer and Shugar (1966)

Additional References for Synthesis

Brown (1955); Fox et al. (1959); Johns (1911); Szer and Shugar (1966)

Synthesis

1. Ueda and Fox (1964)

2. Szer and Shugar (1968a)

(29% yield)   (75% yield)

13. 5-METHYLCYTOSINE   5-MeCyt

$C_5H_7N_3O$

M.W. 125.13

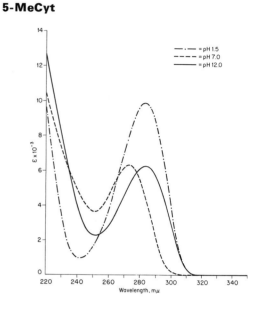

Physical Properties

M.P. (HCl salt) 299–301° (dec.)   Hitchings et al. (1949)
M.P. (picrate) 288–290°   Fox et al. (1959)
pK (basic)   4.6   Shugar and Fox (1952)
   (acidic) 12.4   Shugar and Fox (1952)

Additional References for Synthesis

Fox et al. (1959); Wheeler and Johnson (1904)

Synthesis

1. Hitchings et al. (1949)

(95 % yield)

NH2
CH3
N
O
HOC−CH2−S
N
(79 % yield)
conc. HCl
reflux
NH2
N
CH3
O
N
H
(67 % yield)

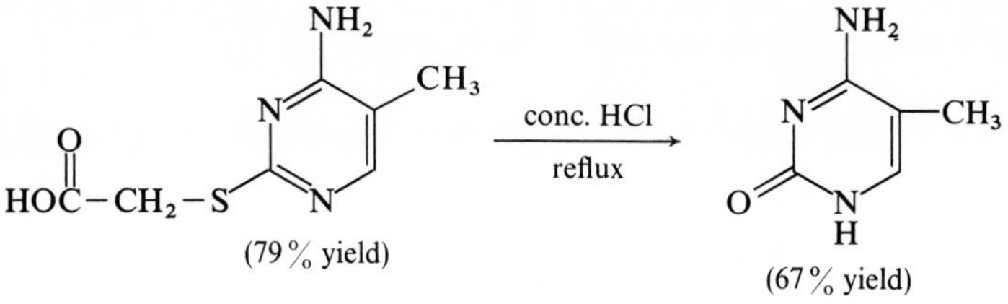

14. 5-HYDROXYMETHYLCYTOSINE   5-HmCyt

$C_5H_7N_3O_2$

M.W. 141.13

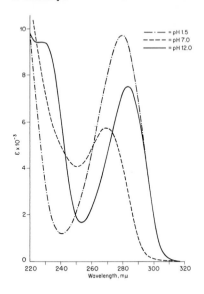

Physical Properties

M.P. $> 200°$ (dec.)   Miller (1955)
pK (basic) 4.3   Fissekis et al. (1964)
   (acidic) ~ 13 Fissekis et al. (1964)

Synthesis

1. Miller (1955)

LiAlH₄

(85% yield)

HCl

(30% yield)

15. *N⁴*-ACETYLCYTOSINE   4-AcCyt

$C_6H_7N_3O_2$

M.W. 153.14

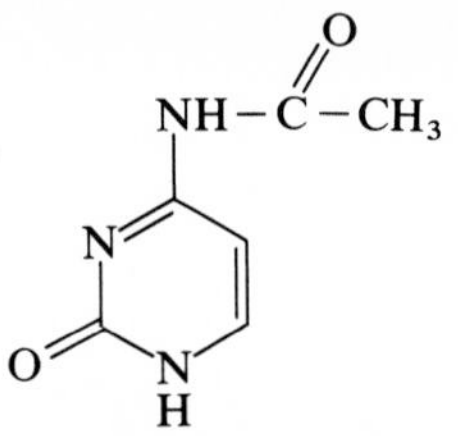

pH	λ_{max}	ε_{max} (x 10^{-3})	λ_{min}
7	244.5	14.2	226
	293	4.9	270

Brown et al. (1956).

Physical Properties

M.P. 326–328°   Codington et al. (1958)

Synthesis

1. Brown et al. (1956); Codington et al. (1958); Wheeler and Johnson (1903)

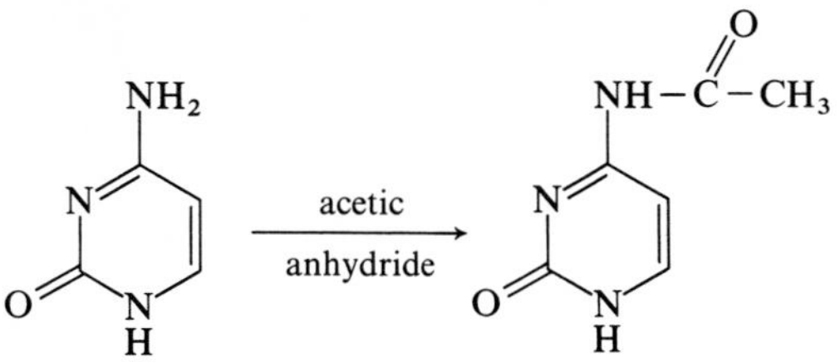

(79% yield)

16. 2-THIOCYTOSINE

$C_4H_5N_3S$

M.W. 127.16

Synthesis

Formed as an intermediate in synthesis of cytosine (**10,** synthesis 5)

17. GUANINE   Gua

$C_5H_5N_5O$

M.W. 151.13

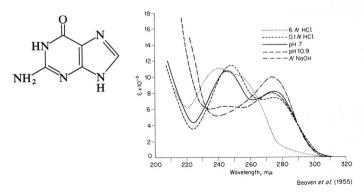

Beaven *et al.* (1955)

Physical Properties

M.P. $>350°$ Traube (1900)

pK (basic) $<0, 3.2$ Taylor (1948)

(acidic) 9.6, 12.4   Taylor (1948)

Synthesis

1. Yamazaki et al. (1967c)

(overall yield, starting
from I, 40%)

2. Traube (1900)

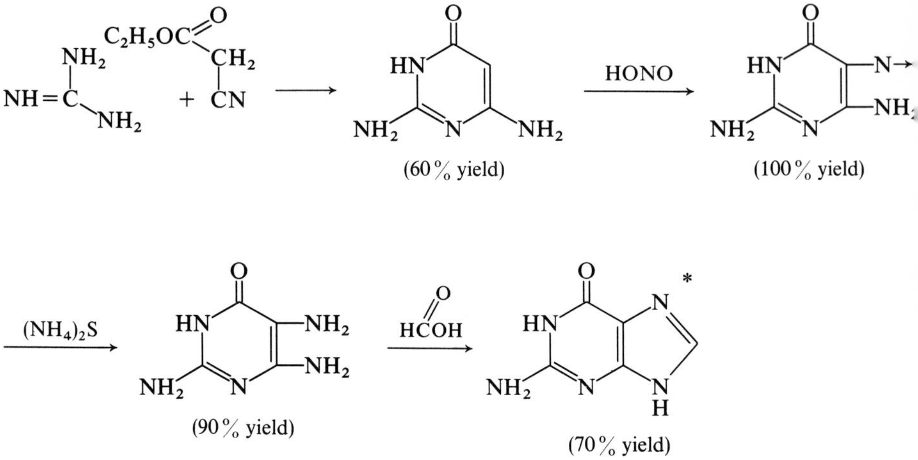

(60% yield)

(100% yield)

(90% yield)

(70% yield)

* See Robins et al. (1953) for an improvement of the last step.

18. 1-METHYLGUANINE   1-MeGua

$C_6H_7N_5O$

M.W. 165.16

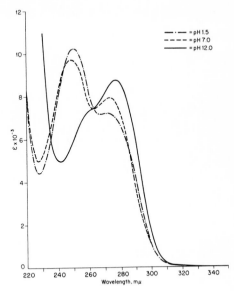

_ . _ . _ = pH 1.5
_ _ _ _ = pH 7.0
_____ = pH 12.0

Physical Properties

M.P. (none) (dec.)	Elion (1962)
pK (basic) ~0, 3.1	Pfleiderer (1961)
(acidic) 10.5	Pfleiderer (1961)

Additional Reference for Synthesis

Traube and Dudley (1910)

Synthesis

1. Elion (1962)

(87% yield)

19. *N²*-METHYLGUANINE   2-MeGua

$C_6H_7N_5O$

M.W. 165.16

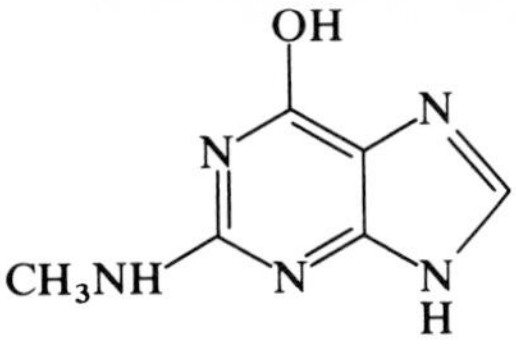

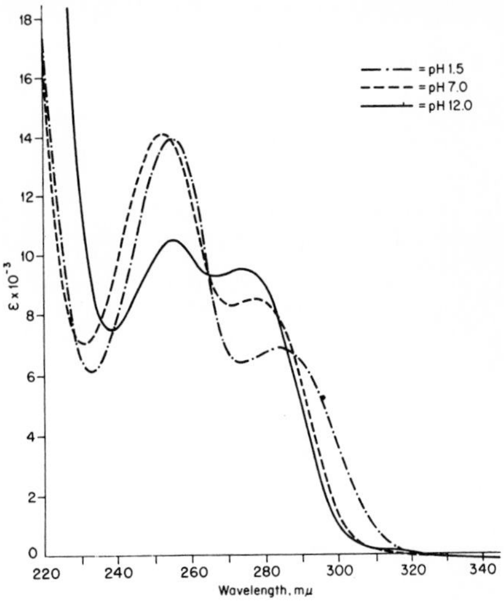

Spectrum legend:
- $-\cdot-\cdot-$ = pH 1.5
- $------$ = pH 7.0
- $———$ = pH 12.0

Physical Properties

pK (basic) 3.3   Shapiro and Gordon (1964)
(acidic) 8.9, 12.8   Shapiro and Gordon (1964)

Synthesis

1. Shapiro et al. (1969)

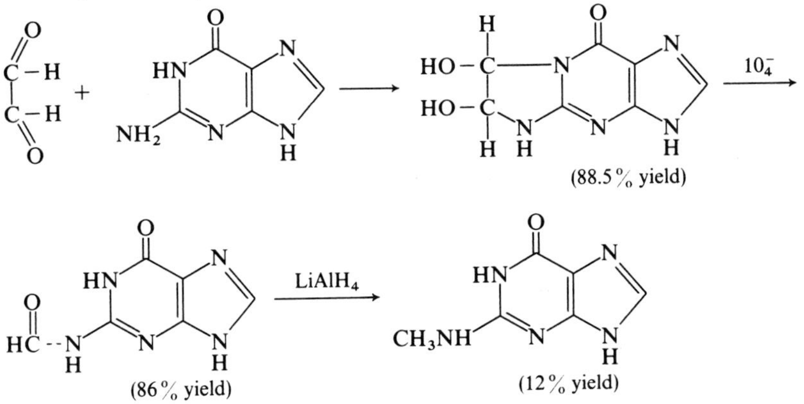

(88.5% yield)

(86% yield)

(12% yield)

20. *N²,N²*-DIMETHYLGUANINE   **2-Me₂Gua**

$C_7H_9N_5O$

M.W. 179.18

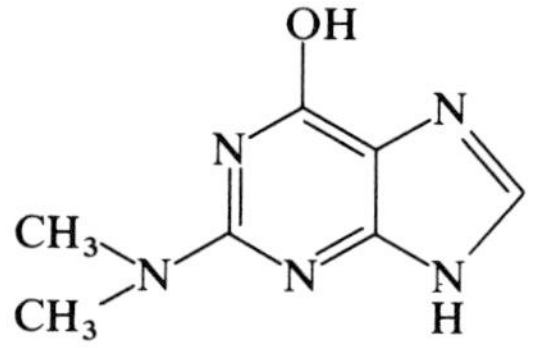

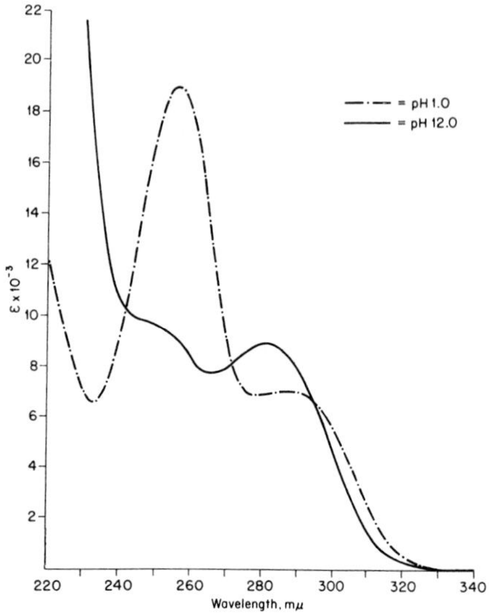

---·--- = pH 1.0
——— = pH 12.0

Additional References for Synthesis

Elion et al. (1956); Gerster and Robins (1965)

Synthesis

1. Elion et al. (1956)

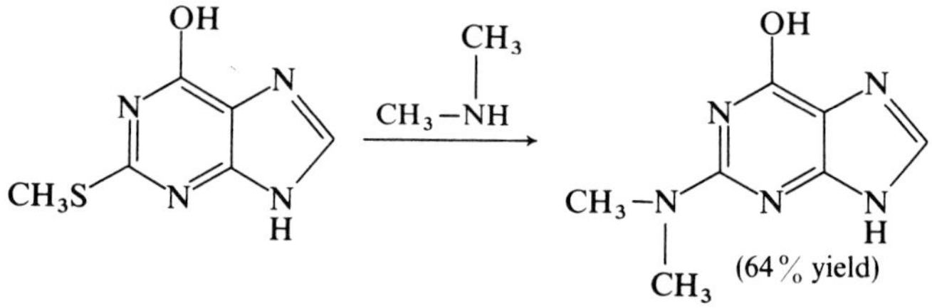

(64% yield)

2. Gerster and Robins (1965)

(70% yield)

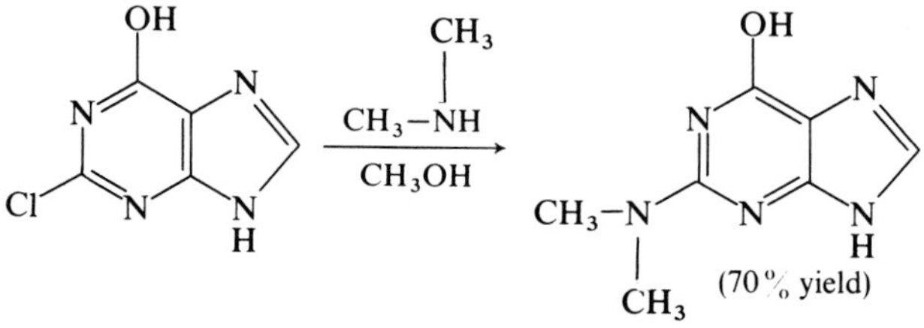

21. 7-METHYLGUANINE   7-MeGua

$C_6H_7N_5O$

M.W. 165.16

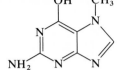

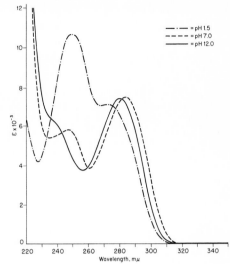

Physical Properties

M.P. none (dec. $> 390°$)   Fischer (1897)
pK (basic) ~ 0, 3.5   Pfleiderer (1961)
    (acidic) 9.95   Pfleiderer (1961)

Additional Reference for Synthesis

Fischer (1897)

Synthesis

1. Jones and Robins (1963)

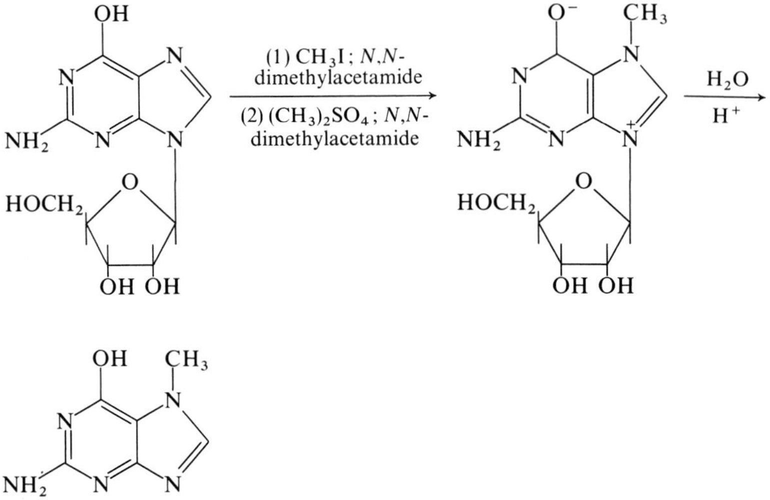

(1) 60% yield; (2) 91% yield

22. HYPOXANTHINE   Hyp

$C_5H_4N_4O$

M.W. 136.11

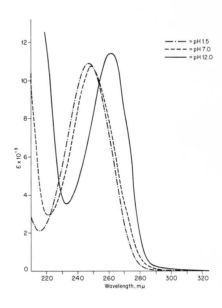

Physical Properties

M.P. $>350°$	Robins et al. (1953)
pK (basic) 2.0	Albert and Brown (1954)
(acidic) 8.8, >12	Bergmann and Dikstein (1955)

Additional References for Synthesis

Baddiley et al. (1959); Gulland and Holiday (1936); Levene and Tipson (1935)

Synthesis

1. Ichikawa et al. (1965)

HCOOH, Zn

K_2CO_3, H_2O, reflux

(43% yield)

2. Richter et al. (1960); Taylor and Cheng (1959)

3. Elion et al. (1952); Robins et al. (1953)

4. Shaw (1950); Shaw and Wooley (1949)

5. Robins et al. (1953)

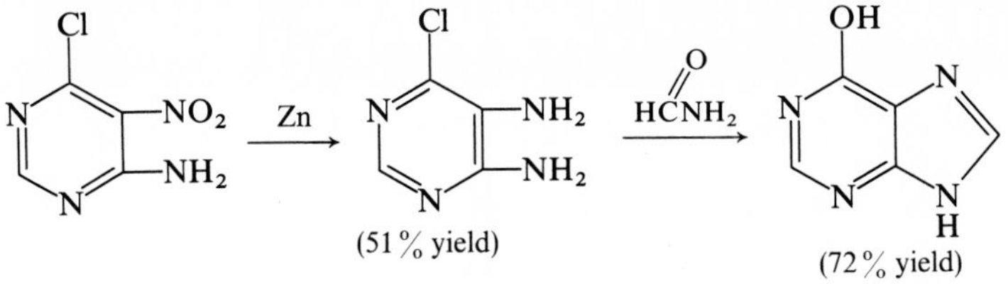

(51% yield)

(72% yield)

6. Taylor et al. (1959a)

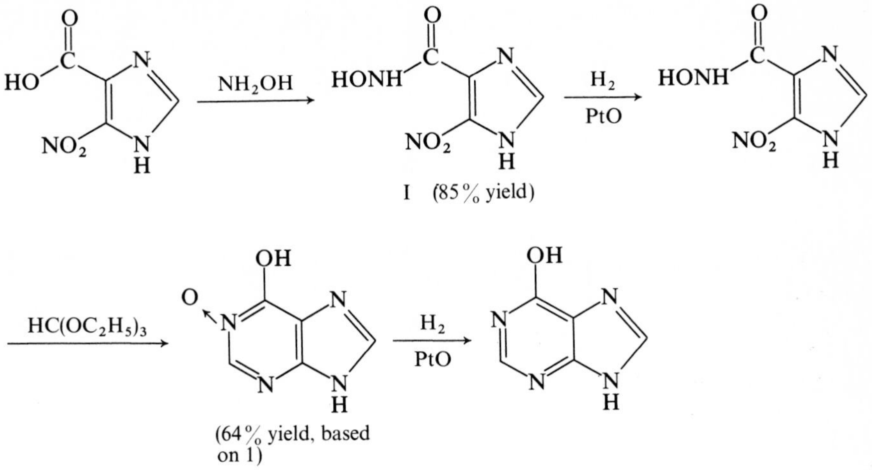

I   (85% yield)

(64% yield, based on 1)

23. 1-METHYLHYPOXANTHINE   1-MeHyp

$C_6H_6N_4O$

M.W. 150.14

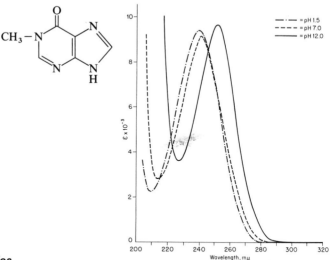

Physical Properties

M.P. 315–320°   Thomas et al. (1968)
pK (basic) approx. 2   Elion (1962)
   (acidic) 8.9, ~13   Elion (1962)

Synthesis

1. Townsend and Robins (1962)

(61% yield)

(87% yield)

(92% yield)

2. Elion (1962)

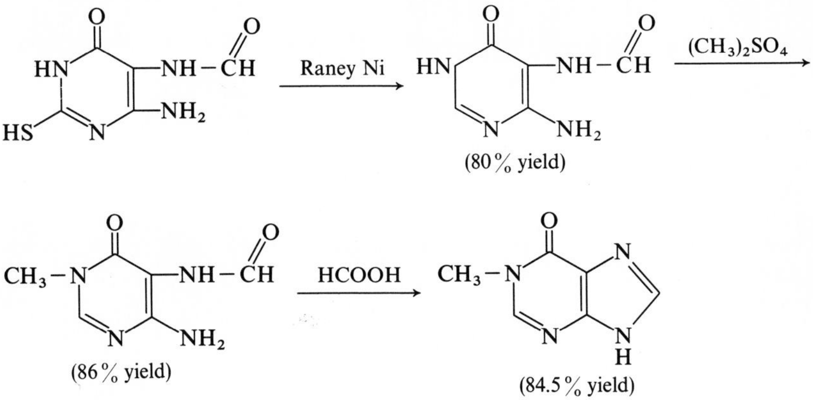

(80% yield)

(86% yield)

(84.5% yield)

3. Thomas et al. (1968)

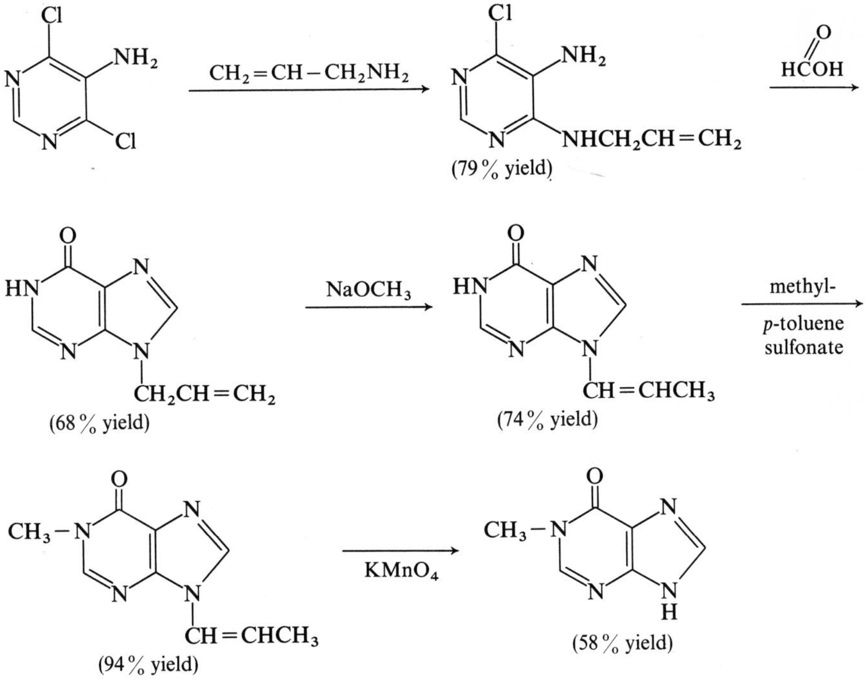

(79% yield)

(68% yield)

(74% yield)

(94% yield)

(58% yield)

24. URACIL Ura

$C_4H_4N_2O_2$

M.W. 112.09

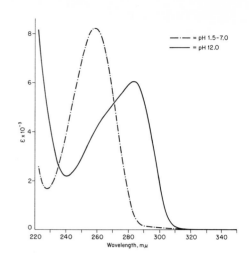

= pH 1.5–7.0
= pH 12.0

Physical Properties

M.P. ~335° (dec.) Levene and Bass (1931)
pK (acidic) 9.5, >13 Shugar and Fox (1952)

Additional Reference for Synthesis

Fischer and Roeder (1901)

Synthesis

1. Davidson and Baudisch (1926)

2. Wheeler and Merriam (1903)

(46% yield) (94% yield)

3. Wempen et al. (1965)

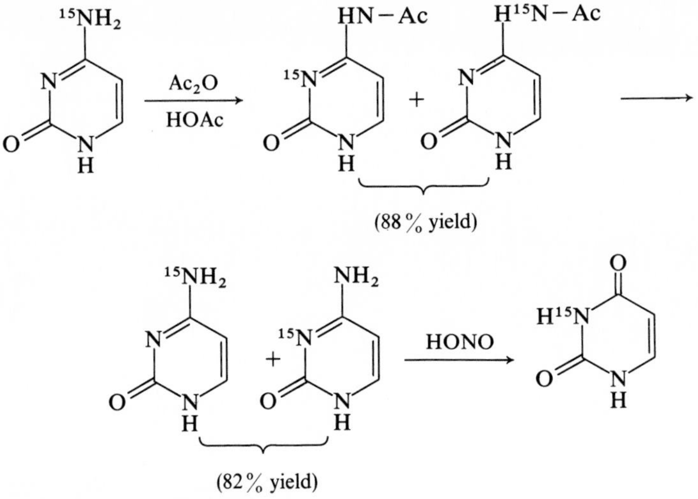

(88 % yield)

(82 % yield)

where Ac = acetyl

25. 3-METHYLURACIL   3-MeUra

$C_5H_6N_2O_2$

M.W. 126.12

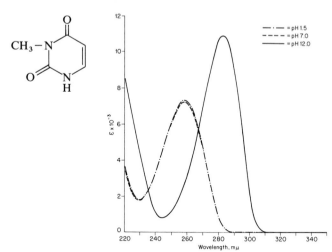

Physical Properties

M.P. 179°   Whitehead (1952); Ueda and Fox (1964)
pK (acidic) 10.0   Shugar and Fox (1952)

Synthesis

1. Whitehead (1952)

(41% yield)

(98% yield)   heat: 255°   (90% yield)

2. D. J. Brown et al. (1955); Johnson and Heyl (1907)

where R = CH_3, C_2H_5

26. 5-METHYLURACIL (THYMINE)   5-MeUra   (Thy)

$C_5H_6N_2O_2$

M.W. 126.11

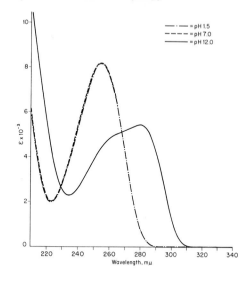

Physical Properties

M.P. 325–335° (dec.) Fischer and Roeder (1901)
313–314°   Scherp (1946)
pK (acidic) 9.9, > 13 Shugar and Fox (1952)

Additional References for Synthesis

Fischer and Roeder (1901); Levene and Bass (1931)

Synthesis

1. Shaw and Warrener (1958)

2. Wheeler and Merriam (1903)

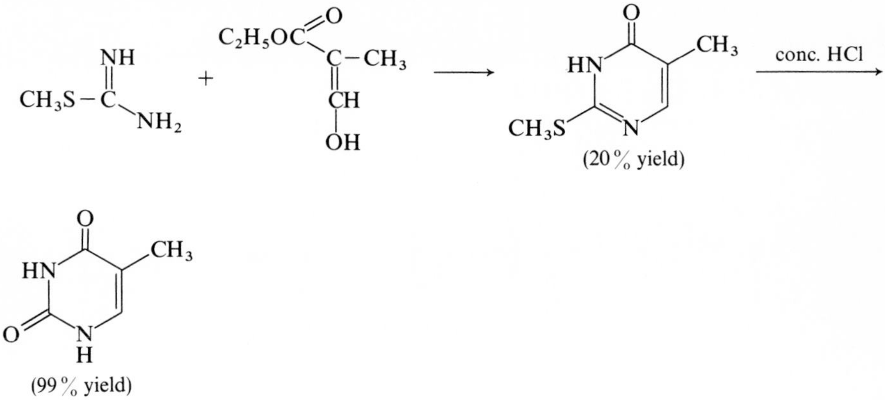

conc. HCl
(20% yield)
(99% yield)

3. Ulbricht (1959)

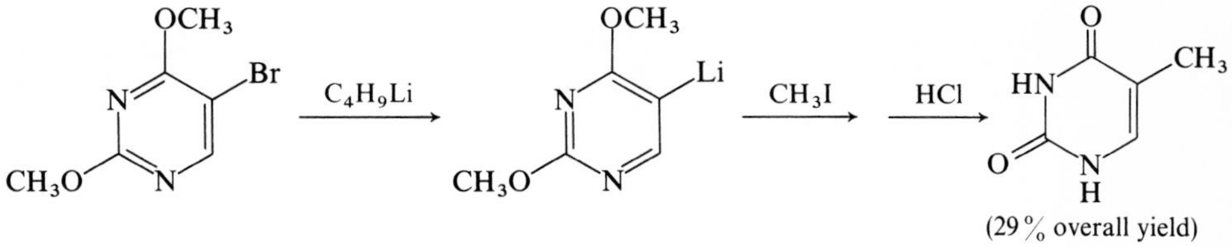

C4H9Li
CH3I
HCl
(29% overall yield)

4. Scherp (1946)

H2
PtO
HCl
(90% yield)
H2SO4
(100% yield)
(50% yield)

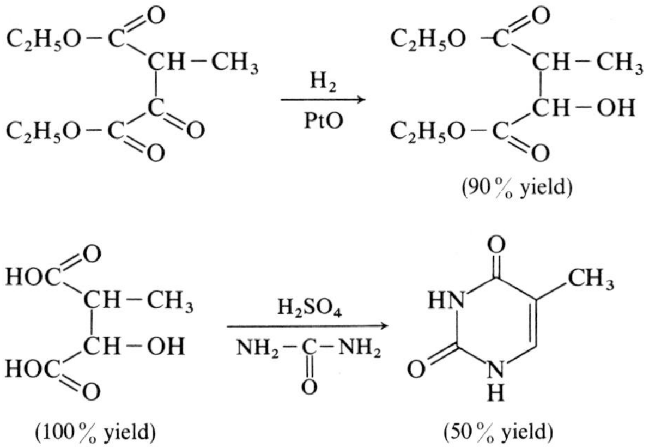

5. Guyot and Mentzer (1958)

27. 4-THIOURACIL   4-SUra

C$_4$H$_4$N$_2$OS

M.W. 128.09

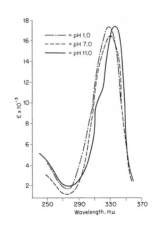

Physical Properties

M.P. 289–290° (dec.)   Fox and Van Praag (1960)

Additional Reference for Synthesis

Wheeler and Liddle (1908)

Synthesis

1. Fox and Van Praag (1960)

P$_2$S$_5$

(50% yield)

28. 5-HYDROXYURACIL   5-OHUra

$C_4H_4N_2O_3$

M.W. 128.09

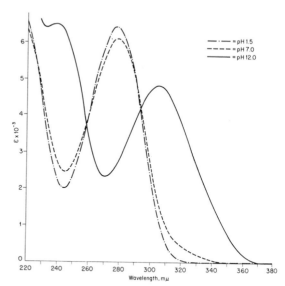

Physical Properties

M.P. none (dec. > 300°)   Wang (1959a)
pK (acidic) 8.0   Dunn, unpublished data

Additional References for Synthesis

Behrend and Roosen (1889); Johnson and McCollum (1906)

Synthesis

1. Wang (1959a)

$$\xrightarrow[\text{H}_2\text{O}]{\text{Br}_2} \qquad (90\% \text{ yield}) \qquad \xrightarrow[\text{H}_2\text{O}]{\text{NaHCO}_3} \qquad (50\% \text{ yield})$$

Preparation of 5-bromouracil is described by Wang (1959b).

29. 5-HYDROXYMETHYLURACIL   5-HmUra

$C_5H_6N_2O_3$

M.W. 142.11

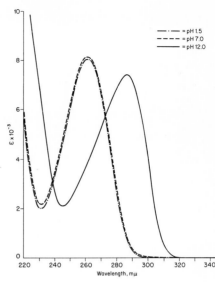

Physical Properties

    M.P. 260–300° (dec.)   Cline et al. (1959)
    pK (acidic) 9.4, ~14   Cohn (1960)

Additional References for Synthesis

    Dornow and Petsch (1954); Johnson and Litzinger (1936)

Synthesis

1. Cline et al. (1959)

(81% yield)

30. 2-THIO-5-(*N*-METHYLAMINOMETHYL)URACIL

$C_6H_9N_3OS$

M.W. 171.22

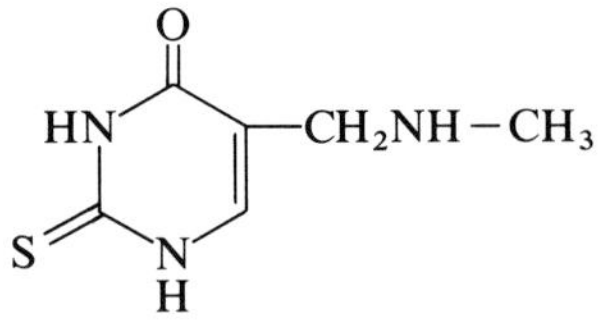

Physical Properties

M.P. (HCl salt) 230–232°   Carbon et al. (1968)

Synthesis

1. Carbon (1960); Carbon et al. (1968)

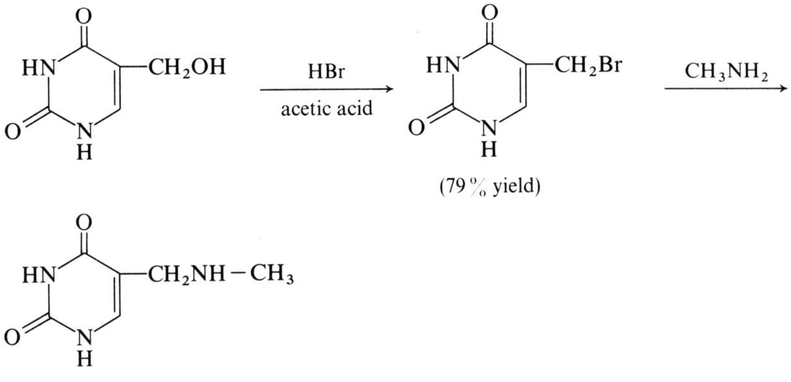

(79% yield)

(40% yield)

31. 5-CARBOXYMETHYLURACIL   5-CMUra

$C_6H_6N_2O_4$

M.W. 170.12

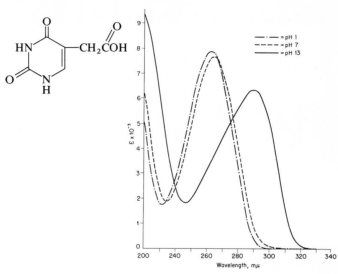

Gray and Lane (1968)

Physical Properties

M.P. 311–311.5°   Gray and Lane (1968)

Synthesis

1. Gray and Lane (1968); Johnson and Speh (1907)

32. 2-THIO-5-CARBOXYMETHYLURACIL, METHYL ESTER

$C_7H_8N_2O_3S$

M.W. 200.15

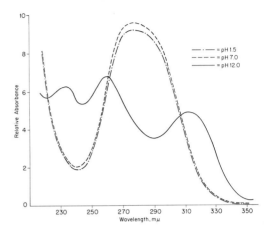

Physical Properties

M.P. 199°   Baczynskyj et al. (1969)

Synthesis

1. Baczynskyj et al. (1969)

Compound I prepared according to Payot and Grob (1954).

33. 5,6-DIHYDROURACIL   H₂Ura

$C_4H_6N_2O_2$

M.W. 114.10

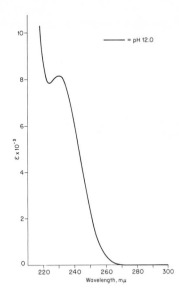

Physical Properties

M.P. 275–276°   Fox and Van Praag (1960)

Additional References for Synthesis

Brown and Johnson (1923); di Carlo et al. (1952); Fischer and Roeder (1901); Green and Cohen (1957); Lengfeld and Stieglitz (1893).

Synthesis

1. Fox and Van Praag (1960)

$$\xrightarrow[\text{Raney Ni}]{H_2}$$

(53 % yield)

34. 5,6-DIHYDRO-5-METHYLURACIL

$C_5H_8N_2O_2$

M.W. 128.13

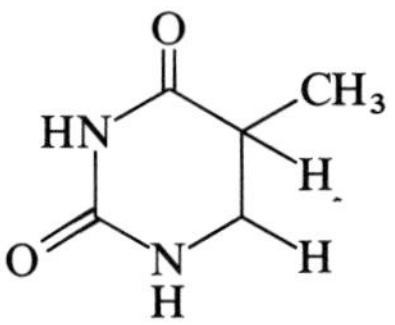

Physical Properties

M.P. 261–263°   Batt et al. (1954)

Ribonucleosides

35. ADENOSINE   Ado   A

$C_{10}H_{13}N_5O_4$

M.W. 267.24

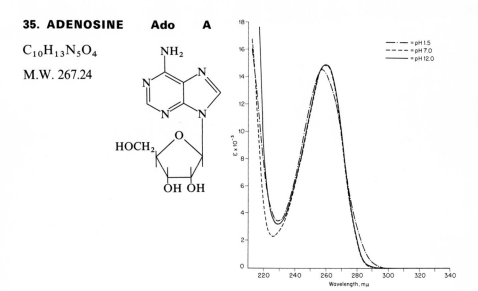

= pH 1.5
= pH 7.0
= pH 12.0

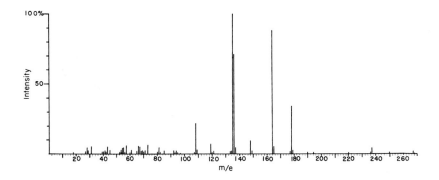

Physical Properties

M.P. 234–235°	Davoll et al. (1948b)
$[\alpha]_D^{11}$ −61.7° (0.706, water)	Davoll et al. (1948b)
pK (basic) 3.51	Harkins and Freiser (1958)
(acidic) 12.5	Levene et al. (1926b)

Additional References for Synthesis

Davoll et al. (1948b); Davoll and Lowy (1951); Kenner et al. (1949); Shimadate et al. (1960)

Adenosine-ribosyl-[^{3}H]was synthesized by Gordon et al. (1958).

9-(α-D-ribofuranosyl)adenine was synthesized by Wright et al. (1958).

9-(β-D-ribopyranosyl)adenine was synthesized by Furukawa and Honjo (1968).

Synthesis

1. Schramm et al. (1967)

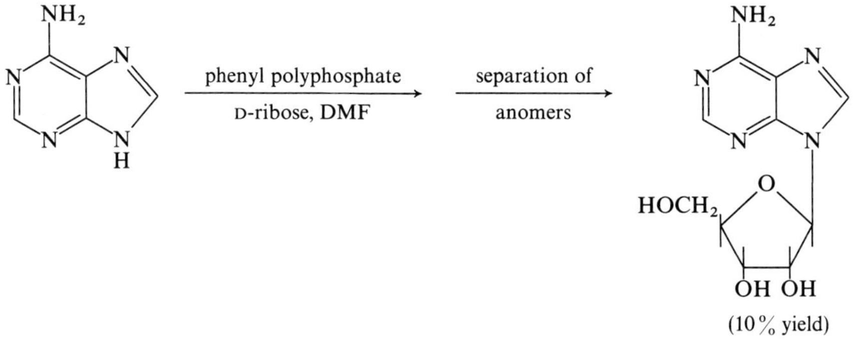
NH2
phenyl polyphosphate
D-ribose, DMF
separation of
anomers
HOCH2
OH OH
(10 % yield)

2. Davoll and Lowy (1952)

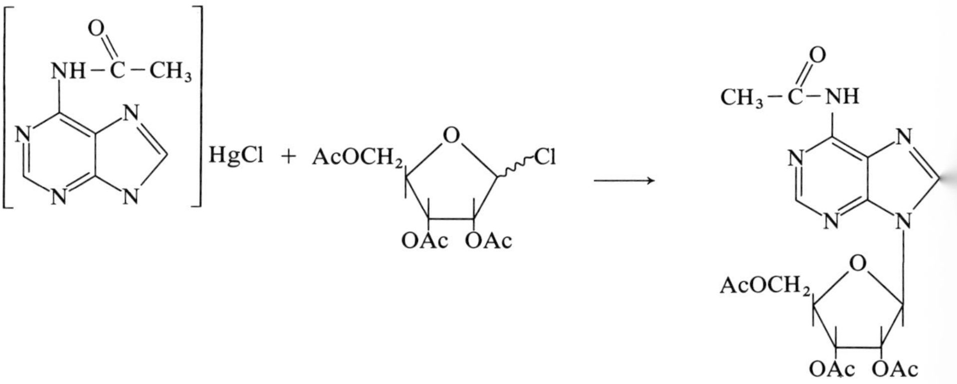
NH−C−CH3
HgCl + AcOCH2
Cl
OAc OAc
CH3−C−NH
AcOCH2
OAc OAc

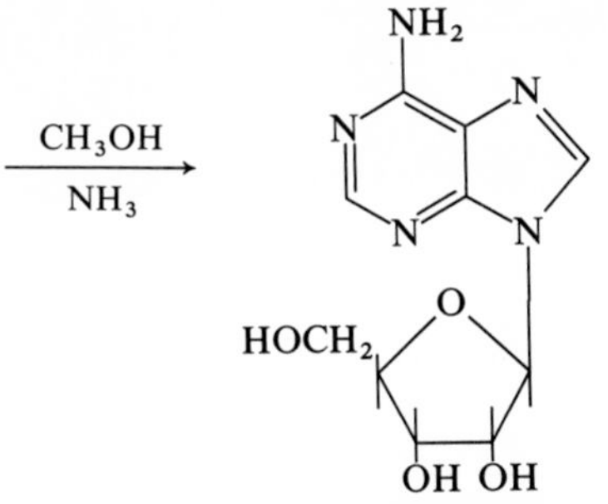

(16% overall yield)

where Ac = acetyl.

3. Fox et al. (1958)

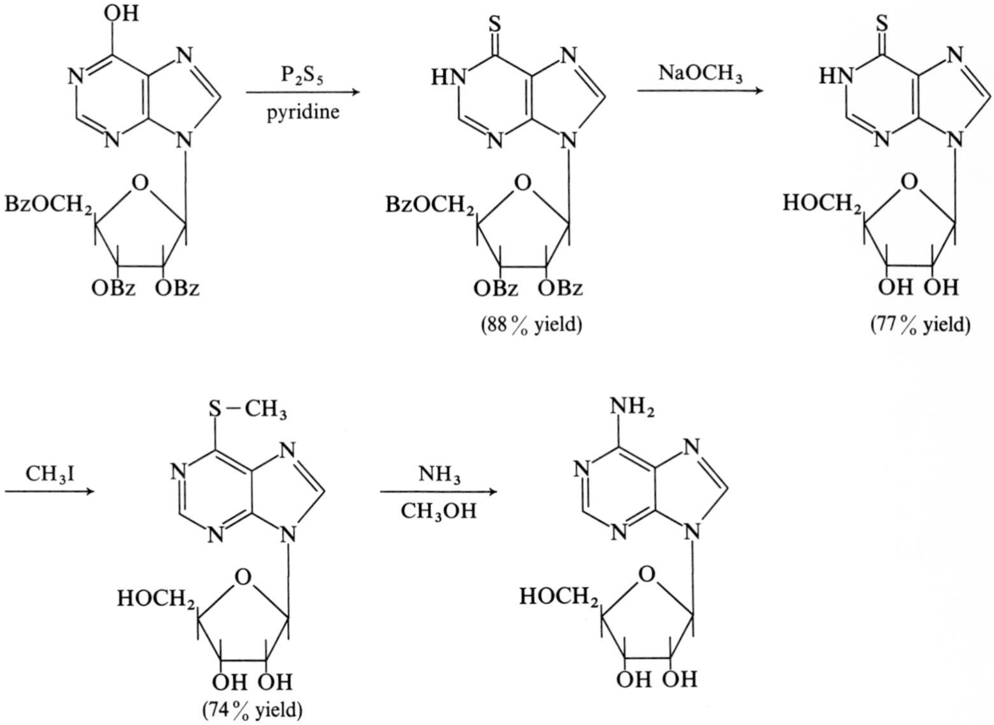

(88% yield)

(77% yield)

(74% yield)

where Bz = benzoyl.

4. Kissman and Weiss (1956); Walsh and Wolfenden (1967)

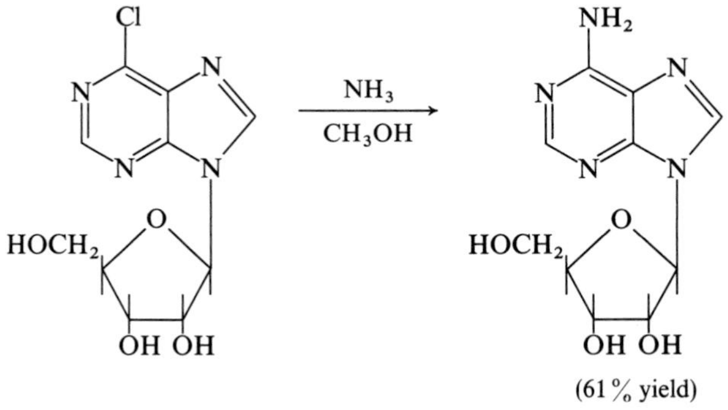

(61 % yield)

5. Shimizu et al. (1967)

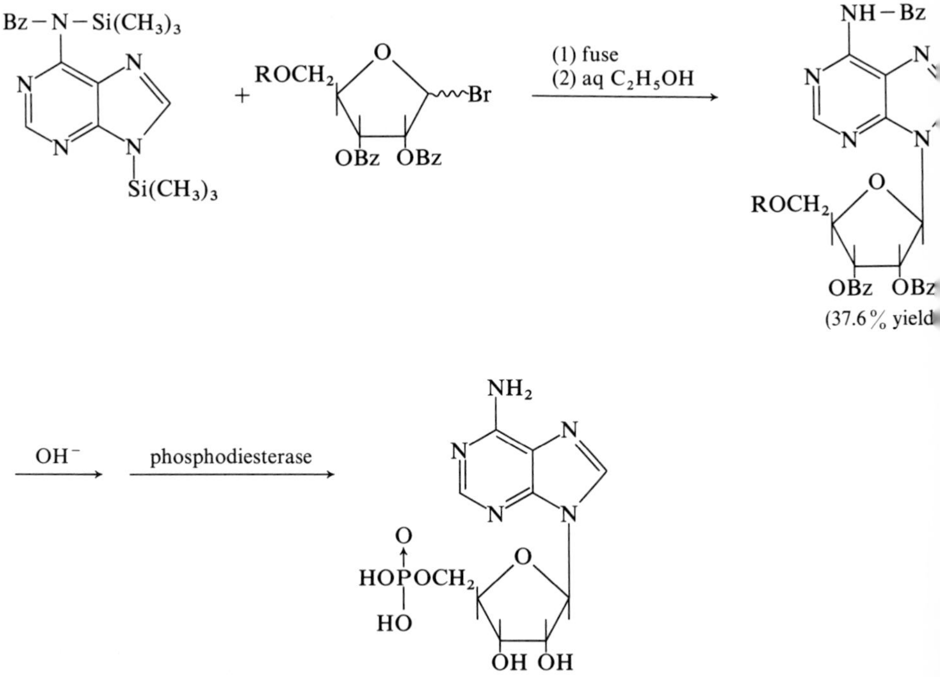

(37.6 % yield)

where Bz = benzoyl and R = $(C_6H_5O)_2-\overset{\overset{\displaystyle O}{\|}}{P}-$.

6. Asai et al. (1967)

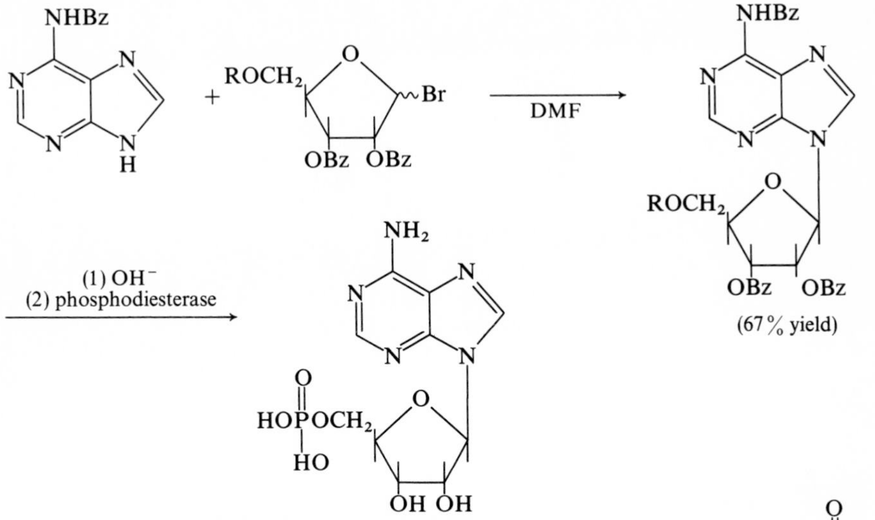

where Bz = benzoyl and R = $(C_6H_5O)_2 - \overset{\overset{O}{\parallel}}{P} -$.

7. Furukawa and Honjo (1968)

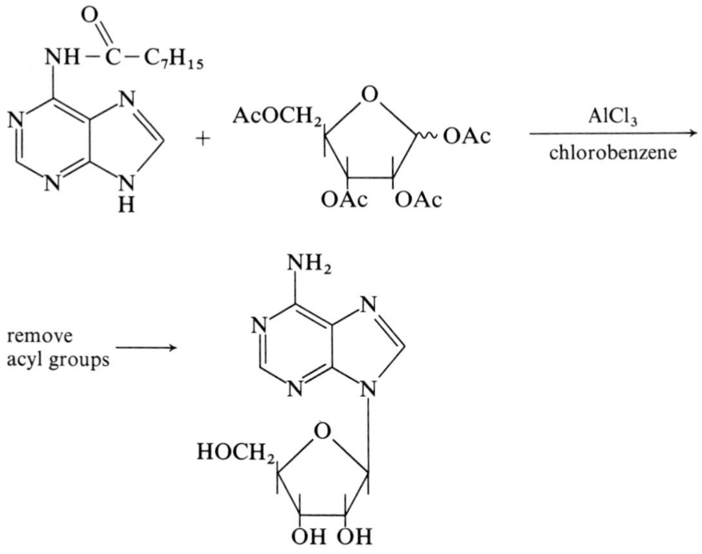

(60% yield)

where Ac = acetyl.

36. 1-METHYLADENOSINE   1-meAdo   m¹A

$C_{11}H_{15}N_5O_4$

M.W. 281.27

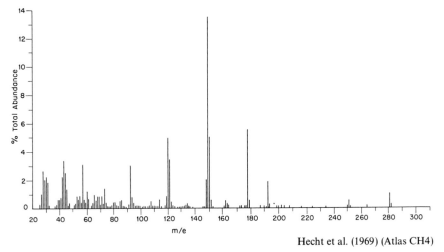

Hecht et al. (1969) (Atlas CH4)

Physical Properties

M.P. 214–217° (dec.)   Jones and Robins (1963)

$[\alpha]_D^{26} -59°$ (2.0, water)   Jones and Robins (1963)

pK (acidic) 8.8   Martin and Reese (1968)

Additional Reference for Synthesis

Bredereck et al. (1948)

Synthesis

1. Jones and Robins (1963)

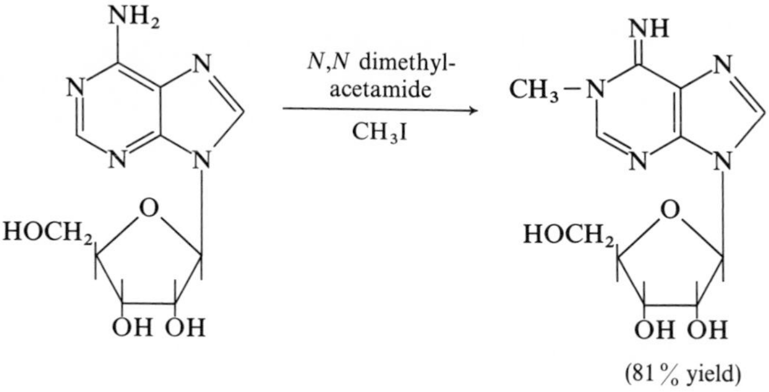

(81 % yield)

1-Methyladenosine rearranges readily in neutral and alkaline solutions to N^6-methyladenosine (Brookes and Lawley, 1960). The kinetics of the rate of the rearrangement at different pH values has been studied by Macon and Wolfenden (1968). N^6-Methyladenosine or its base has often been isolated from nucleic acid samples under conditions that do not preclude the occurrence of the rearrangement during isolation. Therefore analytical values for N^6-methyl- and N^1-methyladenosine are included under this heading. The compound was first isolated as N^6-methyladenine from commercial yeast RNA by Adler et al. (1958) and from the RNA of several organisms by Littlefield and Dunn (1958). N^6-Methyladenine (but not the N^1-methyl derivative) occurs in the DNA of some organisms (Table 1, Chapter 3). This nucleoside appears in the primary sequence of several molecular species of yeast tRNA (Chapter 4).

RNA source	Amount moles/100 moles		References
Carp liver (t)	1.2		Zaitseva et al. (1962)
Mouse adenosarcoma (t)	0.84	(c)	Bergquist and Matthews (1962)
Mouse liver (t)	0.15	(c)	Bergquist and Matthews (1962)
Pig liver	0.95	(a,c)	Dunn (1963)
Pigeon (t)	0.7		Zaitseva et al. (1962)
Rabbit liver (whole RNA)	0.06	(c)	Littlefield and Dunn (1958)
Rat liver (r)	0.10	(a,c)	Dunn (1959); Littlefield and Dunn (1958)
Rat liver (t)	0.6	(c)	Dunn (1959)
S-180 ascites tumor (t)	0.70	(c)	Bergquist and Matthews (1962)
Euglena gracilis	0.12		Brawerman et al. (1962)
Tobacco leaves	0.08	(a,c)	Bergquist and Matthews (1962)
Wheat germ (t)	0.95	(a,c)	Littlefield and Dunn (1958); Hudson et al. (1965)
Yeast, commercial	0.05		Adler et al. (1958); Littlefield and Dunn (1958)
Yeast (t)	0.90	(a,c)	Dunn and Flack (1967a); Hall (1965)
Aerobacter aerogenes	0.06	(c)	Littlefield and Dunn (1958)
Azotobacter vinelandii (t)	0.4		Zaitseva et al. (1962)
Escherichia coli (t)	0.10	(b,c)	Dunn et al. (1960a)
Proteus vulgaris (t)	0.02		Zaitseva et al. (1962)
Sarcina flava (t)	0.6		Zaitseva et al. (1962)
Staphylococcus aureus	0.04	(c)	Littlefield and Dunn (1958)

37. 2-METHYLADENOSINE   2-meAdo   m²A

$C_{11}H_{15}N_5O_4$

M.W. 281.27

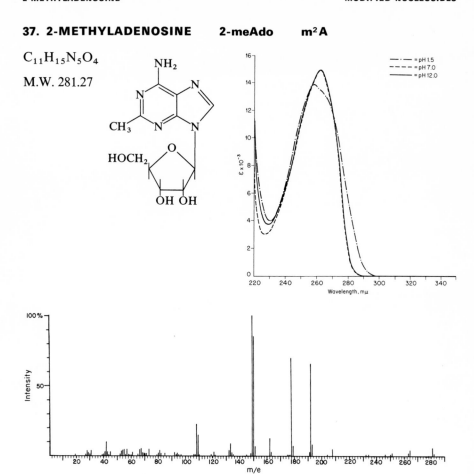

Physical Properties

M.P. (picrate) $>200°$ (dec.)   Davoll and Lowy (1952)

Additional Reference for Synthesis

Littlefield and Dunn (1958) (enzymic synthesis)

Synthesis

1. Davoll and Lowy (1952)

$$\left[\begin{array}{c}\text{HNAc} \\ \text{2-methyladenine structure with } CH_3\end{array}\right] \quad HgCl \quad + \quad AcOCH_2\text{-sugar-Cl (OAc, OAc)} \quad \longrightarrow$$

NAc-purine (CH_3) with AcOCH$_2$ sugar (OAc OAc) $\quad \xrightarrow[\text{NH}_3]{\text{CH}_3\text{OH}} \quad$ NH$_2$-purine (CH_3) with HOCH$_2$ sugar (OH OH)

(11% yield, starting
from 2-methyladenine)

where Ac = acetyl.

2-Methyladenosine was first isolated by Littlefield and Dunn (1958) as the nucleotide and nucleoside from the RNA of several organisms. It also exists as a component of vitamin B_{12} (the 7-α isomer) that occurs in *E. coli* (F. B. Brown et al., 1955; Dion et al., 1954).

RNA source	Amount moles/100 moles		References
Carp liver (t)	0.02		Zaitseva et al. (1962)
S-180 ascites (t)	0.48	(c)	Bergquist and Matthews (1962)
Tobacco leaves (t)	0.04	(c)	Bergquist and Matthews (1962)
Wheat germ	0.02	(c)	Littlefield and Dunn (1958)
Yeast, commercial	0.02	(c)	Littlefield and Dunn (1958)
Aerobacter aerogenes	0.06	(c)	Littlefield and Dunn (1958)
Azotobacter vinelandii (t)	0.18		Zaitseva et al. (1962)
Escherichia coli (t)	0.25	(c)	Dunn et al. (1960a)
Escherichia coli (23S)	0.07		Fellner and Sanger (1968)
Proteus vulgaris (t)	0.1		Zaitseva et al. (1962)
Sarcina flava (t)	0.14		Zaitseva et al. (1962)
Staphylococcus aureus	0.01	(c)	Littlefield and Dunn (1958)

38. N^6-METHYLADENOSINE   6-meAdo m^6A

$C_{11}H_{15}N_5O_4$

M.W. 281.27

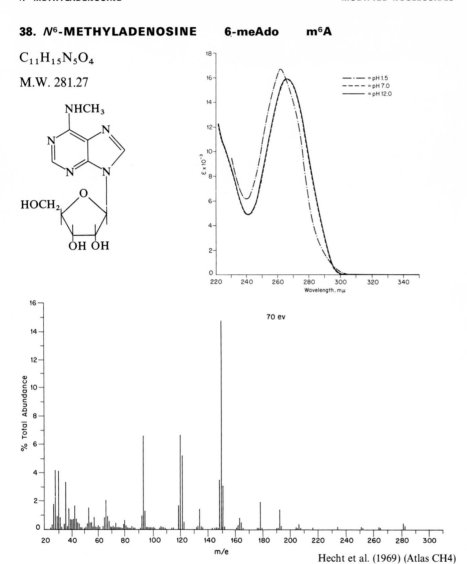

Hecht et al. (1969) (Atlas CH4)

Physical Properties

M.P. 219–221°	Wempen and Fox (1967b)
$[\alpha]_D^{26}$ − 54° (0.6, water)	Johnson et al. (1958)
pK (basic) 4.0	Martin and Reese (1968)

Additional References for Synthesis

Bredereck et al. (1948); Littlefield and Dunn (1958) (enzymic synthesis)

Synthesis

1. Jones and Robins (1963)

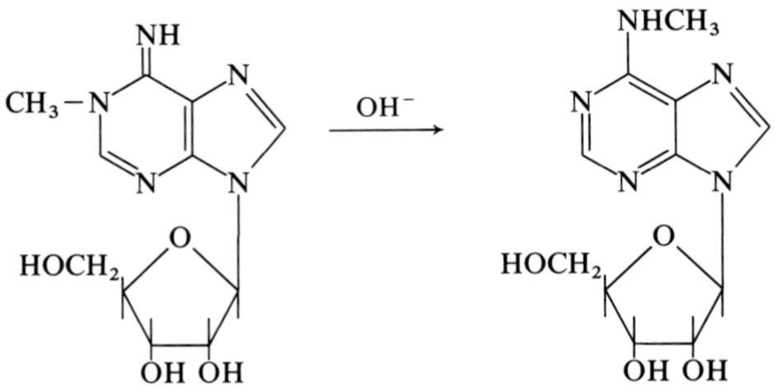

(64% yield)

2. Johnson et al. (1958); Walsh and Wolfenden (1967)

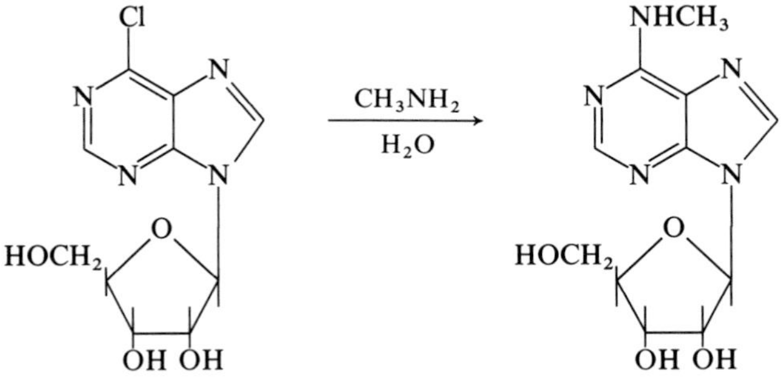

I   (74% yield)

Compound I can be prepared readily according to Gerster et al. (1963) or Wempen and Fox (1967a).

39. *N⁶,N⁶*-DIMETHYLADENOSINE   6-Me₂Ado   m₂⁶A

$C_{12}H_{17}N_5O_4$

M.W. 295.30

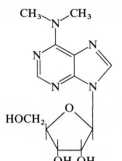

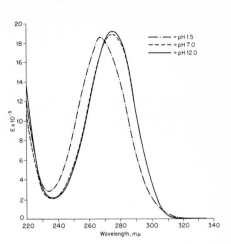

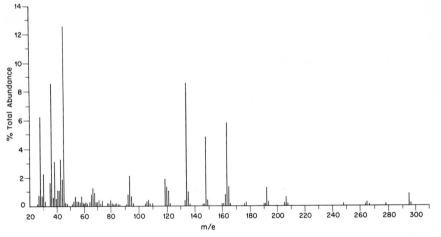

Hecht et al. (1969) (Atlas CH4)

Physical Properties

M.P. 183–184°   Townsend et al. (1964)
$[\alpha]_D^{25}$ −62.6° (2.6, water)   Kissman et al. (1955)
pK (basic) 4.50   Martin and Reese (1968)

Additional References for Synthesis

Andrews and Barber (1958); Kissman et al. (1955); Littlefield and Dunn
(1958) (enzymic synthesis); Weiss et al. (1959)

Synthesis

1. Townsend et al. (1964); Walsh and Wolfenden (1967)

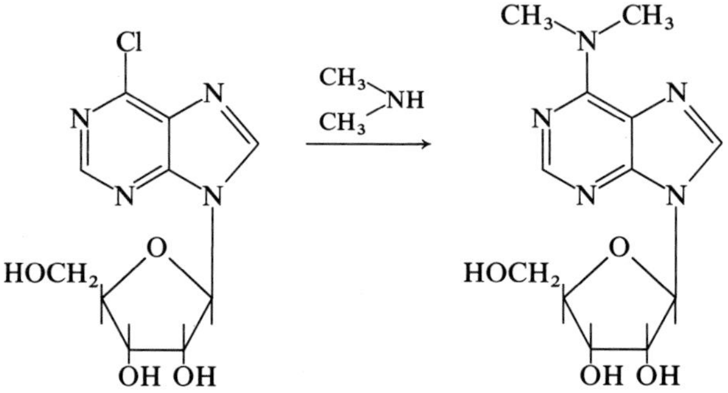

I

Compound I can be prepared readily according to Gerster et al. (1963) or
Wempen and Fox (1967a).

2. Bhat (1968)

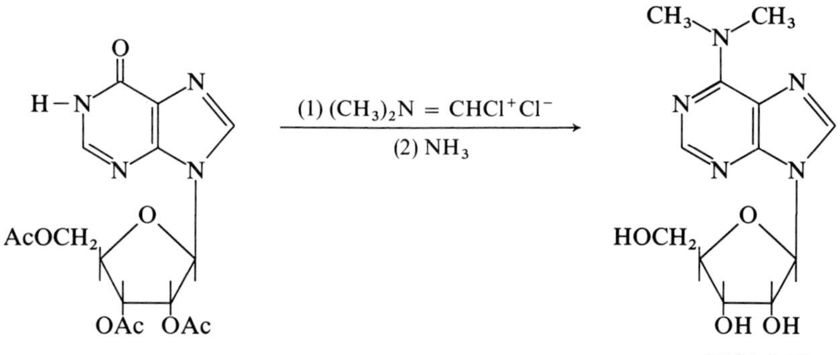

(72% yield)

where Ac = acetyl.

N^6,N^6-*Dimethyladenosine* was first isolated by Littlefield and Dunn (1958) as the nucleotide and nucleoside from the RNA of several sources. It occurs exclusively in the 16S ribosomal RNA of *E. coli* as the dinucleotide, $m_2^6Apm_2^6Ap$ (Dubin and Gunalp, 1967; Fellner and Sanger, 1968; Nichols and Lane, 1966b). The base occurs as a component of the antibiotic puromycin (Waller et al., 1953).

RNA source	Amount, moles/100 moles		References
Mouse adenosarcoma (t)	0.23	(c)	Bergquist and Matthews (1962)
Mouse liver (t)	0.11	(c)	Bergquist and Matthews (1962)
Pigeon liver (t)	0.11		Zaitseva et al. (1962)
Rat liver (r)	0.02	(c)	Dunn (1959); Littlefield and Dunn (1958)
Rat liver (t)	0.02	(c)	Dunn (1959)
S-180 ascites (t)	0.43	(c)	Bergquist and Matthews (1962)
Tobacco leaves (t)	0.07	(c)	Bergquist and Matthews (1962)
Wheat germ	0.01	(c)	Littlefield and Dunn (1958)
Yeast, commercial	0.01	(c)	Littlefield and Dunn (1958)
Aerobacter aerogenes	0.02	(c)	Littlefield and Dunn (1958)
Azotobacter vinelandii (t)	0.3		Zaitseva et al. (1962)
Escherichia coli B	0.15	(c)	Littlefield and Dunn (1958)
Escherichia coli (16S)	0.27		Fellner and Sanger (1968); Nichols and Lane (1966b)
Proteus vulgaris (t)	0.2		Zaitseva et al. (1962)

40. 2'-O-METHYLADENOSINE   *OMeAdo*   **Am**

$C_{11}H_{15}N_5O_4$

M.W. 281.27

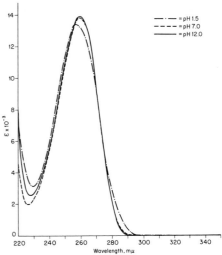

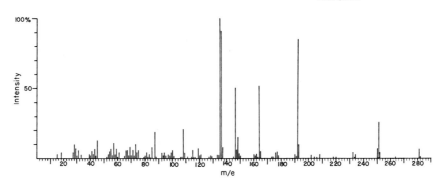

Physical Properties

M.P. 200–202°   Broom and Robins (1965)

$[\alpha]_D^{24} -58.2°$ (1.0, water)   Broom and Robins (1965)

Synthesis

1. Khwaja and Robins (1966)

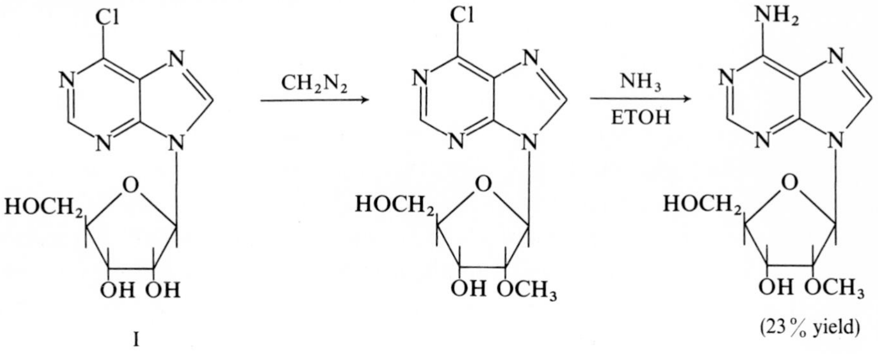

I   (23% yield)

Compound I can be prepared according to Gerster et al. (1963) or Wempen and Fox (1967a).

2. Broom and Robins (1965); Gin and Dekker (1968); Martin et al. (1968)

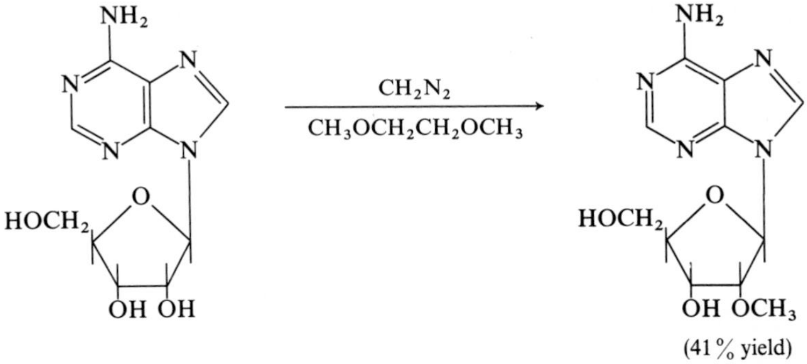

(41% yield)

Methylation by this method produces a mixture of O-methyl derivatives. They are readily separated by chromatography on Dowex-1-X2 [OH⁻] resin (Gin and Dekker, 1968).

2'-O-Methyladenosine and the corresponding methylated derivatives of the three other major ribonucleosides were first isolated by Smith and Dunn (1959a) and characterized as $2'(3')-O$ derivatives. Hall (1963a) identified this group of nucleosides as the 2' isomer; this identification was confirmed by Honjo et al. (1964). Biswas and Myers (1960) identified the 2'-O-methylcytidine in the nucleic acids of the blue-green alga, *Anacystis nidulans.*

Alkali hydrolyzates of wheat germ rRNA and tRNA have yielded alkali-stable dinucleotides composed of all combinations of the four 2'-O-methyl-ribonucleosides and the four major ribonucleosides (Hudson et al., 1965; Lane, 1965; and Singh and Lane, 1964a,b).

The 2'-O-methylribose content of the nucleic acids of plant and mammalian tissues appears to be much higher than that of *E. coli.* This difference is particularly striking in rRNA. The 16S + 28S RNA of L cells and HeLa cells contains about 1 mole % of 2'-O-methylribonucleosides, and these methyl groups account for 80–90% of the total methyl groups in the rRNA (Tamaoki and Lane, 1968; Wagner et al., 1967; Vaughan et al., 1967). *Escherichia coli* rRNA, on the other hand, contains about the same proportion of methyl groups, but only about 10% of them are attached to the $O^{2'}$ position of the nucleoside residues.

The presence of a methyl group at the 2' position of the pyrimidine nucleosides blocks the action of pancreatic ribonucleases. Honjo et al. (1964) found that the 5'-phosphate derivatives of these four 2'-O-methylribonucleosides are not dephosphorylated by the action of bull semen or snake venom 5'-nucleotidases. Gray and Lane (1967) determined that the 2'-O-methyl substituent adjacent to the 3'-phosphoester internucleotide linkage inhibits the action of snake venom diesterase.

RNA source	Amount, moles/100 moles	References
Calf liver (t)	0.19	Hall (1963b, 1964b)
Human liver (r)	0.08	Hall (1964b)
Human liver (t)	0.19	Hall (1964b)
Human tumors	0.05	Hall (1964b)
Mouse Ehrlich ascites	0.17	Hall (1964b)
Rat, Murphy-Sturm lymphosarcoma	0.30	Hall (1964b)
Rat liver (r)	pres.	Smith and Dunn (1959a)
Rat liver (t)	pres.	Smith and Dunn (1959a)
Sheep heart	0.26	Hall (1964b)
Sheep liver	0.21	Hall (1963b, 1964b)

RNA source	Amount, moles/100 moles	References
Chinese cabbage leaves (r)	~0.4	Dunn et al. (1963)
Tobacco leaves	pres.	Smith and Dunn (1959a)
Tobacco leaves (r)	~0.4	Dunn and Flack (1967b)
Wheat germ (r)	0.54	Lane (1965)
Wheat germ (t)	pres.	Smith and Dunn (1959a); Hudson et al. (1965)
Yeast (r)	pres.	Dunn (1961b)
Yeast (t)	0.03	Hall (1963d, 1964b)
Escherichia coli (t)	0.01	Hall (1963b)
Neurospora (r)	pres.	Dunn (1961b)
Neurospora (t)	pres.	Dunn (1961b)

41. N^6-(Δ^2-ISOPENTENYL)ADENOSINE   6-ipAdo   i⁶A

$C_{15}H_{21}N_5O_4$

M.W. 335.36

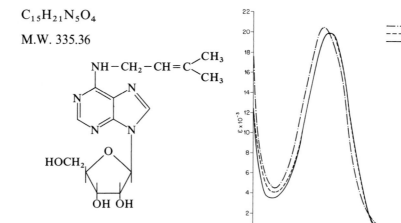

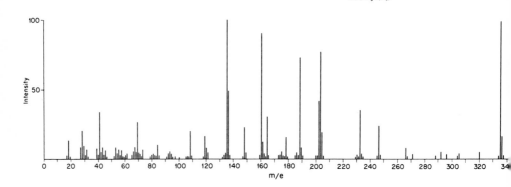

Physical Properties

M.P. 145–147°	Hall et al. (1966)
$[\alpha]_{546}^{25}$ −97° (0.07, ethanol)	Hall et al. (1966)
pK (basic) 3.76	Martin and Reese (1968)

Additional References for Synthesis

Grimm and Leonard (1967); Hall et al. (1966); Hall and Robins (1968b); Leonard et al. (1966)

Synthesis

1. Robins et al. (1967)

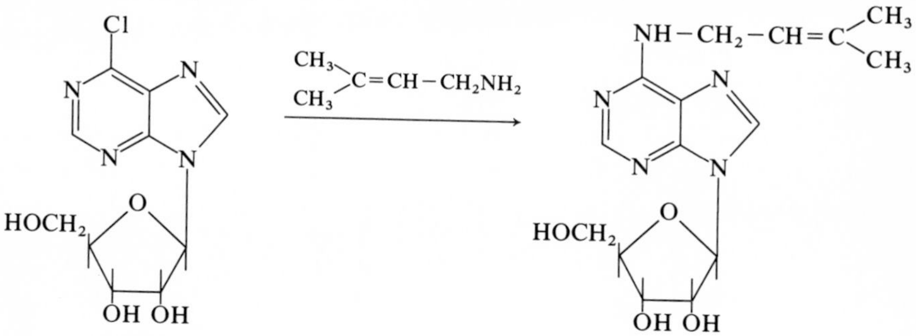

(60% yield)

Compound I can be prepared according to Gerster et al. (1963) or Wempen and Fox (1967a).

2. Grimm et al. (1968)

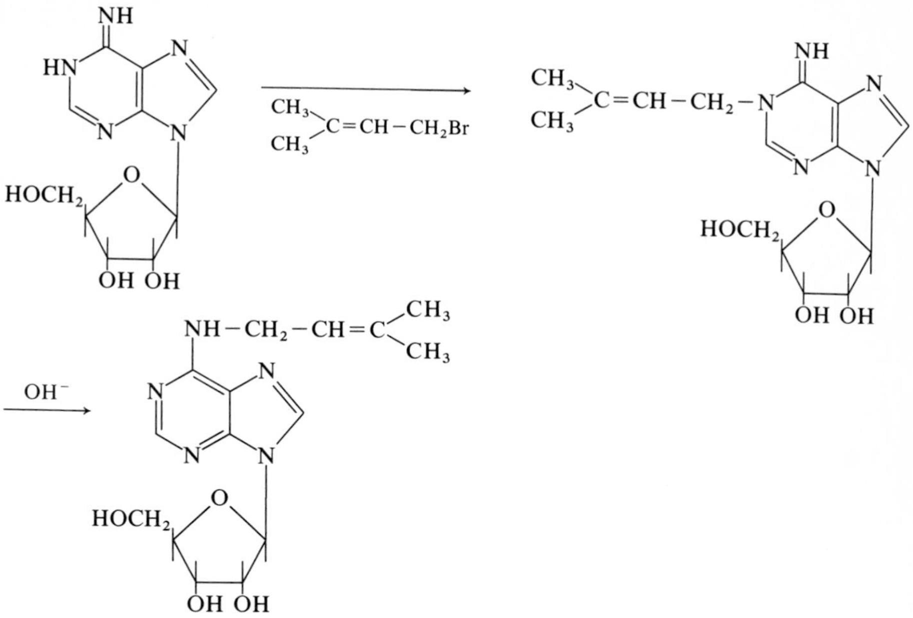

(38% yield overall)

γ,γ-Dimethylallyl bromide can be synthesized according to Hall and Fleysher (1968).

N^6-(Δ^2-Isopentenyl)adenosine was isolated from yeast tRNA and identified by Biemann et al. (1966) and Hall et al. (1966). It occurs in the primary sequence adjacent to the anticodon of yeast and rat liver tRNA^{Ser} and yeast tRNA^{Tyr} (see Chapter 4).

The following data were obtained in our laboratory by means of hydrolysis of tRNA by whole snake venom and bacterial alkaline phosphatase. More recently we found that this procedure does not give complete release of the N^6-(Δ^2-isopentenyl)adenosine (Fittler et al., 1968a). Consequently, the values reported may be low.

RNA source	Amount, moles/100 moles	References
Calf liver (t)	0.05	Robins et al. (1967)
Chick embryo (r)	not detected	Robins et al. (1967)
Chick embryo (t)	0.03	Robins et al. (1967)
Human liver (r)	not detected	Robins et al. (1967)
Human liver (t)	0.05	Robins et al. (1967)
Rat liver (t)	pres.	Staehelin et al. (1968)
Immature corn kernels (t)	not detected	Hall et al. (1967a)
Immature peas (t)	0.003	Hall et al. (1967a)
Spinach leaves (t)	0.02	Hall et al. (1967a)
Tobacco pith cells, grown in culture	pres.	Chen and Hall (1969)
Yeast (t)	0.065	Robins et al. (1967)
Escherichia coli B (t)	not detected	Fittler et al. (1968a)
Lactobacillus acidophilus (t)	pres.	Fittler et al. (1968a)
Lactobacillus plantarum (t)	pres.	Fittler et al. (1968a)
Corynebacterium fascians	pres.*	Matsubara et al. (1968)

* A positive identification of this nucleoside was not made.

42. N^6-(*cis*-4-HYDROXY-3-METHYLBUT-2-ENYL)ADENOSINE

$C_{15}H_{21}N_5O_5$

M.W. 351.36

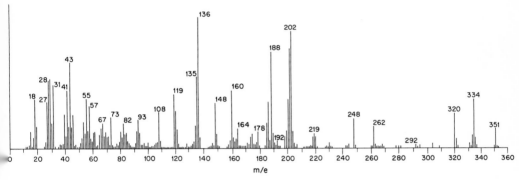

pH	λ_{max}	ε_{max} (x 10^{-3})	λ_{min}
1	265	20.4	—
7	268	20.0	—
12	268	20.0	—

Hall et al. (1967a)

Hall et al. (1967a) (Perkin-Elmer RMU-60)

N^6-(*cis-4-Hydroxy-3-methylbut-2-enyl*)adenosine was isolated and identified as the nucleoside by Hall et al. (1967b), starting from plant tissue tRNA. This compound has not been detected in the tRNA of yeast or mammalian tissue. The *trans* isomer of the compound occurs in plant tissue in a relatively unbound form (perhaps as the nucleotide or nucleoside) (Letham, 1966a,b; Miller, 1965). The chemistry and the biological significance of this tRNA constituent are discussed in Chapter 7.

The values reported here may be low for the same reason that the values for N^6-(Δ^2-isopentenyl)adenosine are low, although we have no specific data on the rate of release of this compound.

RNA source	Amount, moles/100 moles	References
Calf liver (t)	not detected	Hall, unpublished data
Chick embryo (t)	not detected	Hall, unpublished data
Immature corn kernels	0.01	Hall et al. (1967a)
Immature peas (t)	0.05	Hall et al. (1967a)
Mature corn kernels (seed corn)	pres.	Hall, unpublished data
Spinach leaves	0.01	Hall et al. (1967a)
Yeast (t)	not detected	Hall, unpublished data
Escherichia coli B (t)	not detected	Hall, unpublished data
Lactobacillus acidophilus (t)	not detected	Hall, unpublished data

43. *N*⁶-(Δ²-ISOPENTENYL)-2-METHYLTHIOADENOSINE   m²SiA

$C_{16}H_{23}N_5O_4S$

M.W. 381.147

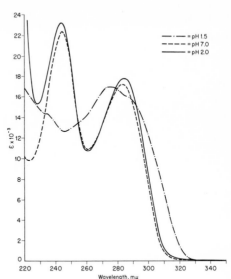

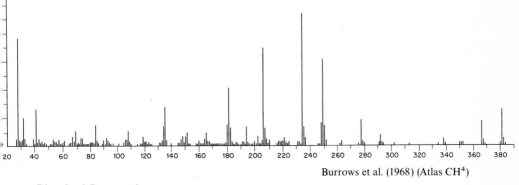

Burrows et al. (1968) (Atlas CH⁴)

Physical Properties

M.P.194–195°   Burrows et al. (1968).

Synthesis

1. Burrows et al. (1968)

where Bz = benzoyl.

N⁶-(Δ²-Isopentenyl)-2-methylthioadenosine has been isolated from *E. coli* tRNA (Burrows et al., 1968). It has been identified in a specific sequence of *E. coli* Su$_{III}^{+}$ tRNA^{Tyr} adjacent to the presumed anticodon (Goodman et al., 1968; Harada et al., 1968).

44. *N*-[9-(β-D-RIBOFURANOSYL)PURIN-6-YLCARBAMOYL]-L-THREONINE   (ThrCo)⁶A

$C_{15}H_{20}N_6O_8$

M.W. 412.36

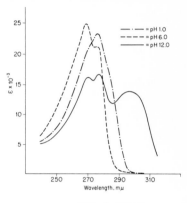

Chheda (1969)

A high-resolution spectrum was obtained on the corresponding free base (Schweizer et al., 1969). The molecular ion did not appear; prominent fragments occurred at m/e of 244.0698, 162.0408, 161.0330, 135.0547, and 100.0399.

M.P. 204–206°   Chheda (1969)

N-[9-(β-D-*Ribofuranosyl*)*purin-6-ylcarbamoyl*]-L-*threonine* [(*N*-(nebularin-6-ylcarbamoyl)-L-threonine] was first isolated from yeast tRNA and characterized as an N^6-(aminoacyl)adenosine by Hall (1964a). Subsequently, Chheda et al. (1969a) and Schweizer et al. (1969) described the isolation and characterization of a major, and perhaps the only, member of this type of derivative (see Chapter 7). This compound has been identified in the primary sequence of yeast tRNA^{Ile} (Takemura et al., 1968).

RNA source	Amount, moles/100 moles	References
Calf liver (t)	0.19	Chheda et al (1969a)
	0.25	D. Magrath (personal comm.)
Rat liver (t)	pres.	Chheda et al. (1969a)
Yeast (r)	not detected	Chheda et al. (1969a)
Yeast (t)	0.28	Chheda et al. (1969a)
	0.34	D. Magrath (personal comm.)
Escherichia coli (r)	not detected	Chheda et al. (1969a)
Escherichia coli (t)	0.07	Chheda et al. (1969a)

Synthesis

Chheda (1969)

where Ac = acetyl.

Compound I can be prepared according to Giner-Sorolla and Bendich (1958).

45. 2'(3')-O-RIBOSYLADENOSINE   ORibAdo

$C_{15}H_{21}N_5O_8$

M.W. 399.36

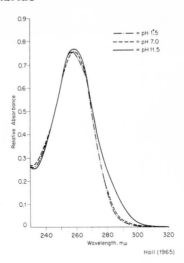

Hall (1965)

Reference for Synthesis

Lis and Passarge (1969)

2'(3')-O-Ribosyladenosine was isolated from samples of yeast tRNA and characterized by Hall (1965). More recently, Chambon et al. (1966), Sugimura et al. (1967), and Hasegawa et al. (1967) described the presence of a polymer in animal cells that appears to consist of repeating units of the phosphate of this nucleoside. The polymer is insoluble in acid; it is reasonable to assume that, if it were present in yeast, it would be co-isolated together with the tRNA. Therefore the source of 2'(3')-O-ribosyladenosine may not have been the tRNA but rather this polymer.

46. CYTIDINE   Cyd   C

$C_9H_{13}N_3O_5$

M.W. 243.22

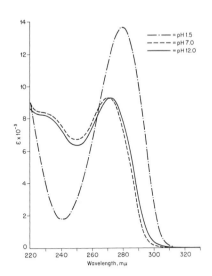

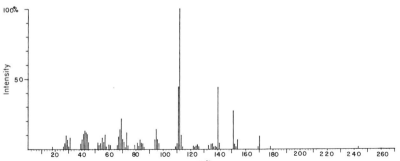

The trimethylsylyl derivative gives a molecular ion. McCloskey et al. (1968)

Physical Properties

M.P. (sulfate) 224–225° (dec.)	Elmore (1950)
$[\alpha]_D^{16} + 34.2°$ (2.0, water)	Elmore (1950)
pK (basic) 4.1	Fox and Shugar (1952)
(acidic) 12.3	Levene et al. (1926a)

Additional References for Synthesis

Cytidine-[2-^{14}C]-ribosyl-[^{3}H] was synthesized by Codington et al. (1958). A mixture of α- and β-phenyl esters of cytidine 5'-phosphate was synthesized by Asai et al. (1967), using a direct condensation method.

Synthesis

1. Fox et al. (1957)

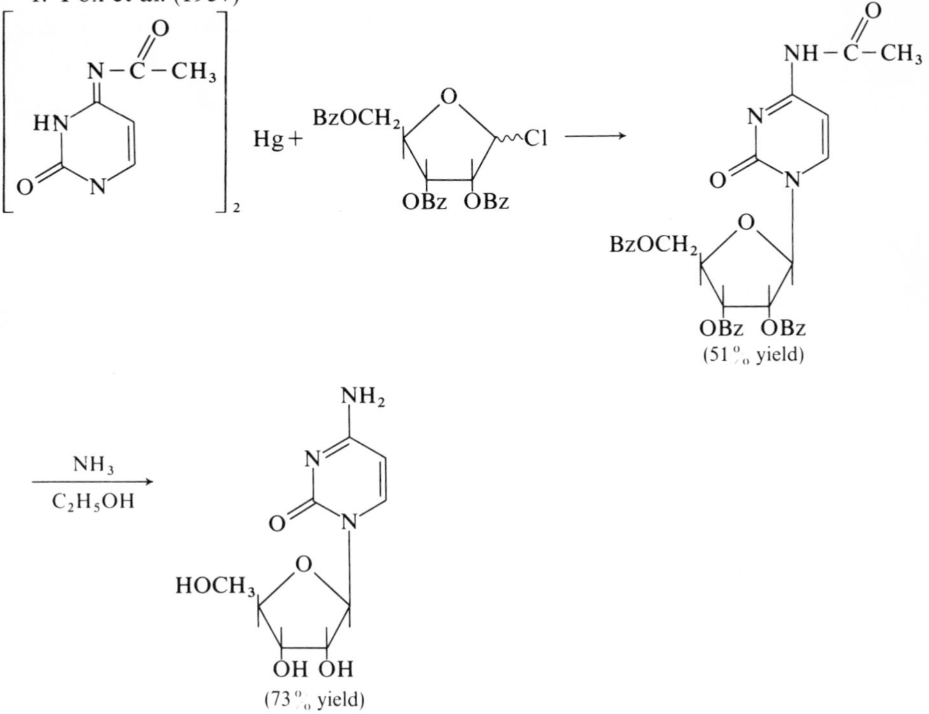

(51 % yield)

(73 % yield)

where Bz = benzoyl.

2. Nishimura and Iwai (1964); Nishimura et al. (1964)

where Bz = benzoyl.

3. Howard et al. (1947)

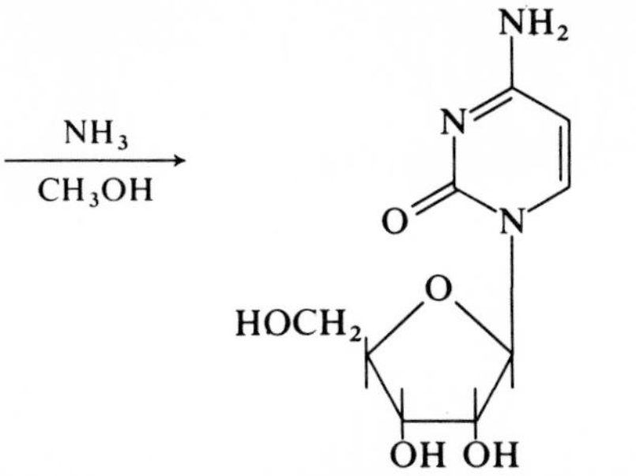

(10% yield, based on I)

where Ac = acetyl.

Cytidine can be prepared in better yield by using 2,4-dimethoxypyrimidine as the starting material (Prystaš and Šorm, 1968).

4. Ueda and Nishino (1968)

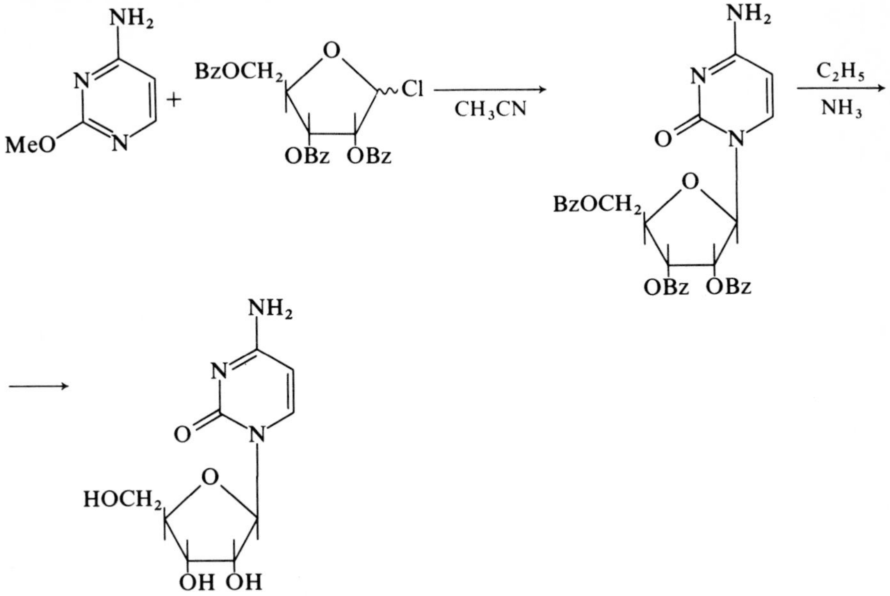

where Bz = benzoyl.

5. Fox et al. (1959)

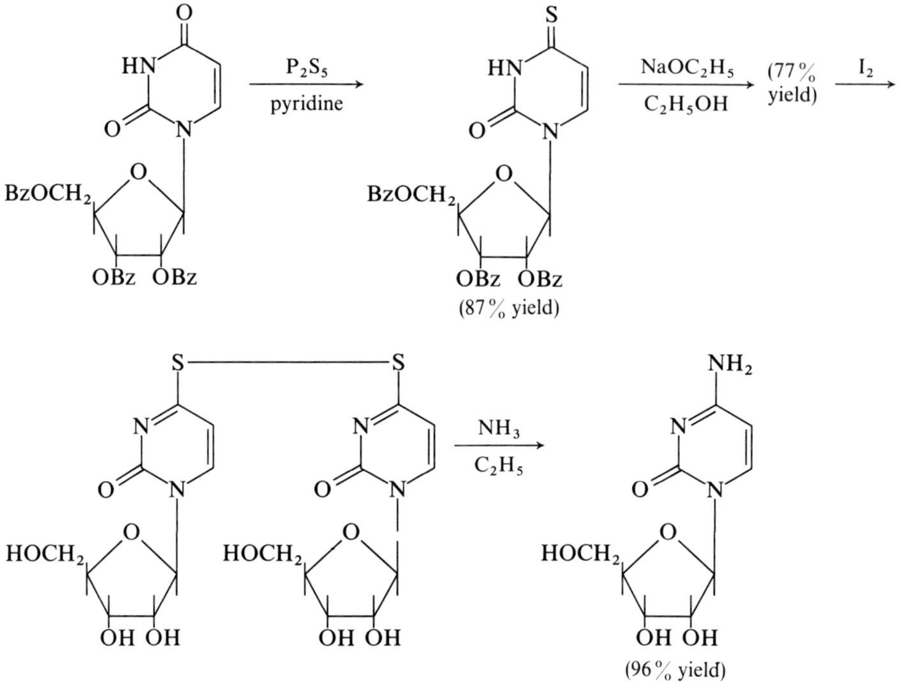

where Bz = benzoyl.

6. Shimizu et al. (1967)

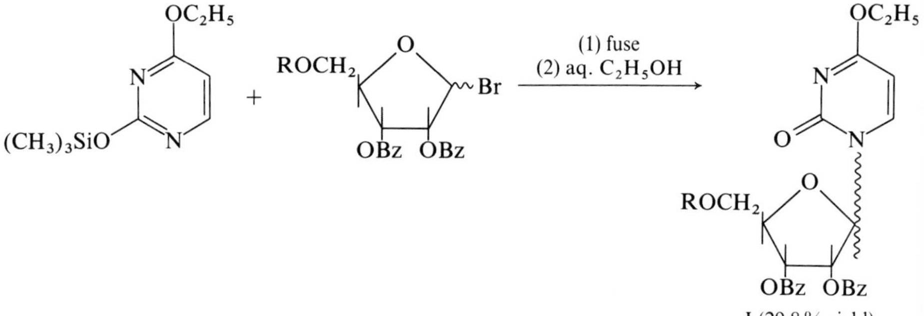

I (29.8% yield)

$$\xrightarrow[\text{C}_2\text{H}_5\text{OH}]{\text{NH}_3}$$

$$\xrightarrow{\text{H}_2/\text{Pt}}$$

(27.4% yield, based on I)

where R = $(\text{C}_6\text{H}_5\text{O})_2\overset{\overset{\displaystyle O}{\|}}{\text{P}}-$.

47. 3-METHYLCYTIDINE   3-MeCyd   m³C

$C_{10}H_{15}N_3O_5$

M.W. 257.24

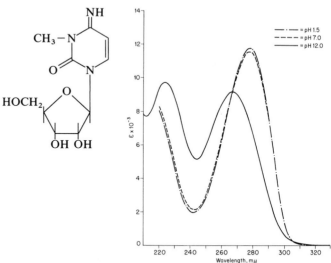

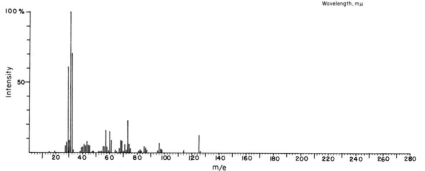

Physical Properties

M.P. 193–194° (methosulfate)   Brookes and Lawley (1962)
pK (basic) 8.7   Ueda and Fox (1963)
   (acidic) > 12   Ueda and Fox (1963)

Additional Reference for Synthesis

Bredereck et al. (1948)

Synthesis

(1) Brookes and Lawley (1962); (2) Haines et al. (1964)

[(1) 86% yield; (2) 40% yield]

3-Methylcytidine was first detected as the nucleoside in yeast tRNA (Hall, 1963a). It has been identified in a specific sequence of rat liver tRNA^{Ser} (Staehelin et al., 1968).

RNA source	Amount, moles/100 moles	References
Rat liver (t)	pres.	Staehelin et al. (1968)
Yeast (t)	0.1	Hall (1963a); Dunn and Flack (1967a)

48. $N^4,O^{2'}$-DIMETHYLCYTIDINE $O,4$-Me$_2$Cyd   m^{4}Cm

$C_{11}H_{17}N_3O_5$

M.W. 271.27

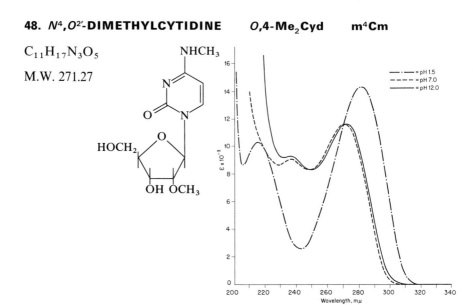

Spectra obtained on N^4-methylcytidine

Physical Properties

pK (basic) 3.9   Nichols and Lane (1966a)

Synthesis

1. The synthesis of N^4-methylcytidine has been reported by Fox et al. (1959).

(93% yield; M.P. 202–203°)

2. Szer and Shugar (1966)

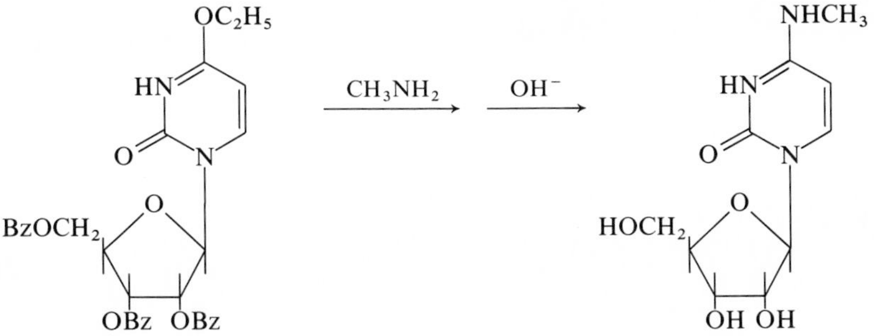

[86% yield; M.P. 237° (dec.)]

where Bz = benzoyl.

3. Ziff and Fresco (1968)

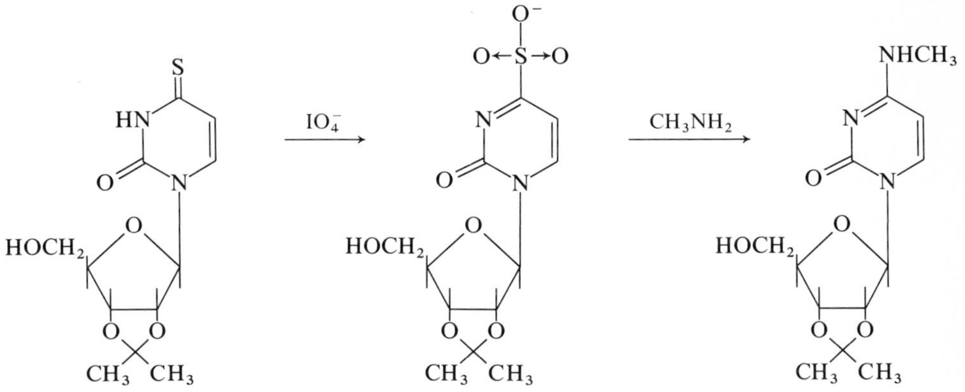

$N^4,O^{2'}$ Dimethylcytidine was isolated from an alkaline hydrolysate of *E. coli* ribosomal RNA by Nichols and Lane (1966a,b; 1968b). It represents the only identified methylated nucleoside containing a methyl group attached to both the sugar and base positions. It occurs in the 16S RNA of *E. coli* at a level of about 0.05 mole %, which corresponds to about 1 per 2000 nucleotides (Nichols and Lane, 1967). The chain length of the 16S component is about 1500 (Fikus et al., 1962). In all probability there is one such residue per 16S molecule (Fellner and Sanger, 1968; Nichols and Lane, 1968a,b).

49. 5-METHYLCYTIDINE   5-MeCyd   m⁵C

$C_{10}H_{15}N_3O_5$

M.W. 257.24

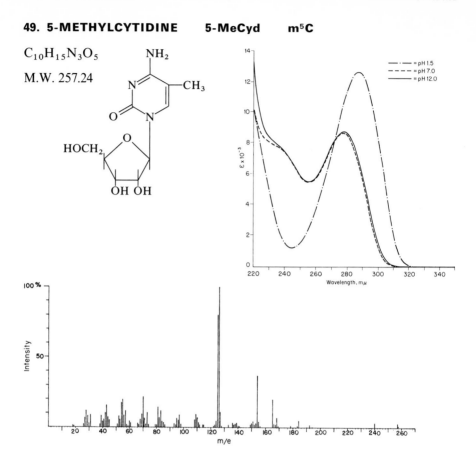

Physical Properties

M.P. 210–211° (dec.)	Fox et al. (1959)
$[\alpha]_D^{23} -3°$ (2.5, 1 N NaOH)	Fox et al. (1959)
pK (basic) 4.3	Fox et al. (1959)
(acidic) > 13	Fox et al. (1959)

Synthesis

1. Fox et al. (1959)

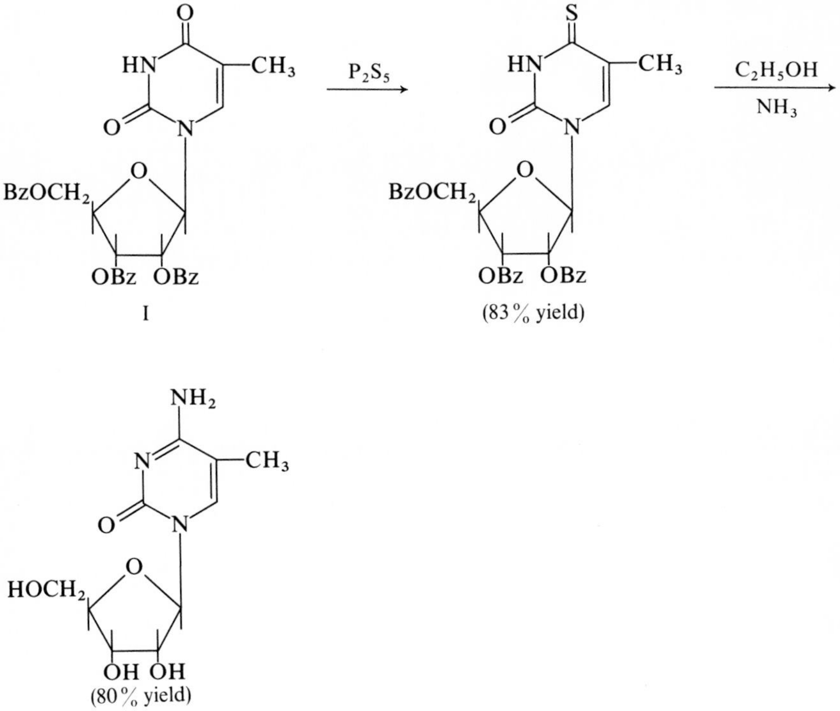
P_2S_5
C_2H_5OH
NH_3
I
(83% yield)
(80% yield)

where Bz = benzoyl.

Compound I can be prepared according to Wempen and Fox (1967a).

5-Methylcytidine was originally isolated from the RNA of *E. coli* by Amos and Korn (1958) and was positively identified by Dunn (1960) in the RNA of several species. It has been identified in the primary sequences of several molecular species of yeast tRNA (Chapter 4).

RNA source	Amount, moles/100 moles		References
Pig liver (t)	1.2	(c)	Dunn (1960)
Rat liver (r)	0.02	(c)	Dunn et al. (1960b)
Rat liver (t)	2.0	(c)	Dunn (1960)
Chinese cabbage leaf (r)	0.03	(c)	Dunn et al. (1963)
Chinese cabbage leaf (t)	0.8	(c)	Dunn et al. (1963); Dunn and Flack (1967b)
Tobacco leaf (t)	0.75	(c)	Dunn and Flack (1967b)
Wheat germ (t)	1.37		Hudson et al. (1965)
Yeast (r)	0.02	(c)	Dunn (1961b)
Yeast (t)	1.2	(c)	Dunn and Flack (1967a)
Aerobacter aerogenes	0.03	(c)	Dunn (1960)
Escherichia coli	1.5		Amos and Korn (1958)
Escherichia coli (16S)	0.13		Fellner and Sanger (1968)
Escherichia coli (23S)	0.07		Fellner and Sanger (1968)
Neurospora (r)	0.02	(c)	Dunn (1961b)
Neurospora (t)	0.8	(c)	Dunn (1961b), also unpublished data

50. 2'-O-METHYLCYTIDINE   *O*MeCyd   **Cm**

$C_{10}H_{15}N_3O_5$

M.W. 257.24

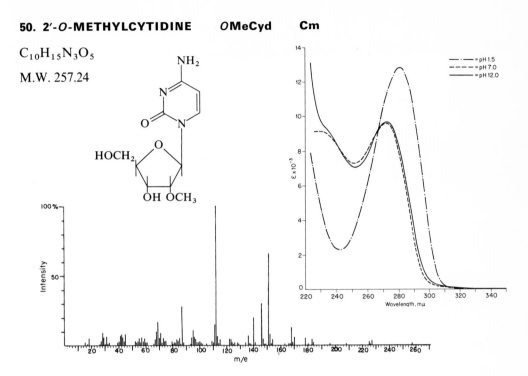

Physical Properties

M.P. 252–253° (dec.)	Martin et al. (1968)
$[\alpha]_D^{21}$ +54° (1.1, water)	Furukawa et al. (1965)
pK (basic) 4.2	Furukawa et al. (1965)

Synthesis

1. Furukawa et al. (1965)

where Tr = trityl.

2. Martin et al. (1968)

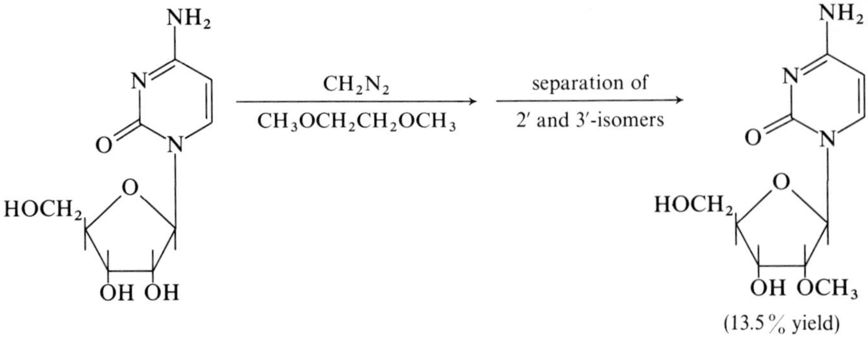

NH2
N
O
N
HOCH2
O
OH OH
CH2N2
CH3OCH2CH2OCH3
separation of
2′ and 3′-isomers
NH2
N
O
N
HOCH2
O
OH OCH3
(13.5% yield)

2′-O-Methylcytidine has been identified in the primary sequence of yeast $tRNA^{Phe}$(Table 1, Chapter 4). See 2′-*O*-methyladenosine for general remarks.

RNA source	Amount, moles/100 moles	References
Calf liver (t)	0.11	Hall (1963b, 1964b)
Human liver (r)	0.09	Hall (1964b)
Human liver (t)	0.34	Hall (1964b)
Human tumor	0.06	Hall (1964b)
Mouse Ehrlich ascites	0.14	Hall (1964b)
Rat liver	pres.	Morisawa and Chargaff (1963)
Rat, Murphy-Sturm lymphosarcoma	0.20	Hall (1964b)
Rat liver (r)	pres.	Dunn (1959)
Rat liver (t)	pres.	Dunn (1959)
Sheep liver	0.26	Hall (1963b)
Chinese cabbage leaf (r)	pres.	Dunn et al. (1963)
Wheat germ (r)	0.34	Lane (1965)
Wheat germ (t)	0.4	Hudson et al. (1965)
Yeast (r)	pres.	Dunn (1961b)
Yeast (t)	0.28	Hall (1963d, 1964b); Morisawa and Chargaff (1963)
Anacystis nidulans	0.05	Biswas and Myers (1960)
Escherichia coli (r)	0.03	Nichols and Lane (1967)
Escherichia coli (t)	0.06	Hall (1963b, 1964b)
Neurospora (r)	pres.	Dunn (1961b)

51. *N*⁴-ACETYLCYTIDINE   4-AcCyd   ac⁴C

$C_{11}H_{15}N_3O_6$

M.W. 285.25

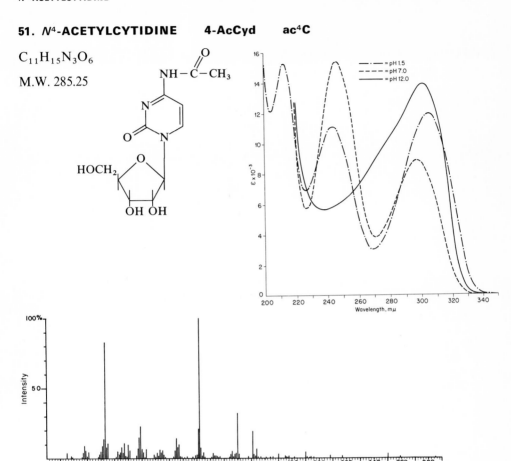

Physical Properties

M.P. 208–209°   Van Montagu and Stockx (1965)

$[\alpha]_D^{23}$ +60.1° (1.0, water)   Van Montagu and Stockx (1965)

pK (basic) < 1.5   Van Montagu and Stockx (1965)

Additional References for Synthesis

Mizuno et al. (1965); Watanabe and Fox (1966)

Synthesis

(1) Van Montagu and Stockx (1965); (2) Saski and Mizuno (1967)

[(2) 67% yield]

N⁴-Acetylcytidine was identified in a specific sequence of yeast and rat liver tRNA^Ser (Zachau et al., 1966b; Staehelin et al., 1968).

52. 2-THIOCYTIDINE

$C_9H_{13}N_3O_4S$

M.W. 259.28

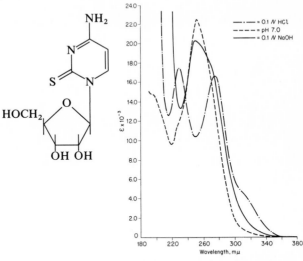

J. Carbon, private communication

Physical Properties

M.P. 208–209°   Ueda et al. (1966)

$[\alpha]_D^{25}$ +64.2 (1.8, H_2O)   Lee and Wigler (1968)

Synthesis

1. Ueda et al. (1966)

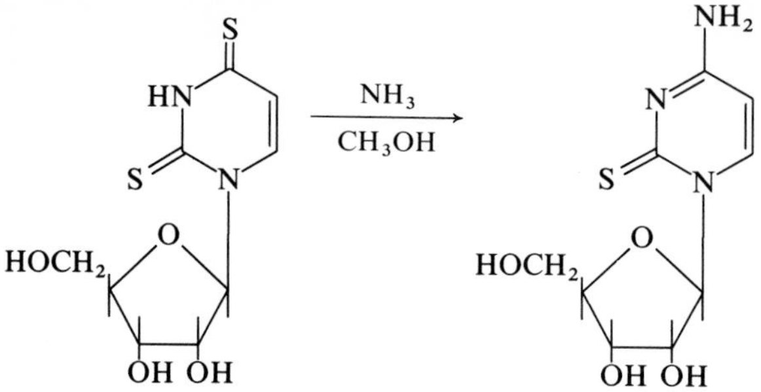

$\xrightarrow{\text{NH}_3}$ CH_3OH

(15% yield, starting from I)

where Bz = benzoyl.

Compound I can be prepared according to Fox et al. (1959).

2. Ueda and Nishino (1968)

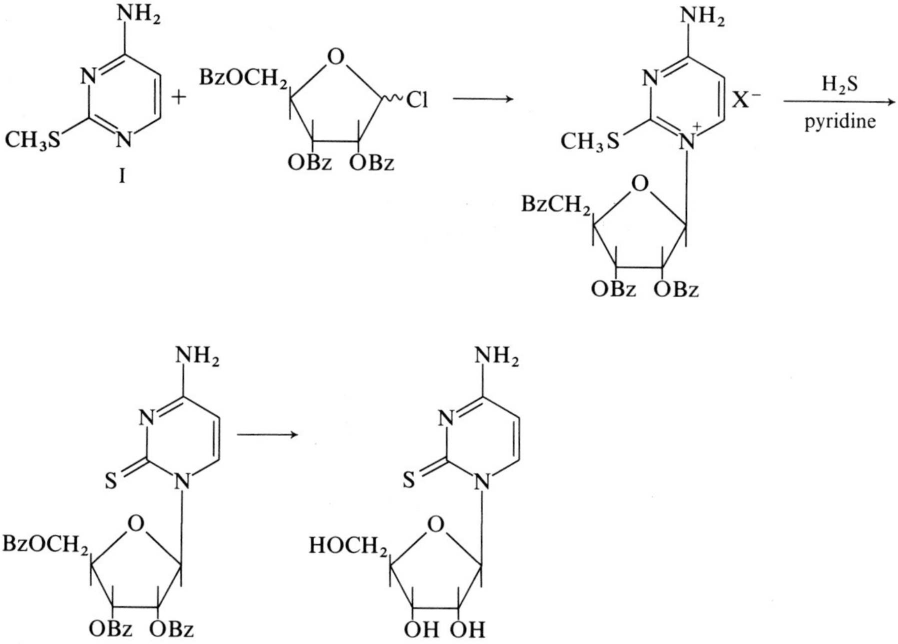

I

$\xrightarrow{\text{H}_2\text{S}}$ pyridine

(40% yield from I)

where Bz = benzoyl.

3. Lee and Wigler (1968)

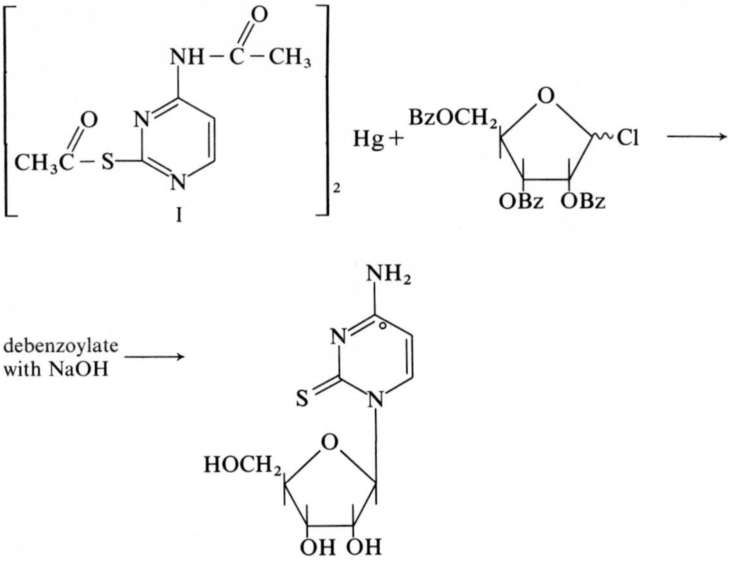

debenzoylate
with NaOH $\longrightarrow$

(yield 31% based on I)

where Bz = benzoyl.

2-Thiocytidine has been detected in an alkaline hydrolysate of *E. coli* tRNA (Carbon et al., 1968).

53. GUANOSINE   Guo   G

$C_{10}H_{13}N_5O_5$

M.W. 283.24

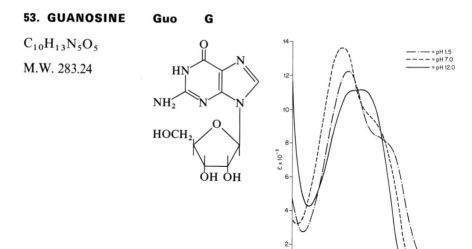

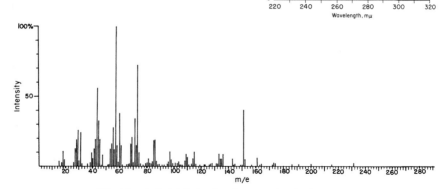

The trimethylsylyl derivative gives a molecular ion. McCloskey et al. (1968)

Physical Properties

M.P. > 235° (dec.)	Davoll and Lowy (1951)
$[\alpha]_D^{26} -72°$ (1.4, 0.1 N NaOH)	Davoll and Lowy (1951)
pK (basic) 1.6	Levene and Bass (1931)
(acidic) 9.2, 12.3	Levene and Bass (1931)

Additional References for Synthesis

Davoll and Lowy (1951); Levene and Bass (1931); Yamazaki et al. (1967c)

Synthesis

1. Yamazaki et al. (1967a)

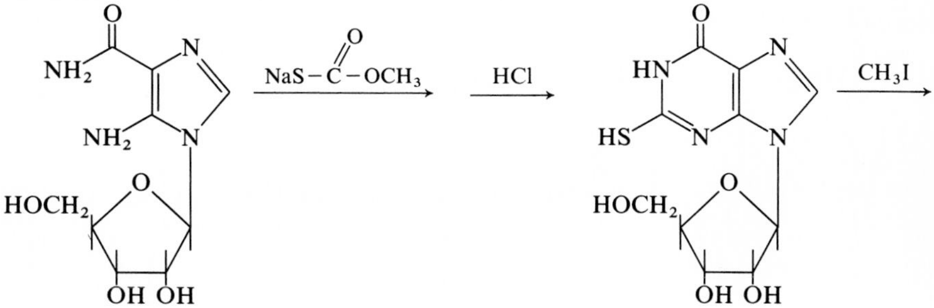

I

II (64% yield)

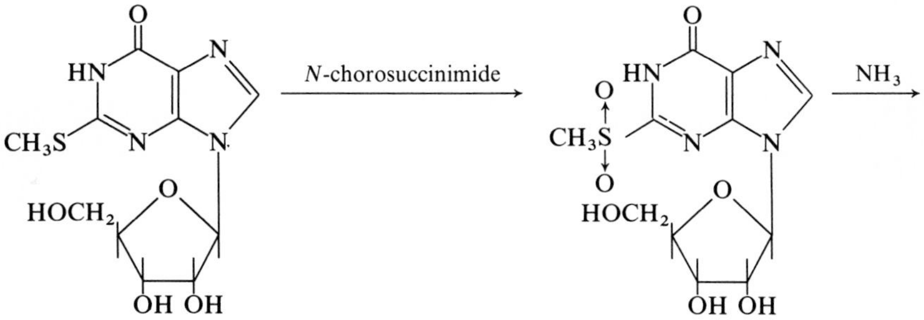

III (73% yield)

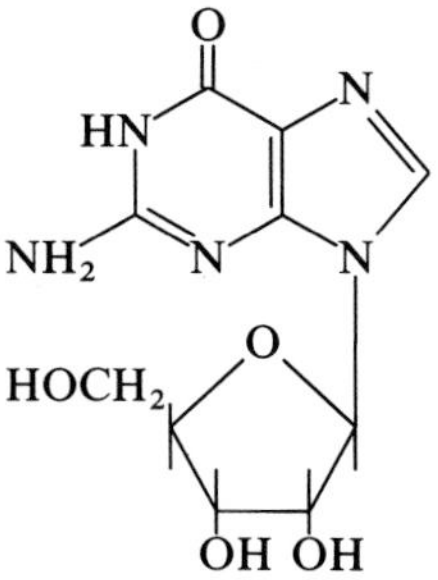

(56% yield from III)

Compound I can be readily obtained by a fermentation process (Shiro et al., 1962).

2. Davoll et al. (1948a)

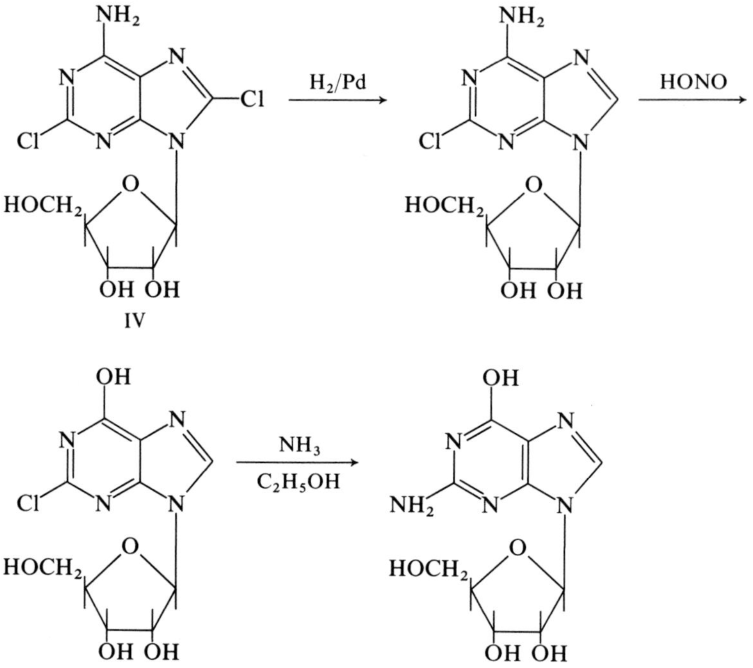

IV

(7% yield, based on IV)

3. Furukawa and Honjo (1968)

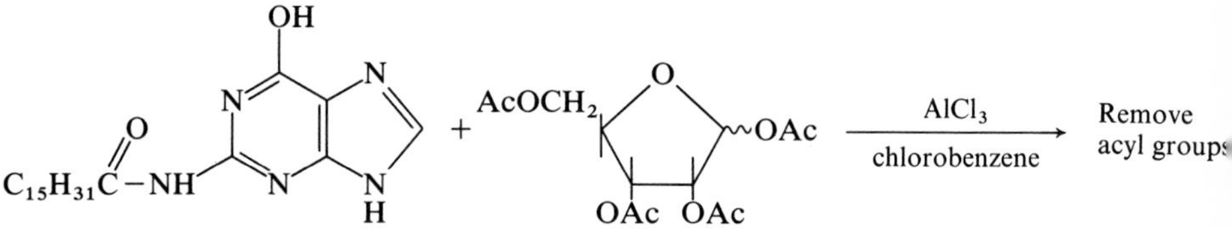

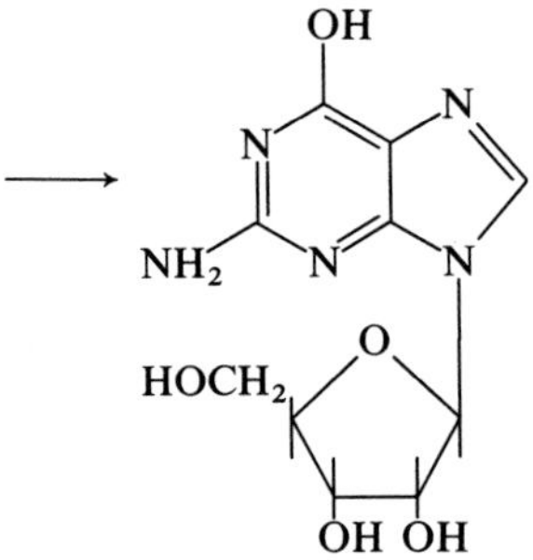

(59% yield)

where Ac = acetyl.

54. 1-METHYLGUANOSINE   1-MeGuo   m¹G

$C_{11}H_{15}N_5O_5$

M.W. 297.27

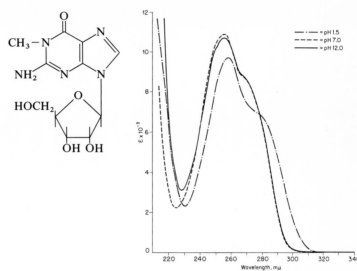

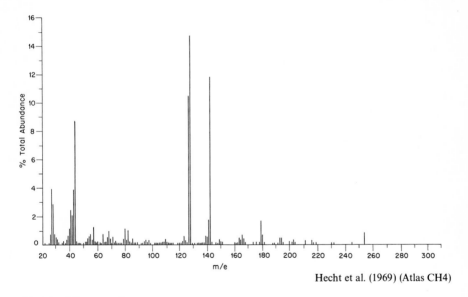

Hecht et al. (1969) (Atlas CH4)

Physical Properties

M.P. 225–227° (dec.)   Broom et al. (1964)

Additional References for Synthesis

Smith and Dunn (1959b) (enzymic synthesis); Broude et al. (1967) (synthesis via diazomethane)

Synthesis

1. Broom et al. (1964)

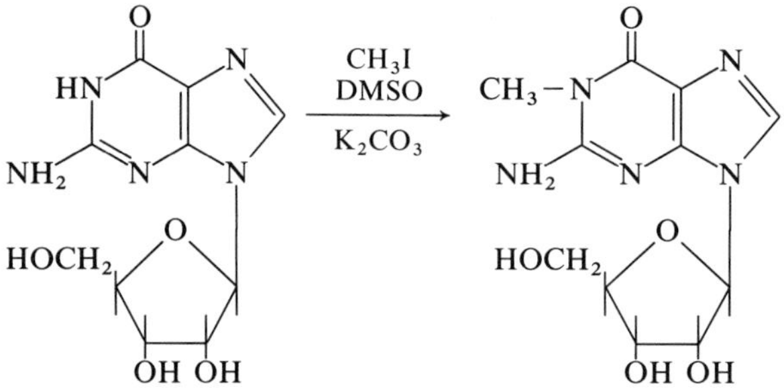

(53% yield)

2. Bredereck et al. (1948)

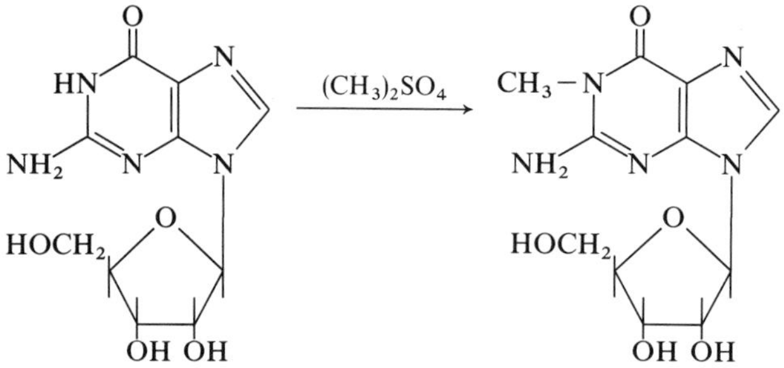

(50% yield)

1-Methylguanosine was first isolated as the free base from commercial yeast RNA by Adler et al. (1958), and as the nucleoside from the RNA of several organisms by Smith and Dunn (1959b). It has been identified in the primary sequences of yeast tRNA^{Ala} and tRNA^{Val} (Table 1, Chapter 4).

RNA source	Amount, moles/100 moles		References
Carp liver (t)	0.7		Zaitseva et al. (1962)
Frog liver (t)	0.7		Zaitseva et al. (1962)
Mouse adenosarcoma (t)	0.07–0.20	(c)	Bergquist and Matthews (1962)
Mouse liver (t)	0.45	(c)	Bergquist and Matthews (1962)
Mouse spleen (t)	0.13	(c)	Bergquist and Matthews (1962)
Pigeon liver (t)	0.7		Zaitseva et al. (1962)
Rabbit liver (r)	0.03	(c)	Bergquist and Matthews (1962)
Rabbit liver (t)	0.94	(c)	Bergquist and Matthews (1962)
Rat liver (r)	0.02	(c)	Dunn (1959); Smith and Dunn (1959b)
Rat liver (t)	0.8	(c)	Dunn (1959); Sluyser and Bosch (1962); Smith and Dunn (1959b)
Hamster liver (t)	11.0		Mittelman et al. (1969)
S-180 ascites (t)	0.6	(c)	Bergquist and Matthews (1962)
Chinese cabbage leaves (t)	0.50	(c)	Dunn et al. (1963); Dunn and Flack (1967b)
Sugar beet leaves	0.05	(c)	Smith and Dunn (1959b)
Tobacco leaves (t)	0.50	(c)	Dunn and Flack (1967b)
Wheat germ (t)	0.73		Hudson et al. (1965)
Yeast (t)	1.00	(c)	Dunn and Flack (1967a)
Aerobacter aerogenes	0.02	(c)	Smith and Dunn (1959b)
Azotobacter vinelandii (t)	0.24		Zaitseva et al. (1962)
Escherichia coli (t)	0.14	(c)	Dunn et al. (1960a)
Escherichia coli (23S)	0.07		Fellner and Sanger (1968)
Neurospora (t)	0.5	(c)	Dunn (1961b), also unpublished data
Proteus vulgaris (t)	0.4		Zaitseva et al. (1962)
Sarcina flava (t)	0.2		Zaitseva et al. (1962)

55. *N²*-METHYLGUANOSINE   2-MeGuo   m²G

$C_{11}H_{15}N_5O_5$

M.W. 297.27

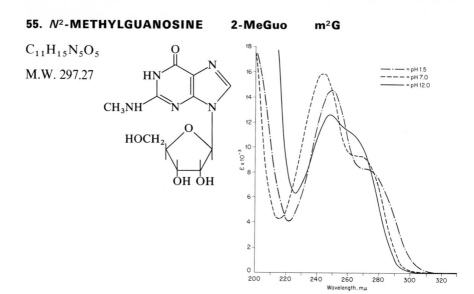

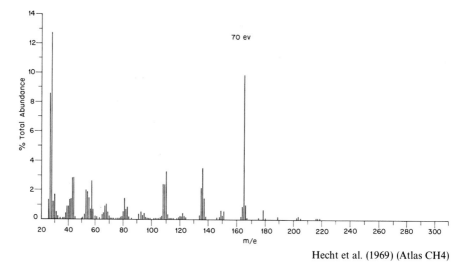

70 ev

Hecht et al. (1969) (Atlas CH4)

Physical Properties

M.P. none (dec. > 200°)	Gerster and Robins (1965)
$[\alpha]_D^{26} -34.6°$ (1.0 DMS/E*)	Gerster and Robins (1965)
pK (basic) 2.3†	Smith and Dunn (1959b)
(acidic) 9.7†	Smith and Dunn (1959b)

* Dimethylsulfoxide-ethanol (50-50).
† Determined from electrophoretic mobility.

Additional Reference for Synthesis

Smith and Dunn (1959b) (enzymic synthesis)

Synthesis

1. Yamazaki et al. (1967a)

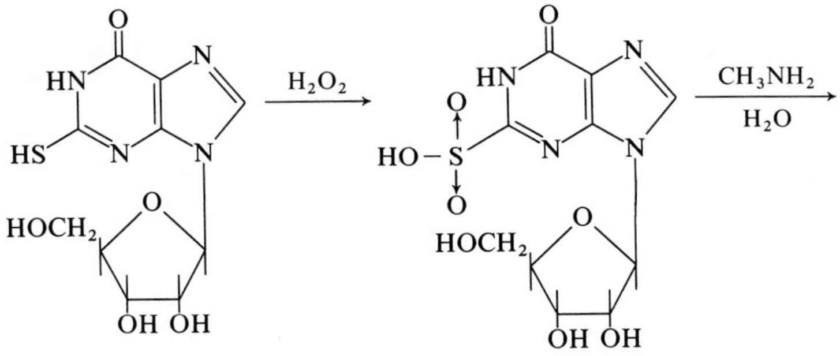

Intermediate II in
synthesis of guanosine

V

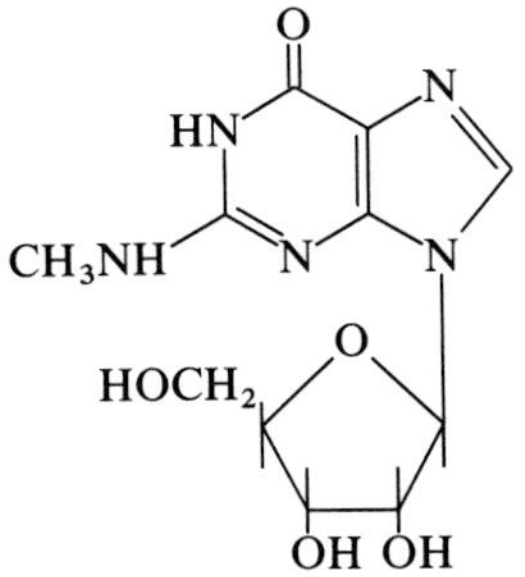

(62.4% yield from II)

2. Gerster and Robins (1965, 1966)

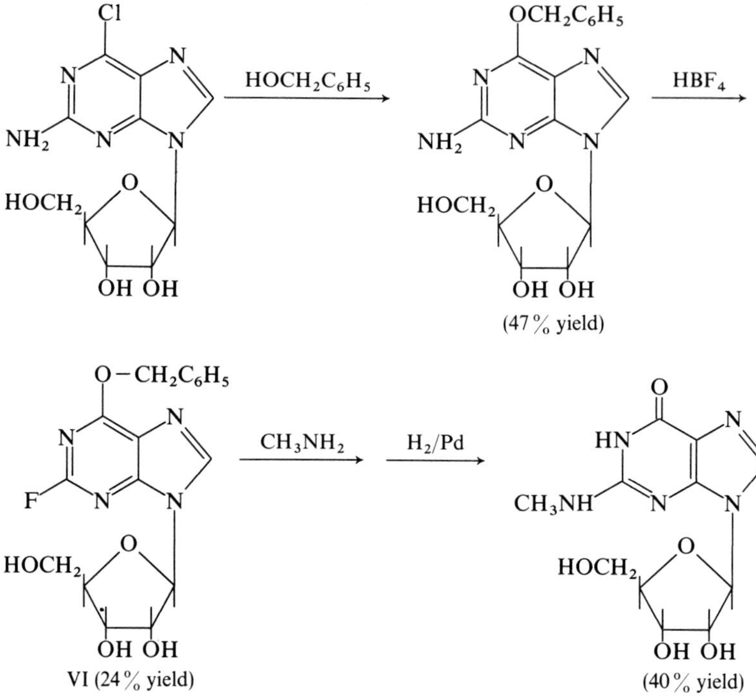

HOCH2C6H5
HBF4
(47% yield)
CH3NH2
H2/Pd
VI (24% yield)
(40% yield)

N^2-Methylguanosine was first isolated as the free base from commercial yeast by Adler et al. (1958), and as the nucleoside from the RNA of several species by Smith and Dunn (1959b). It has been identified in the primary sequence of yeast $tRNA^{Phe}$ and $tRNA^{Tyr}$ (Chapter 4).

RNA source	Amount, moles/100 moles		References
Mouse adenosarcoma (t)	2.5	(c)	Bergquist and Matthews (1962)
Mouse liver (t)	1.6	(c)	Bergquist and Matthews (1962)
Mouse spleen (t)	0.64	(c)	Bergquist and Matthews (1962)
Rabbit liver (r)	0.01	(c)	Bergquist and Matthews (1962)
Rabbit liver (t)	0.5	(c)	Bergquist and Matthews (1962)
Rat liver (r)	0.02	(c)	Dunn (1959); Smith and Dunn (1959b)
Rat liver (t)	0.4	(c)	Dunn (1959); Sluyser and Bosch (1962); Smith and Dunn (1959a)
S-180 ascites (t)	1.2	(c)	Bergquist and Matthews (1962)
Chinese cabbage leaves (t)	0.2	(c)	Dunn et al. (1963)
Euglena gracilis	0.09		Brawerman et al. (1962)
Tobacco leaves (t)	0.40	(c)	Bergquist and Matthews (1962)
Wheat germ (t)	0.29		Hudson et al. (1965)
Yeast (t)	0.35	(c)	Dunn and Flack (1967a); Dunn (1961b), also unpublished data
Aerobacter aerogenes	0.01	(c)	Smith and Dunn (1959b)
Escherichia coli (16S)	0.27		Fellner and Sanger (1968)
Escherichia coli (23S)	0.13		Fellner and Sanger (1968)
Neurospora (t)	0.10		Dunn (1961b), also unpublished data

56. *N²,N²*-DIMETHYLGUANOSINE   2-Me₂Guo   m₂²G

$C_{12}H_{17}N_5O_5$

M.W. 311.30

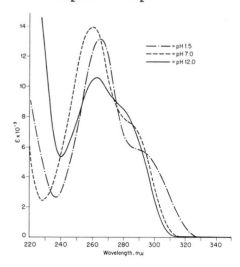

Physical Properties

M.P. 242° (dec.)	Gerster and Robins (1965)
$[\alpha]_D^{26}$ −35.6° (1.1, DMS/Et*)	Gerster and Robins (1965)
pK (basic) 2.5†	Smith and Dunn (1959b)
(acidic) 9.7†	Smith and Dunn (1959b)

* Dimethylsulfoxide-ethanol (50-50).
† Determined from electrophoretic mobility.

Additional References for Synthesis

Bredereck et al. (1948); Smith and Dunn (1959b) (enzymic synthesis)

Synthesis

1. Yamazaki et al. (1967a)

Intermediate V
in synthesis of N^2-methylguanosine

(52% yield from II in
synthesis of guanosine)

2. Gerster and Robins (1965, 1966)

Intermediate VI
in synthesis of N^2-methylguanosine

(48% yield)

N^2,N^2-*Dimethylguanosine* was first isolated from the RNA of several organisms by Smith and Dunn (1959b).

RNA source	Amount, moles/100 moles		References
Mouse adenosarcoma (t)	0.64	(c)	Bergquist and Matthews (1962)
Mouse liver (t)	0.28	(c)	Bergquist and Matthews (1962)
Mouse spleen (t)	0.17	(c)	Bergquist and Matthews (1962)
Rabbit liver (r)	0.01	(c)	Bergquist and Matthews (1962)
Rabbit liver (t)	0.6	(c)	Bergquist and Matthews (1962)
Rat liver (r)	0.02	(c)	Dunn (1959); Smith and Dunn (1959b)
Rat liver (t)	0.65	(c)	Dunn (1959); Sluyser and Bosch (1962); Smith and Dunn (1959b)
S-180 ascites (t)	0.65	(c)	Bergquist and Matthews (1962)
Chinese cabbage leaf (t)	0.55	(c)	Dunn and Flack (1967b)
Sugar beet leaves	0.025	(c)	Smith and Dunn (1959b)
Tobacco leaves (t)	0.55	(c)	Dunn and Flack (1967b)
Wheat germ (t)	0.55		Hudson et al. (1965)
Yeast (t)	0.70	(c)	Dunn and Flack (1967a); Dunn (1961b), also unpublished data
Neurospora (t)	0.65	(c)	Dunn (1961b), also unpublished data

57. 7-METHYLGUANOSINE   7-MeGuo   m⁷G

$C_{11}H_{15}N_5O_5$

M.W. 297.27

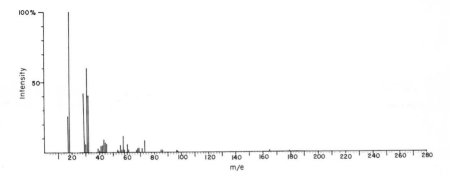

pH	λ_{max}	ε_{max} (x 10^{-3})	λ_{min}
3	257	10.7	230
7	258	8.5	238
	281	7.4	—
9	282	8.0	242

Physical Properties

M.P. 165° (hemihydrate)   Lawley and Brookes (1963)
$[\alpha]_D^{27}$ −33.5° (0.4, water)   Jones and Robins (1963)
pK (acidic) 7.1   Lawley and Brookes (1963)

Additional References for Synthesis

Haines et al. (1962); Lawley and Brookes (1963)

Synthesis

Jones and Robins (1963)

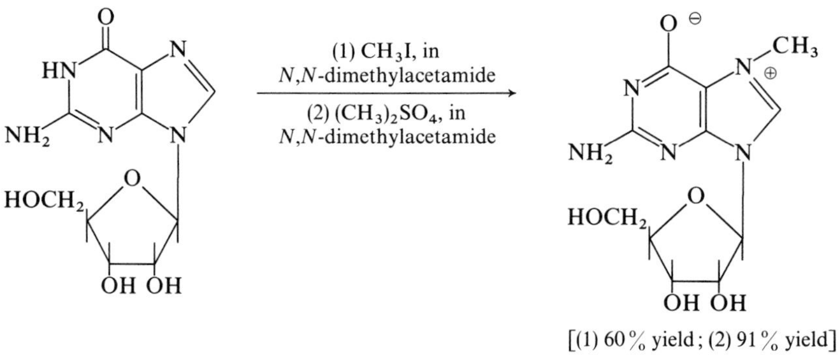

(1) CH$_3$I, in
N,N-dimethylacetamide

(2) (CH$_3$)$_2$SO$_4$, in
N,N-dimethylacetamide

[(1) 60% yield; (2) 91% yield]

7-Methylguanosine was isolated as the free base from acid hydrolyzates of the tRNA of pig liver and yeast by Dunn (1963). At neutral and alkaline pH values, this nucleoside undergoes ring scission to form 2-amino-4-hydroxy-5-(*N*-methyl)formamido-6-ribosylaminopyrimidine.

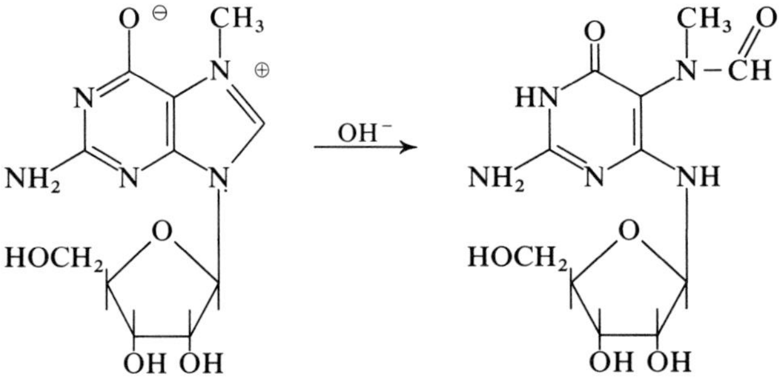

OH$^-$

Because of the ease of this reaction, isolation procedures involving enzymatic processes at slightly alkaline pH values result in conversion to the ring-opened form. Hall (1967) first reported the isolation of the ring-opened form from an enzymatic hydrolyzate of yeast tRNA. 7-Methylguanosine occurs in yeast tRNA[Phe] (Chapter 4).

Conversion to the ring-opened form by treatment with alkali is not quantititative and therefore cannot be used as a reliable quantitative assay for 7-methylguanosine. Other products appear to be formed; Dunn (private communication) finds that analysis for 7-methylguanosine based on the ring-opened base gives results 25% lower than that based on isolation of 7-methylguanylic acid.

RNA source	Amount, moles/100 moles		References
Pig liver (t)	0.20	(c)	Dunn (1963)
Rat liver (r)	pres.		Craddock et al. (1968)
Rat liver (t)	pres.		Craddock et al. (1968)
Chinese cabbage leaves (t)	0.30	(c)	Dunn et al. (1963); Dunn and Flack (1967b)
Tobacco leaf (t)	0.20	(c)	Dunn and Flack (1967b)
Yeast (t)	0.35	(c)	Dunn and Flack (1967b); Dunn (1961b), also unpublished data
Escherichia coli (t)	0.50	(c)	Dunn and Flack, unpublished data
Escherichia coli (16S)	0.13		Fellner and Sanger (1968)
Escherichia coli (23S)	0.07		Fellner and Sanger (1968)

58. 2′-*O*-METHYLGUANOSINE *O*MeGuo Gm

$C_{11}H_{15}N_5O_5$

M.W. 297.27

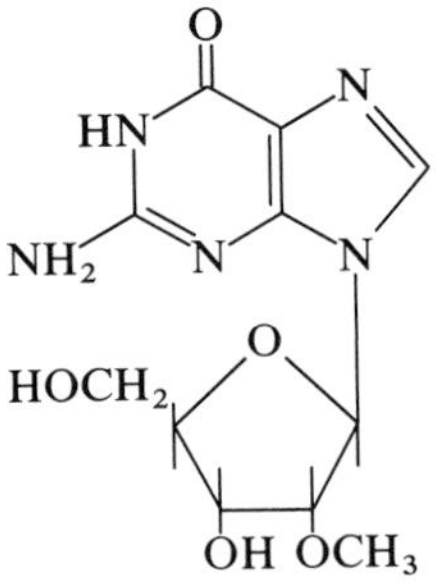

pH	λ_{max}	ε_{max} (x 10^{-3})	λ_{min}
1	256	10.7	---
11	258	9.8	---

Physical Properties

M.P. 218–220° Khwaja and Robins (1966)

$[\alpha]_D^{22}$ −38.4° (0.6, water) Khwaja and Robins (1966)

Synthesis

1. Khwaja and Robins (1966)

CH_2N_2
CH_3OH
NaHS
H_2O_2
aq. ammonia
(57% overall yield)
(50% yield)

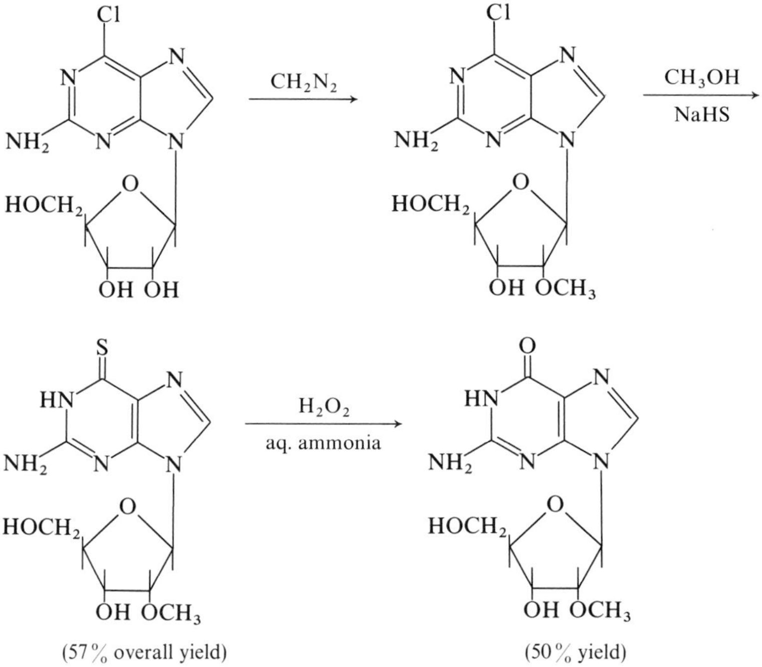

2'-O-Methylguanosine has been identified in the primary sequence of several molecular species of yeast tRNA (Chapter 4). See 2'-*O*-methyladenosine for general remarks.

RNA source	Amount, moles/100 moles	References
Calf liver (t)	0.16	Hall (1963b, 1964b)
Human liver (r)	0.15	Hall (1964b)
Human liver (t)	0.30	Hall (1964b)
Human tumor	0.12	Hall (1964b)
Mouse Ehrlich ascites	0.23	Hall (1964b)
Rat liver	pres.	Morisawa and Chargaff (1963)
Rat liver (r)	pres.	Dunn (1959)
Rat liver (t)	pres.	Dunn (1959)
Rat, Murphy-Sturm lymphosarcoma	0.30	Hall (1964b)
Sheep heart	0.38	Hall (1964b)
Sheep liver	0.27	Hall (1963b, 1964b)
Chinese cabbage leaf (r)	pres.	Dunn et al. (1963)
Wheat germ (r)	0.38	Lane (1965)
Wheat germ (t)	0.47	Hudson et al. (1965)
Yeast (r)	pres.	Dunn (1961b)
Yeast (t)	0.25	Hall (1963d, 1964b); Morisawa and Chargaff (1963)
*Escherichia coli**	0.10	Hall (1963b, 1964b)
Escherichia coli (r)	0.02	Nichols and Lane (1967)
Neurospora (r)	pres.	Dunn (1961b)

* 2'-*O*-Methylguanosine is the most common O^2-methyl nucleoside in *E. coli* RNA (see also Sanger et al., 1965).

59. INOSINE   Ino   I

$C_{10}H_{12}N_4O_5$

M.W. 268.23

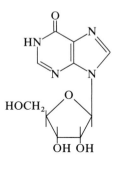

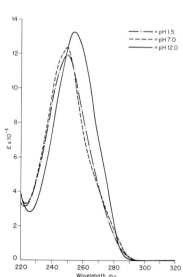

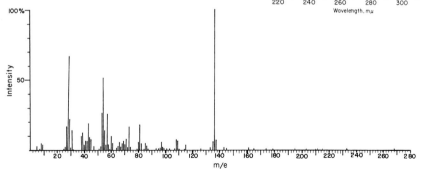

Physical Properties

M.P. 218°	Levene and Bass (1931)
$[\alpha]_D^{22} -52.6°$ (2.5, water)	Levene et al. (1926b)
pK (basic) 1.2	Beaven et al. (1955)
(acidic) 8.8, 12.3	Beaven et al. (1955)

Additional Reference for Synthesis

Levene and Bass (1931)

Synthesis

1. Yamazaki et al. (1967b)

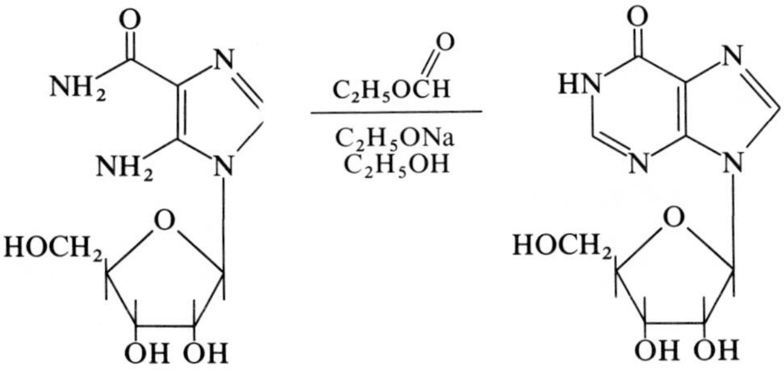

I   (74% yield)

Compound I can be readily obtained by a fermentation process (Shiro et al., 1962).

2. Baddiley et al. (1959)

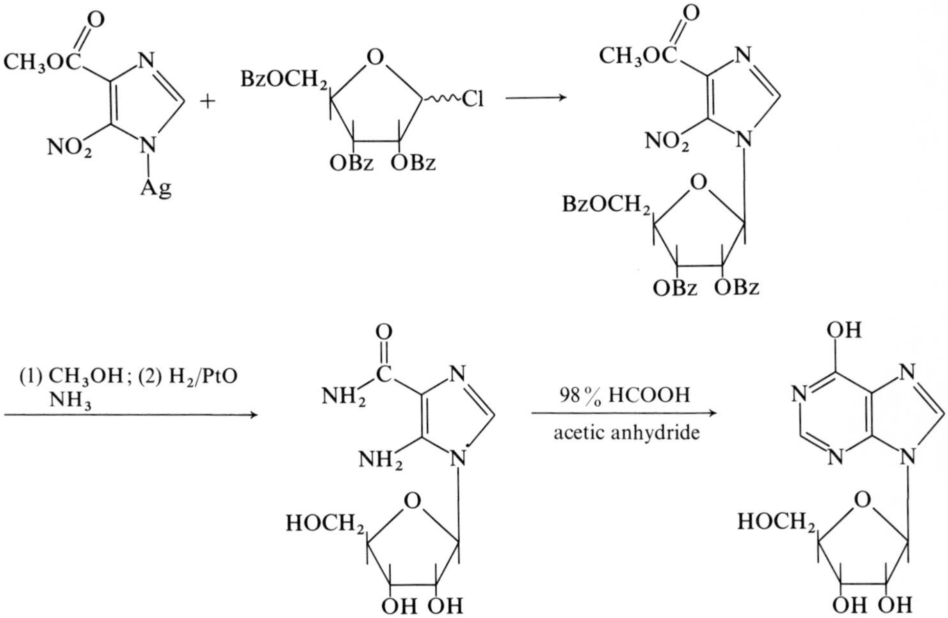

where Bz = benzoyl.   (no yield reported)   (poor yield)

3. Shimizu et al. (1967)

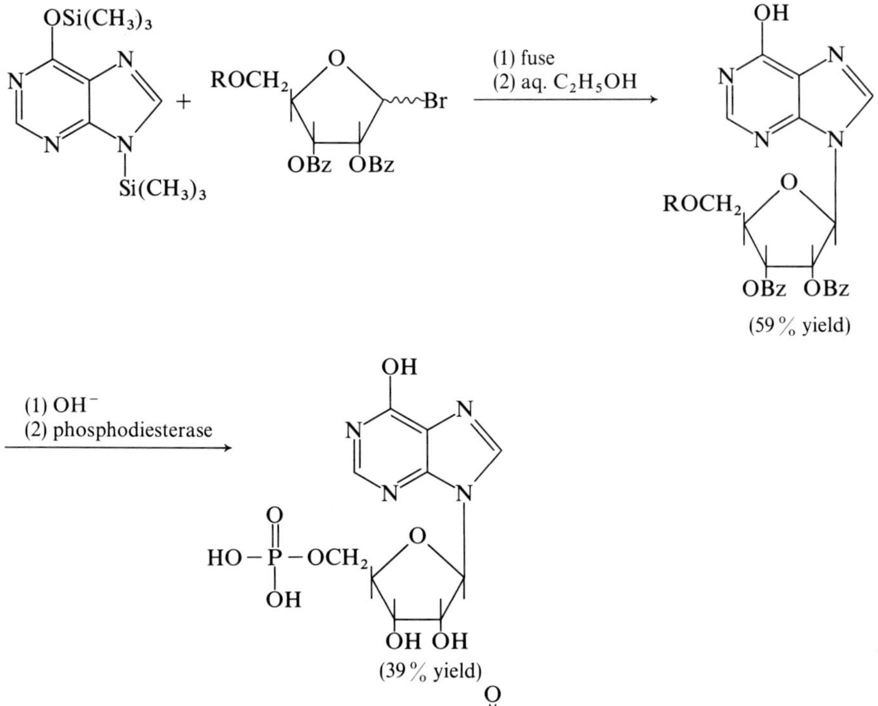

(59% yield)

(39% yield)

where Bz = benzoyl, and R = $(C_6H_5O)_2 - \overset{\displaystyle O}{\underset{\displaystyle \|}{P}} -$.

4. Asai et al. (1967)

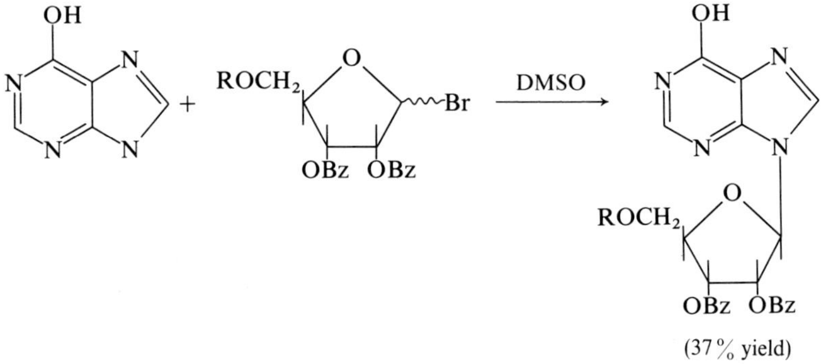

(37% yield)

$$\text{(1) OH}^-$$
$$\xrightarrow{\text{(2) phosphodiesterase}}$$

where Bz = benzoyl, and R = $(C_6H_5O)_2-\overset{\displaystyle O}{\overset{\displaystyle \|}{P}}-$.

Inosine was identified in yeast tRNA (Dunn, 1961b; Hall, 1963c). It occurs as the third letter of the anticodon of yeast tRNA^{Ala}, tRNA^{Ser}, and tRNA^{Val} (see Chapter 4). It has not been found in any other position of the tRNA molecule, and this fact may be related to its ability to pair with more than one base residue (wobbling).

RNA source	Amount, moles/100 moles	References
Pig liver (t)	0.15 (c)	Dunn (1963), also unpublished data
Tobacco leaf (t)	0.10	Dunn and Flack (1967b)
Wheat germ (t)	0.09	Hudson et al. (1965)
Yeast (t)	0.33	Hall (1963c)

60. 1-METHYLINOSINE   1-MeIno   m¹I

$C_{11}H_{14}N_4O_5$

M.W. 282.25

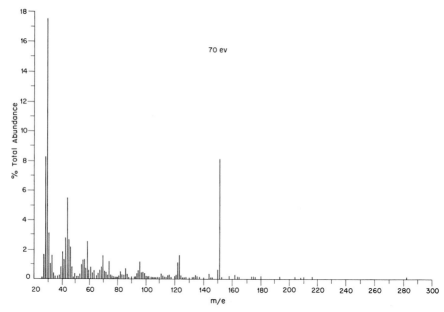

pH	λ_{max}	ε_{max} (x 10^{-3})	λ_{min}
2	250	10.4	223
6	251	10.4	226
12	249	10.7	—

70 ev

Hecht et al. (1969) (Atlas CH4)

Physical Properties

M.P. 210–212°   Jones and Robins (1963)

$[\alpha]_D^{28}$ −49.2° (0.5, water)   Jones and Robins (1963)

Additional Reference for Synthesis

Miles (1961)

Synthesis

Jones and Robins (1963)

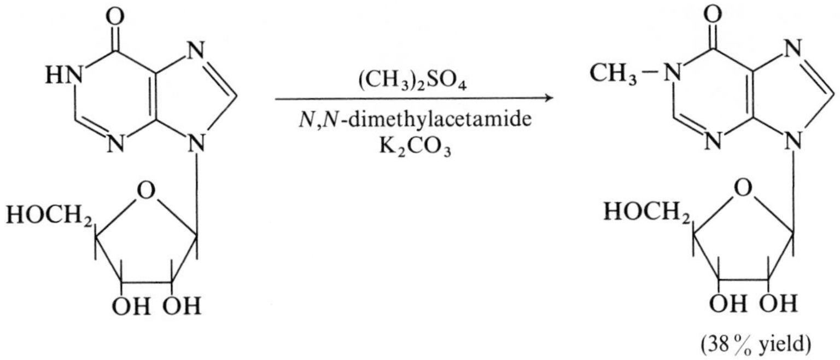

(38% yield)

1-Methylinosine occurs in yeast tRNA at a level of about 0.05 mole % (Hall, 1963c). It has been identified in the primary sequence of yeast tRNA[Ala] (Chapter 4).

RNA source	Amount, moles/100 moles	References
Yeast (t)	0.045	Hall (1963c); Holley et al. (1965)

61. URIDINE   Urd   U

$C_9H_{12}H_2O_6$

M.W. 244.20

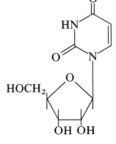

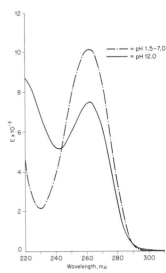

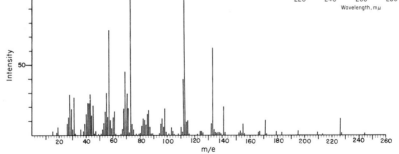

m/e

Physical Properties

M.P. 163–165°   Elmore (1950)

$[\alpha]_D^{16}$ +9.6° (2.0, water)   Elmore (1950)

pK (acidic) 9.2, 12.5   Fox and Shugar (1952)

Synthesis

1. Nishimura and Iwai (1964); Nishimura et al. (1964); Wempen and Fox (1967a)

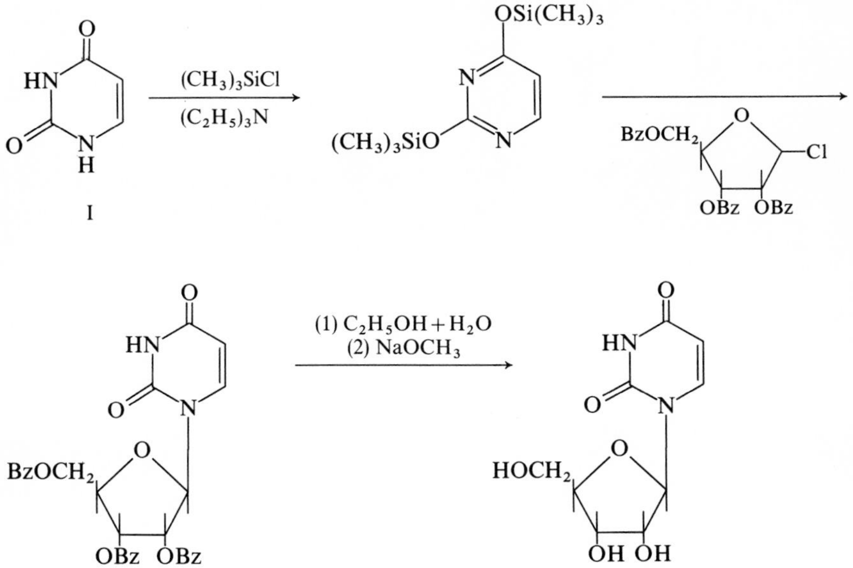

(41% yield, based on I)

(86% yield)

Note. About 25% of the product is the α anomer. This can be easily removed by fractional crystallization at the benzoylated stage.

where Bz = benzoyl.

2. Shaw et al. (1958)

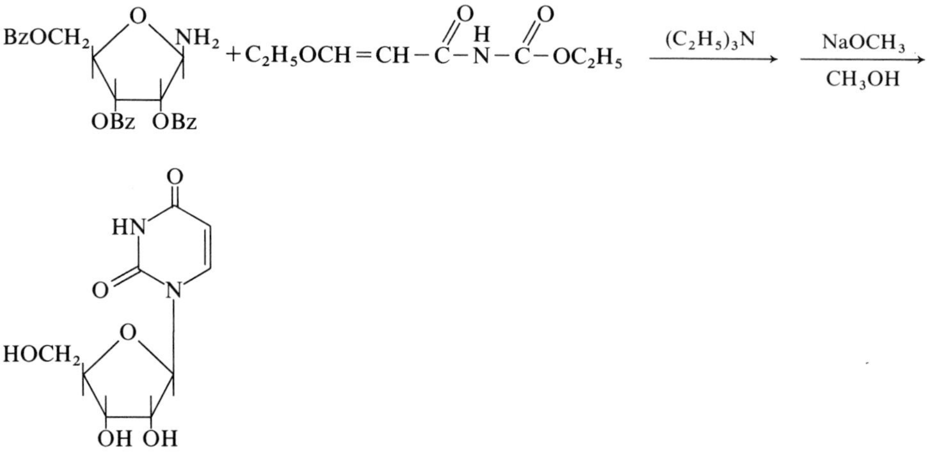

(40% yield overall)

where **Bz** = benzoyl.

3. Shaw et al. (1958)

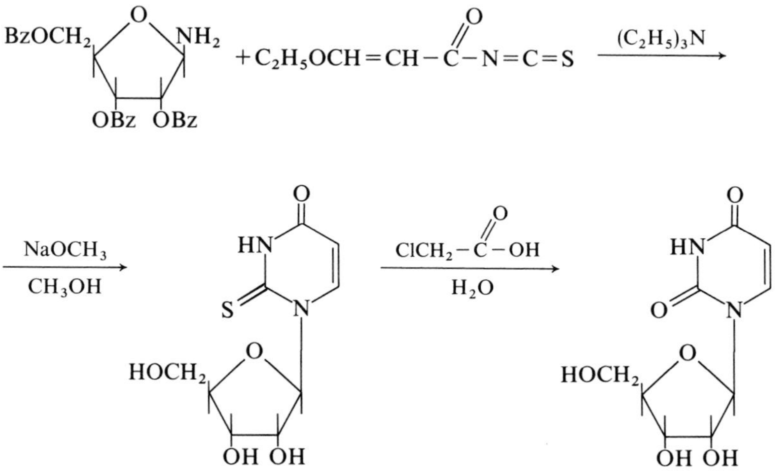

(15% yield)

where **Bz** = benzoyl.

4. Prystaš and Šorm (1968)

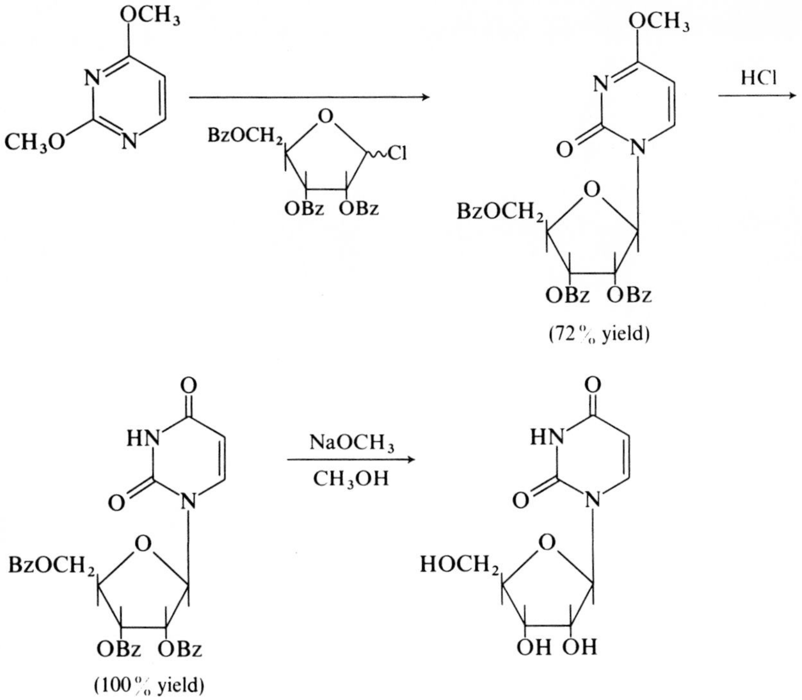

(72 % yield)

(100 % yield)

where Bz = benzoyl.

62. 3-METHYLURIDINE   3-MeUrd   m³U

$C_{10}H_{14}N_2O_6$

M.W. 258.23

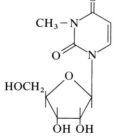

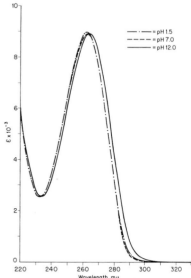

= pH 1.5
= pH 7.0
= pH 12.0

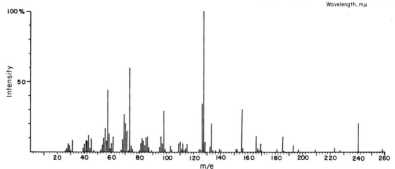

Physical Properties

M.P. 119–120°   Miles (1956)

$[\alpha]_D^{26}$ +20.1° (water)   Miles (1956)

Additional References for Synthesis

Levene and Tipson (1934); Scannell et al. (1959); Visser et al. (1953); Szer and Shugar (1968b)

Synthesis

Haines et al. (1964); Miles (1956); Thedford et al. (1965)

(74% yield)

3-Methyluridine was detected as the nucleoside in yeast tRNA (0.01 mole %) and human liver tRNA (0.03 mole %) (Hall, 1963a). Since 3-methylcytidine readily undergoes deamination at slightly alkaline pH values, the possibility exists that this nucleoside could have been formed from 3-methylcytidine during the isolation procedure. Dunn and Flack (1967a) report the failure to detect any 3-methyluridylic acid in mild acid hydrolyzates of yeast tRNA, although >0.003% should have been detectable.

The attachment of a methyl at the 3 position of uridylic acid derivatives markedly changes the properties of the compound. For example, 3-methyluridylyl-3′ derivatives are resistant to the action of pancreatic ribonuclease (Szer and Shugar, 1961; Thedford et al., 1965). Methyl groups at position 3 of uridine residues of polyuridylic acid eliminate all vestiges of a secondary structure (Szer and Shugar, 1961); this fact is in distinct contrast to the effect of methyl groups at C-5.

63. 5-METHYLURIDINE   5-MeUrd   m⁵U or T

$C_{10}H_{14}N_2O_6$

M.W. 258.23

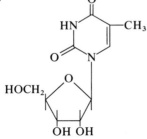

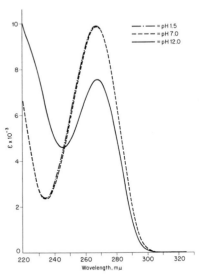

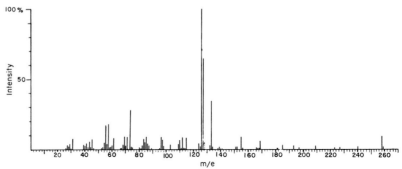

Physical Properties

M.P. 183–185°	Fox et al. (1956)
$[\alpha]_D^{31}$ − 10° (2.0, water)	Fox et al. (1956)
pK (acidic) 9.7	Fox et al. (1956)

Additional References for Synthesis

Littlefield and Dunn (1958) (enzymic synthesis); Reichard (1955) (enzymic synthesis). The α- and β-ribopyranosyl forms of 5-methyluridine have been synthesized by Farkas et al. (1964) and Naito et al. (1961).

Synthesis

1. Fox et al. (1956); Wempen and Fox (1967a)

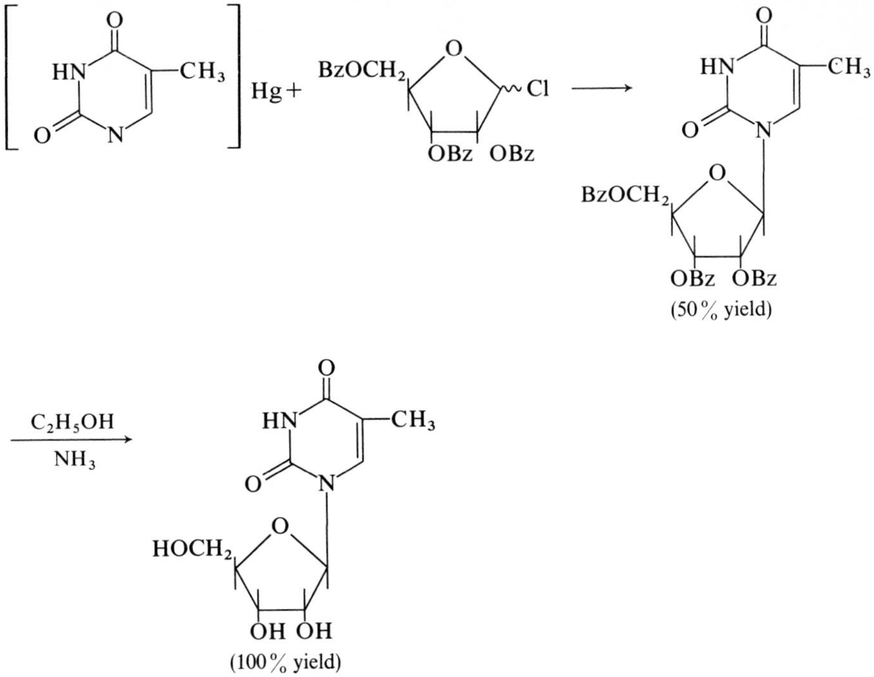

(50% yield)

(100% yield)

where Bz = benzoyl.

2. Cline et al. (1959)

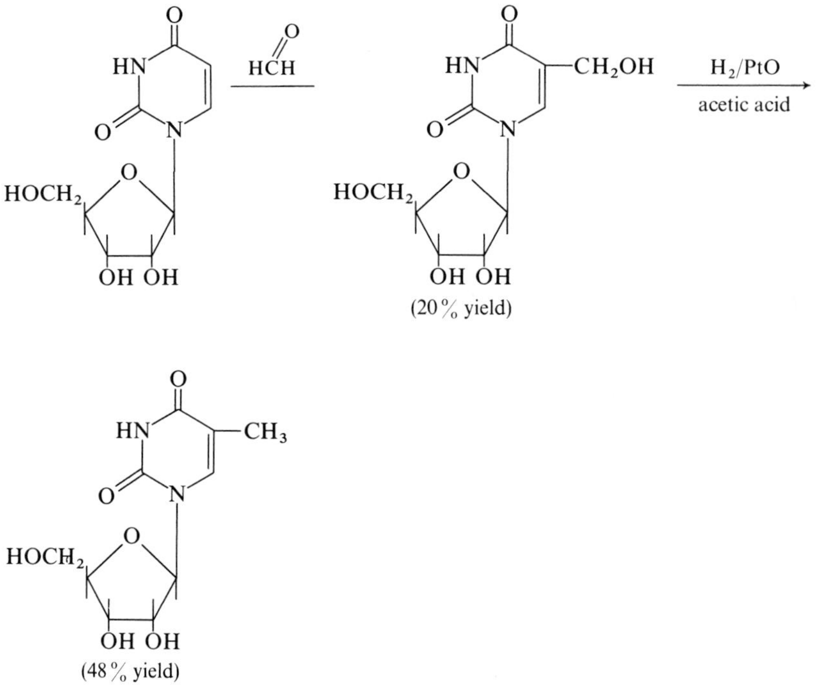

(20% yield)

(48% yield)

3. Shimizu et al. (1967)

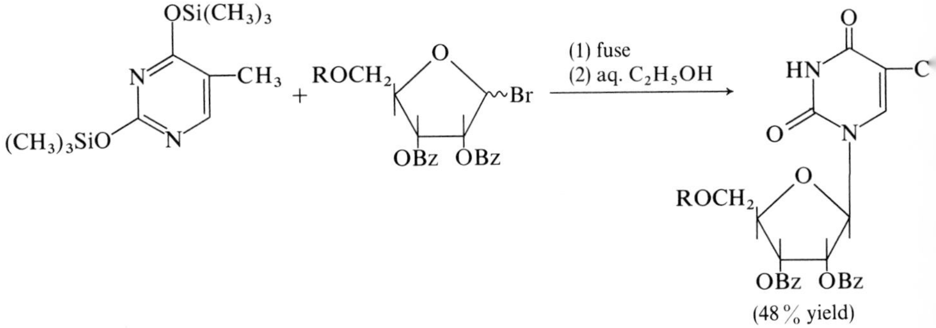

(48% yield)

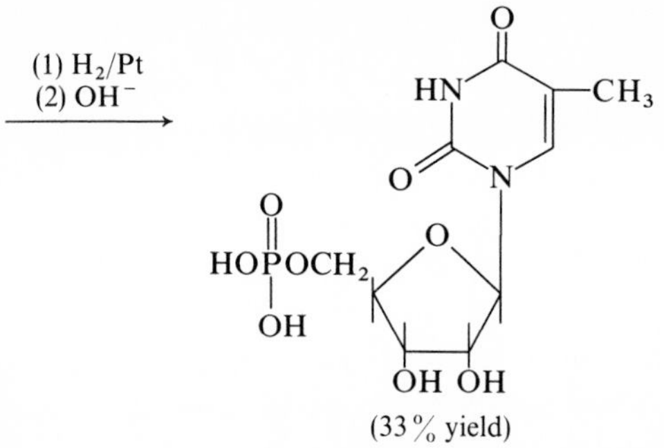

(1) H₂/Pt
(2) OH⁻ →

(33% yield)

where Bz = benzoyl, and R = $(C_6H_5O)_2 - \overset{\overset{\displaystyle O}{\|}}{P} -$.

4. Budowsky et al. (1968a)

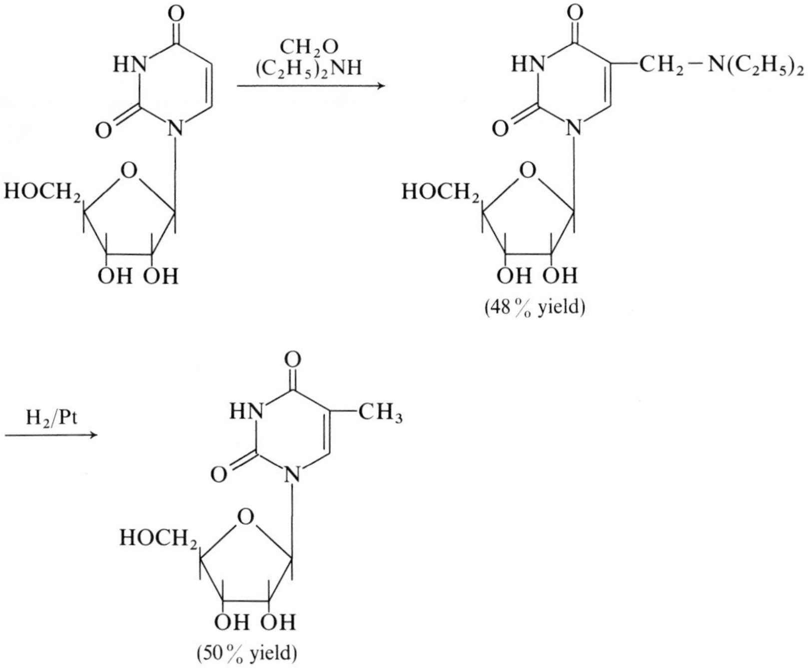

$\xrightarrow[\text{(C}_2\text{H}_5)_2\text{NH}]{\text{CH}_2\text{O}}$

(48% yield)

$\xrightarrow{\text{H}_2/\text{Pt}}$

(50% yield)

5-Methyluridine was originally identified in the RNA of several sources by Littlefield and Dunn (1958). A 5-methyl group attached to uridylic acid residues in polynucleotides can influence considerably the stacking pattern of adjacent nucleotides. This influence is indicated by the substantial increase in the hyperchromicity of poly-5-methyluridylic acid over that of polyuridylic acid (Griffin et al., 1958), and by a relatively high melting point, $T_m = 36°$ (Shugar and Szer, 1962). The presence of the 5-methyl group in uridylyl-3' oligoribonu-cleotide derivatives has no effect on the normal action of pancreatic ribonuclease (Thedford et al., 1965). 5-Methyluridine has been found in every molecular species of tRNA for which the sequence has been determined.

RNA source	Amount, moles/100 moles		References
Human kidney (t)	0.09		Price et al. (1963)
Rat liver (t)	0.40		Price et al. (1963)
Chinese cabbage leaf (t)	0.70	(c)	Dunn and Flack (1967b)
Tobacco leaf (t)	0.75	(c)	Dunn and Flack (1967b)
Wheat germ	0.8	(c)	Littlefield and Dunn (1958); Hudson et al. (1965)
Yeast (t)	1.2	(c)	Dunn and Flack (1967a); Dunn (1961b), also unpublished data
Aerobacter aerogenes	0.24	(c)	Littlefield and Dunn (1958)
Azotobacter vinelandii (t)	1.7		Zaitseva et al. (1962)
Escherichia coli (t)	1.1	(c)	Dunn et al. (1960a)
Escherichia coli (23S)	0.13		Fellner and Sanger (1968)
Neurospora (t)	1.1	(c)	Dunn (1961b), also unpublished data
Sarcina flava (t)	1.8		Zaitseva et al. (1962)
Staphylococcus aereus	0.2	(c)	Littlefield and Dunn (1958)

64. 2'-O-METHYLURIDINE　　OMeUrd　　Um

$C_{10}H_{14}N_2O_6$

M.W. 258.23

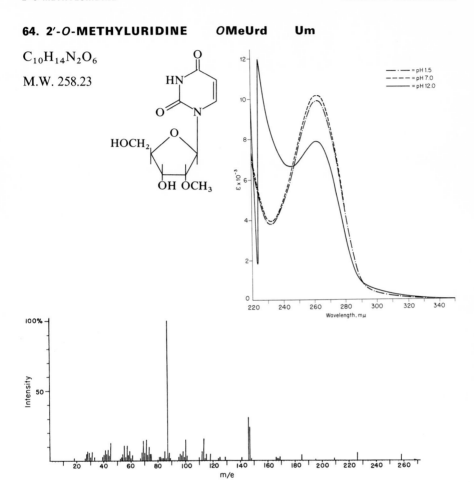

Physical Properties

M.P. 159°　　　　　　　　　　　Furukawa et al. (1965)
$[\alpha]_D^{20}$ +41° (1.6, water)　　Furukawa et al. (1965)
pK (acidic) ~9.3　　　　　　　Furukawa et al. (1965)

Synthesis

1. Furukawa et al. (1965)

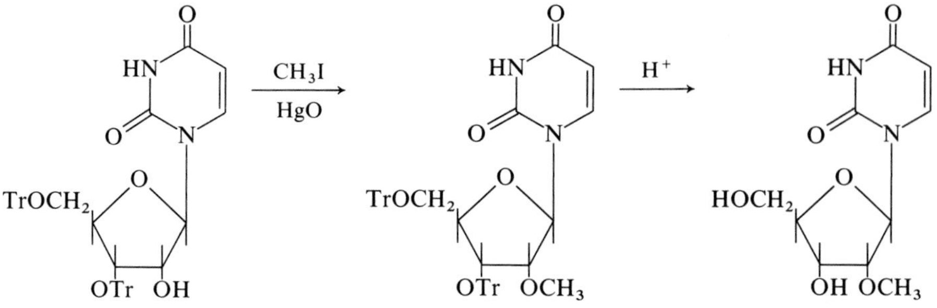

where Tr = trityl.

2. Martin et al. (1968)

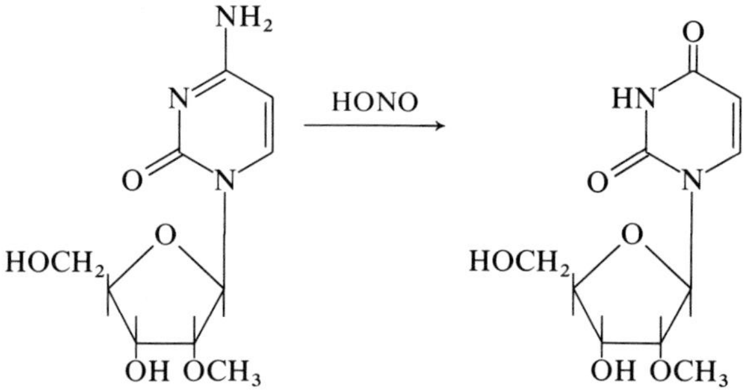

(65% yield)

2'-O-Methyluridine occurs in yeast tRNA$_{I,II}^{Ser}$ (Table 1, Chapter 4). See *2'-O-*Methyladenosine for general remarks.

RNA source	Amount, moles/100 moles		References
Calf liver (t)	0.21		Hall (1963b,1964b)
Human liver (r)	0.04		Hall (1964b)
Human liver (t)	0.22		Hall (1964b)
Human tumor	0.11		Hall (1964b)
Mouse, Ehrlich ascites	0.20		Hall (1964b)
Rat liver (r)	0.18		Dunn (1959) (value unpublished)
Rat liver (t)	0.30	(c)	Dunn (1959) (value unpublished)
Rat, Murphy-Sturm lymphosarcoma	0.21		Hall (1964b)
Sheep liver	0.31		Hall (1963b, 1964b)
Chinese cabbage leaves (r)	0.35		Dunn et al. (1963)
Wheat germ (r)	0.47		Lane (1965)
Wheat germ (t)	0.30		Hudson et al. (1965)
Yeast (r)	0.07		Dunn (1961b)
Escherichia coli	0.06		Hall (1963d, 1963b, 1964b)
Escherichia coli (23S)	0.07		Fellner and Sanger (1968); Nichols and Lane (1967)
Neurospora (r)	0.16		Dunn (1961b)
Neurospora (t)	0.07	(c)	Dunn (1961b)

65. 5-(β-D-**RIBOFURANOSYL)URACIL** (**PSEUDOURIDINE**)

$C_9H_{12}N_2O_6$

M.W. 244.20

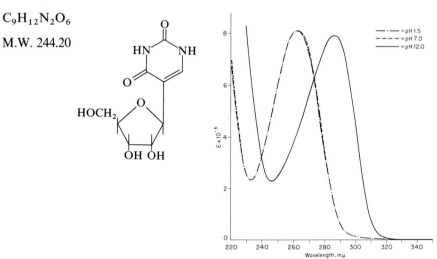

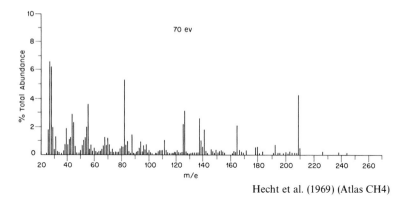

Hecht et al. (1969) (Atlas CH4)

Physical Properties

M.P. 223–224°   Shapiro and Chambers (1961)
$[\alpha]_D$-3.0° (1.0, water)   Yu and Allen (1959)
pK (acidic) 8.9, > 13   Chambers (1966)

Additional References for Synthesis

Cohn (1967); Cohn et al. (1963) (isolation from urine)

Synthesis

1. Shapiro and Chambers (1961)

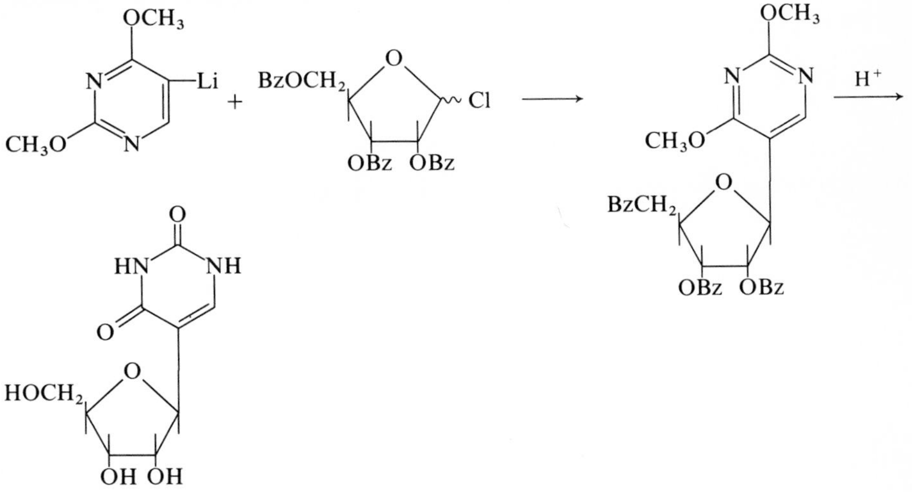

(0.27% yield)

where Bz = benzoyl.

2. Brown et al. (1968)

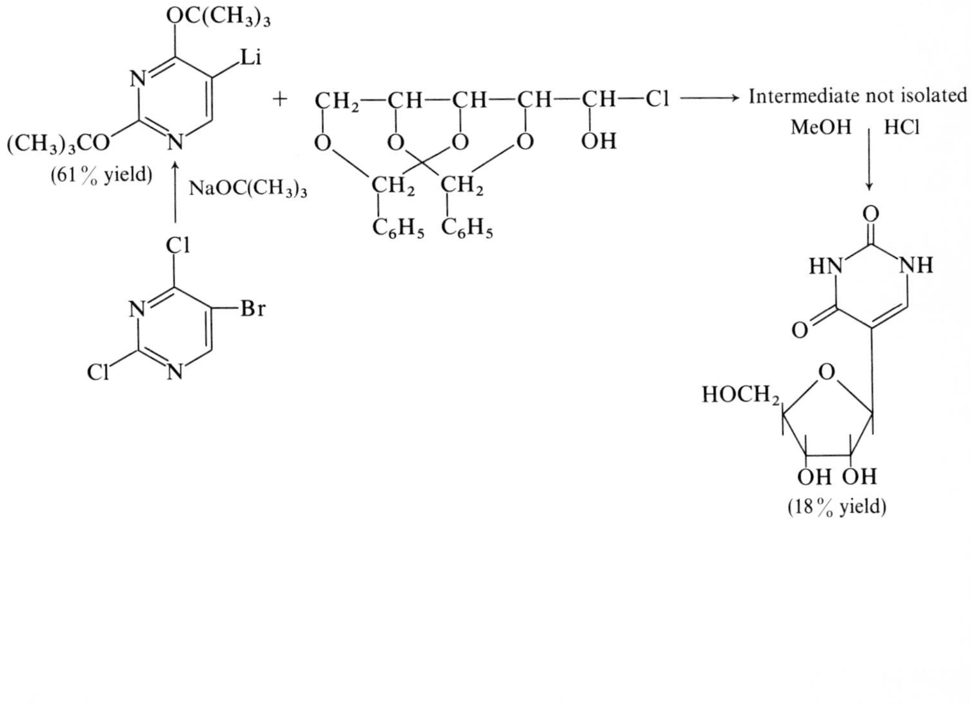

(18% yield)

5-(β-ᴅ-Ribofuranosyl)uracil appeared originally in the elution profile of an RNA hydrolyzate (Cohn and Volkin, 1951) and later became known as the "fifth" ribonucleotide (Davis and Allen, 1957). It was identified by Cohn (1960) as a 5-ribosyluracil, and the β-ᴅ-ribofuranosyl configuration was established by Michelson and Cohn (1962). This nucleoside is the most predominant of the modified nucleosides. In yeast and rat liver tRNA, it accounts for about 25% of the uridine residues. 5-(β-ᴅ-Ribofuranosyl)uracil has been identified in every known sequence of tRNA (Chapter 4). The chemistry of this nucleoside is described in Chapter 8, p. 394.

RNA source	Amount, moles/100 moles		References
Carp liver (t)	5.0		Zaitseva et al. (1962)
Frog liver (t)	5.0		Zaitseva et al. (1962)
Pig liver (t)	3.2	(c)	Dunn (1963)
Pigeon liver (t)	4.0		Zaitseva et al. (1962)
Rat liver (r)	1.6	(c)	Dunn et al. (1960b)
Rat liver (t)	4.3	(c)	Dunn (1959)
Chinese cabbage leaf (r)	1.4	(c)	Dunn et al. (1963)
Chinese cabbage leaf (t)	2.9	(c)	Dunn and Flack (1967b)
Tobacco leaf (t)	2.6	(c)	Dunn and Flack (1967b)
Wheat germ (t)	2.7		Hudson et al. (1965)
Yeast (r)	0.5	(c)	Dunn (1961b)
Yeast (t)	4.5		Monier et al. (1960); Dunn (1961b)
Azotobacter vinelandii (t)	4.2		Zaitseva et al. (1962)
Escherichia coli B (r)	0.15	(c)	Dunn et al. (1960a)
Escherichia coli B (t)	2.0	(c)	Dunn et al. (1960a)
Escherichia coli 16S	0.12		Dubin and Gunalp (1967)
Escherichia coli 23S	0.30		Dubin and Gunalp (1967); Fellner and Sanger (1968)
Neurospora (r)	1.2	(c)	Dunn (1961b)
Neurospora (t)	4.3	(c)	Dunn (1961b)
Proteus vulgaris (t)	4.4		Zaitseva et al. (1962)
Sarcina flava (t)	4.0		Zaitseva et al. (1962)

66. 5-(2'-O-METHYL-β-D-RIBOFURANOSYL)URACIL   OMeψrd   ψm

$C_{10}H_{14}N_2O_6$

M.W. 258.23

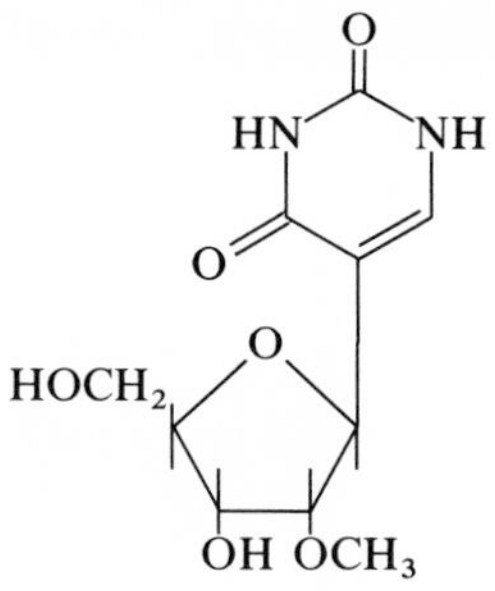

pH	λ_{max}	ε_{max} (x 10^{-3})	λ_{min}
1	261	—	—
7	261	—	—
13	281	—	—

5-(2'-O-Methylribosyl)uracil was first isolated as the nucleoside from an enzymic digest of yeast tRNA (Hall, 1964b); however, the amount obtained did not permit a rigorous identification. Evidence for its existence rests solely on the spectral and chromatographic behavior of the isolated sample. A similar compound was isolated as the nucleotide by Hudson et al. (1965) from an alkaline digest of tRNA of wheat germ. Its chromatographic and spectral properties conform to those expected for 5-(2'-O-methylribosyl)uracil. It has been identified in a specific sequence of rat liver tRNA[Ser] (Staehelin et al., 1968).

RNA source	Amount, moles/100 moles	References
Rat liver (t)	pres.	Staehelin et al. (1968)
Wheat germ (t)	0.12	Hudson et al. (1965)
Yeast (t)	pres.	Hall (1964b)

67. 2-THIO-5-CARBOXYMETHYLURIDINE METHYL ESTER   mcm⁵s²U

$C_{12}H_{16}N_2O_7S$

M.W. 332.32

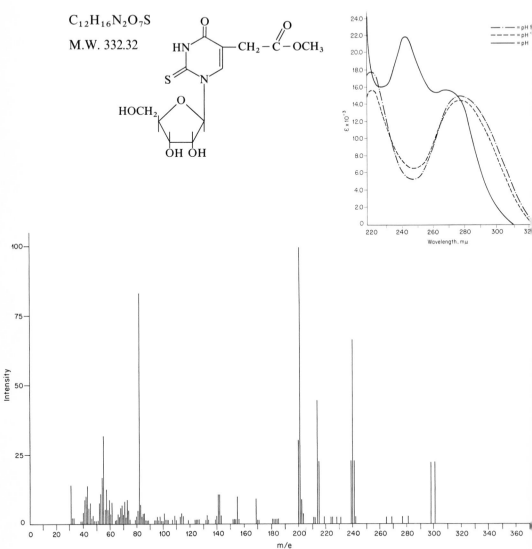

Baczynskyj et al. (1968) (CEC-21-110B)

Physical Properties

M.P. 295°   Baczynskyj et al. (1969); Vorbrüggen, and Strehlke (1969)

$[\alpha]_D$ + 19.8°   (0.5, water) Vorbrüggen and Strehlke (1969).

Synthesis

1. Baczynskyj et al. (1969)

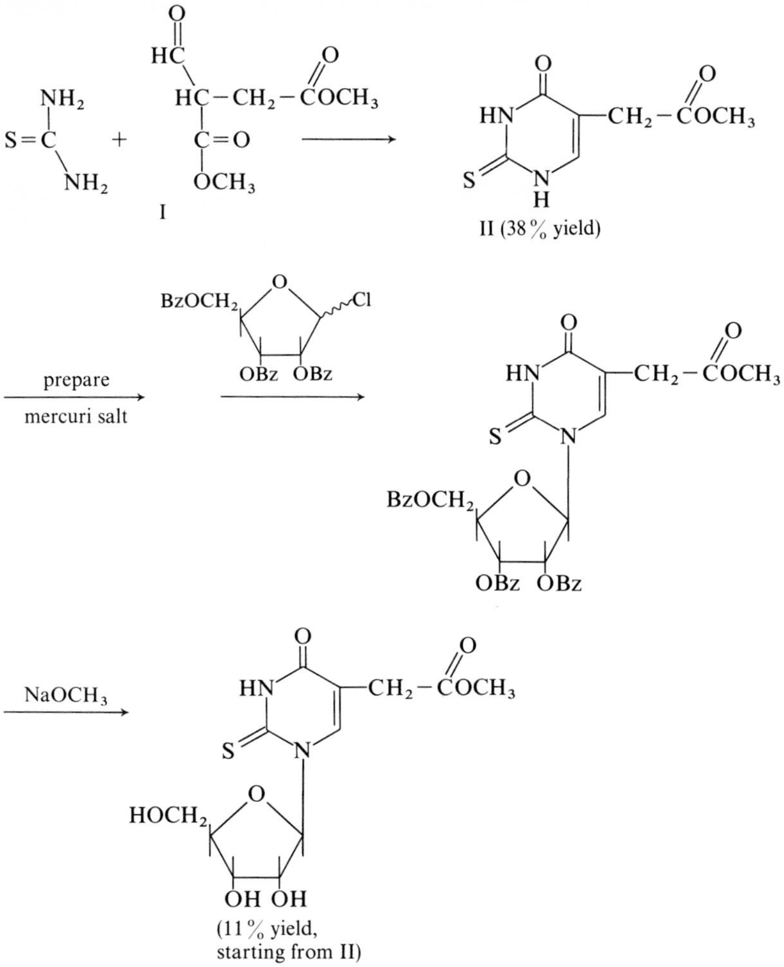

II (38% yield)

where Bz = benzoyl.

Compound I synthesized according to Payot and Grob (1954).

2-Thio-5-carboxymethyluridine methyl ester was detected in a snake venom hydrolyzate of yeast tRNA (Baczynskyj et al., 1968).

68. 4-THIOURIDINE   4-Srd   S⁴U

$C_9H_{12}N_2O_5S$

M.W. 260.20

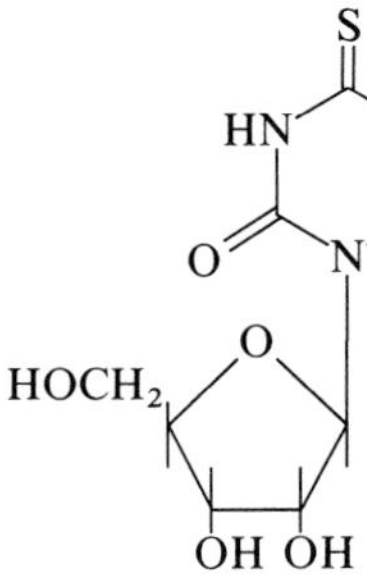

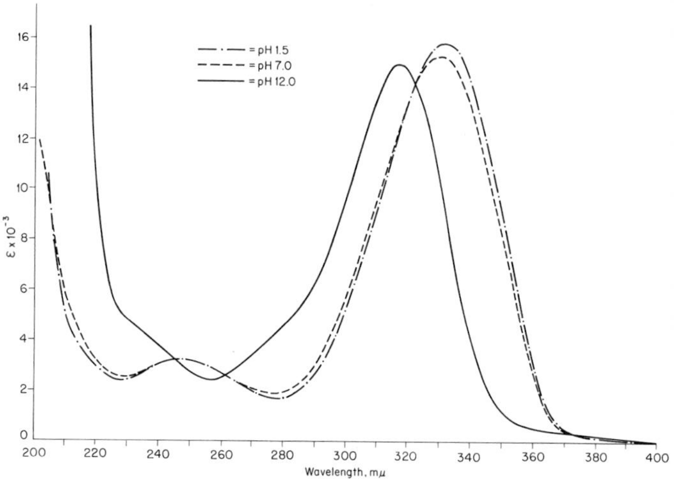

Physical Properties

pK (acidic) 8.2   Lipsett (1965)

Synthesis

Fox et al. (1959)

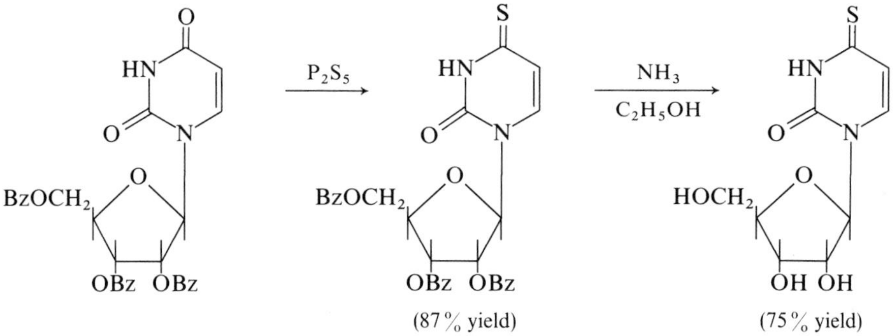

where Bz = benzoyl.

4-Thiouridine was isolated from the tRNA of *E. coli* and identified by Lipsett (1965). It accounts for about 1.5% of the total nucleotides in the microorganism. It has not been detected in the tRNA of yeast or mammalian tissue (Hall, unpublished results), although the sulfur-containing nucleoside, 2-thio-5-carboxymethyluridine methyl ester, occurs in the tRNA of yeast (Baczynskyj et al., 1968). Lipsett and Doctor (1967) purified a species of *E. coli* tRNA^{Tyr} and found two 4-thiouridylate residues per molecule.

RNA source	Amount, moles/100 moles	References
Bacillus subtilis (t)	0.01	Cerutti et al. (1968a)
Escherichia coli B (t)	0.9	Cerutti et al. (1968a)
	0.8	Carbon and David (1968)

69. 5-HYDROXYURIDINE   5-OH-Urd   ho⁵U

$C_9H_{12}N_2O_7$

M.W. 260.21

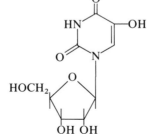

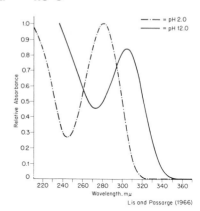

Lis and Passarge (1966)

m/e

Physical Properties

M.P. 242–245°   Roberts and Visser (1952)
pK 7.8   Visser (1968)

Additional References for Synthesis

Levene and La Forge (1912); Ueda (1960)

Synthesis

Roberts and Visser (1952); Visser (1968)

5-Hydroxyuridine (*isobarbituridine*) was identified as the nucleoside in yeast total RNA by Lis and Passarge (1966). No quantitative data were given by the authors.

70. 2-THIO-5-(*N*-METHYLAMINOMETHYL)URIDINE   mnm⁵s²U

$C_{11}H_{17}N_3O_5S$

M.W. 303.33

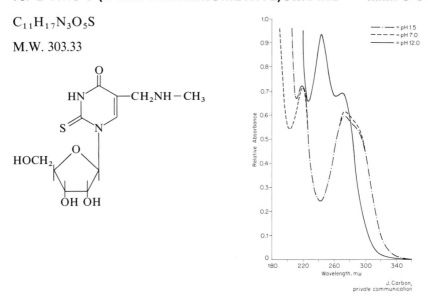

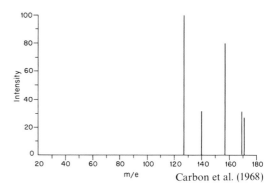

Carbon et al. (1968)

2-Thio-5-(N-methylaminomethyl)uridine has been detected in an alkaline hydrolyzate of *E. coli* tRNA (Carbon et al., 1968).

71. 5-CARBOXYMETHYLURIDINE   cm⁵U

$C_{11}H_{14}N_2O_8$

M.W. 302.24

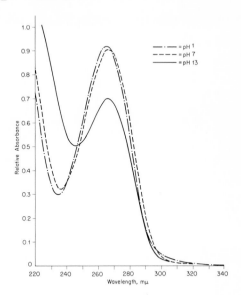

Gray and Lane (1968)

5-Carboxymethyluridine was first detected in wheat embryo tRNA by Gray and Lane (1967, 1968). Gray and Lane (1968) report that very little of this compound can be isolated from a snake venom diesterase digest of tRNA; on the other hand, 5-carboxymethyluridine can be readily obtained from an alkaline digest. This fact led the authors to suggest that in native tRNA the carboxymethyl group is esterified. There is precedence for the existence of an esterified carboxyl group in tRNA; Baczynskyj et al. (1968) identified 2-thio-5-carboxymethyluridine methyl ester in yeast tRNA.

RNA source	Amount, moles/100 moles	References
Wheat embryo (r)	not detected	Gray and Lane (1968)
Wheat embryo (t)	0.15	Gray and Lane (1968)
Yeast (r)	not detected	Gray and Lane (1968)
Yeast (t)	0.34	Gray and Lane (1968)
Escherichia coli (r)	not detected	Gray and Lane (1968)
Escherichia coli (t)	not detected	Gray and Lane (1968)

72. 5,6-DIHYDROURIDINE h₂Urd hU

$C_9H_{14}N_2O_6$

M.W. 246.22

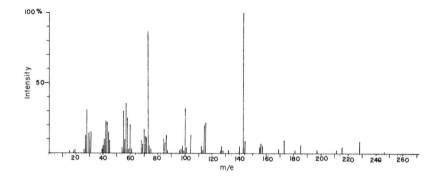

pH	λ_{max}	ε_{max} (x 10^{-3})	λ_{min}
W	208	6.6	—
13	235	—	—

Physical Properties

M.P. 106–108°   Hanze (1967)

$[\alpha]_D^{23}$ −36.8° (2.12, water)   Cerutti et al. (1968b)

Additional Reference for Synthesis

Levene and La Forge (1912)

Synthesis

(1) Green and Cohen (1957); (2) Hanze (1967); (3) Cerutti et al. (1968b)

(85% yield)

5,6-Dihydrouridine was first detected in pancreatic digests of yeast tRNA[Ala] by Madison and Holley (1965). From two to six 5,6-dihydrouridylic acid units occur in each of the yeast tRNA molecules of known sequences (Chapter 4).

RNA source	Amount, moles/100 moles	References
Rat liver (t)	1.1	Cerutti et al. (1968a)
Wheat embryo (t)	1.9	Gray and Lane (1968)
Yeast (t)	3.0	Gray and Lane (1968)
	3.6	Magrath and Shaw (1967)
Bacillus subtilis (t)	1.7	Cerutti et al. (1968a)
Escherichia coli B (t)	2.2	Gray and Lane (1968)
	2.2	Magrath and Shaw (1967)
	3.5	Cerutti et al. (1968a)
Pea bud chromosomes	8.5	Huang and Bonner (1965); Jacobson and Bonner (1968)
Chicken embryo chromosomes	9.6	Huang (1967)
Calf thymus chromosomes	8.5	Jacobson and Bonner (1968)

73. 5,6-DIHYDRO-5-METHYLURIDINE   hT

$C_{10}H_{16}N_2O_6$

M.W. 260.25

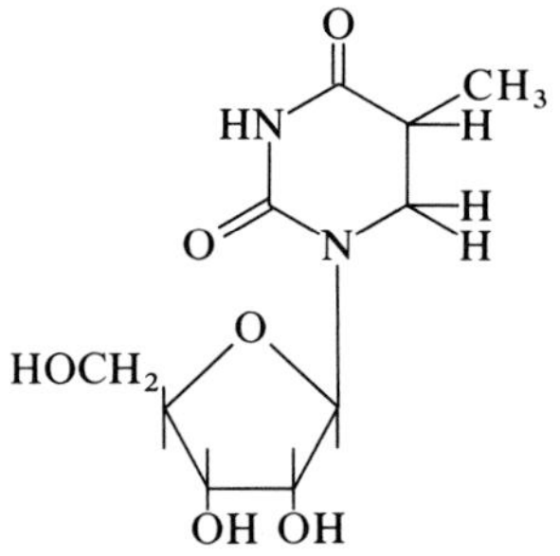

Synthesis

Jacobson and Bonner (1968)

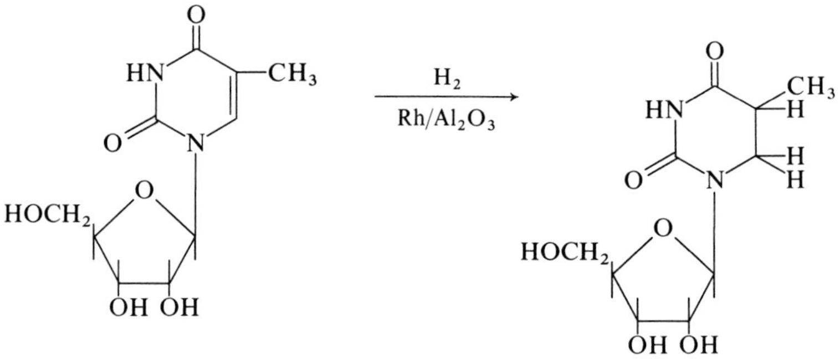

5,6-Dihydro-5-methyluridine was first reported by Jacobson and Bonner (1968) to occur in a species of RNA associated with the chromosomes of ascites tumor cells. It accounts for 8.1 mole % of the nucleosides in this RNA.

Deoxyribonucleosides

74. DEOXYADENOSINE   dAdo   dA

$C_{10}H_{13}N_5O_3$

M.W. 251.24

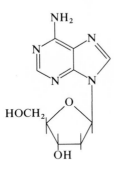

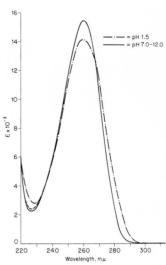

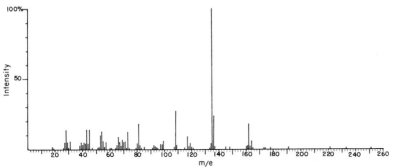

Physical Properties

M.P. 191–192°   Andersen et al. (1952); Pedersen and Fletcher
                        (1960)

$[\alpha]_{589}^{25}$ −26° (0.49, water)   Ness and Fletcher (1960)

pK (basic)   3.8

(acidic) 12.5

Additional References for Synthesis

The pyranosyl form of 2-deoxyadenosine has been synthesized by Nagasawa et al. (1967) and Robins et al. (1964).

α-2-deoxyadenosine has been synthesized by Robins and Robins (1965).

Synthesis

1. Ness and Fletcher (1960)

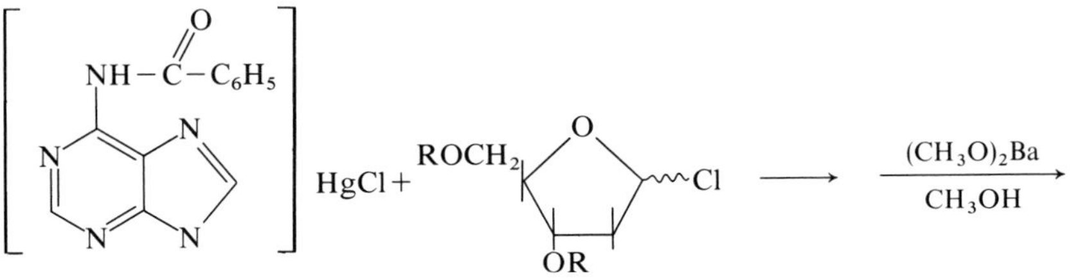

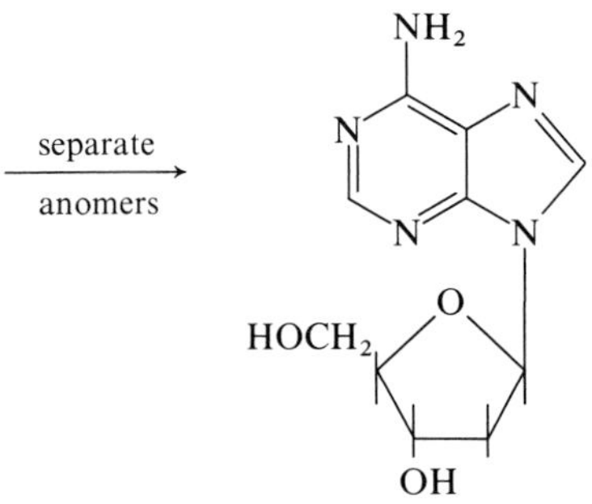

(10% overall yield)   where R = p-nitrobenzoyl.

2. Schramm et al. (1961, 1967)

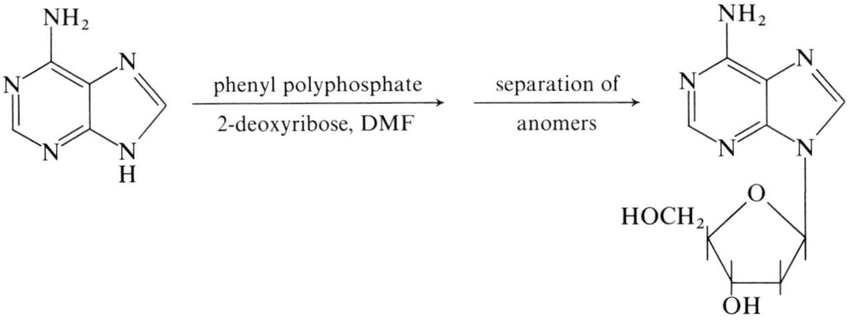

(15–20% yield)

3. Pedersen and Fletcher (1960)

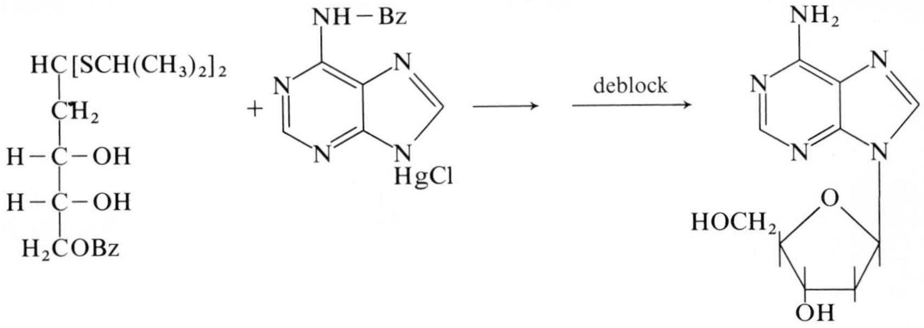

HC[SCH(CH$_3$)$_2$]$_2$

where Bz = benzoyl.

(6.3% overall yield)
(α anomer also obtained)

4. Venner (1960)

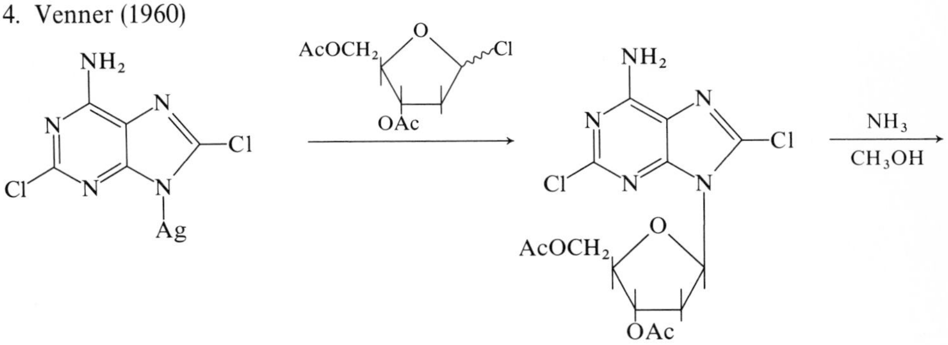

(54.8% yield)

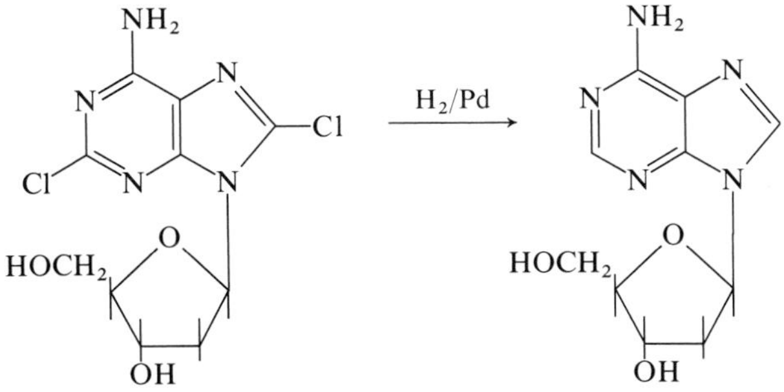

(11.9% yield)　　　　(45.4% yield)

where Ac = acetyl.

5. Anderson et al. (1959)

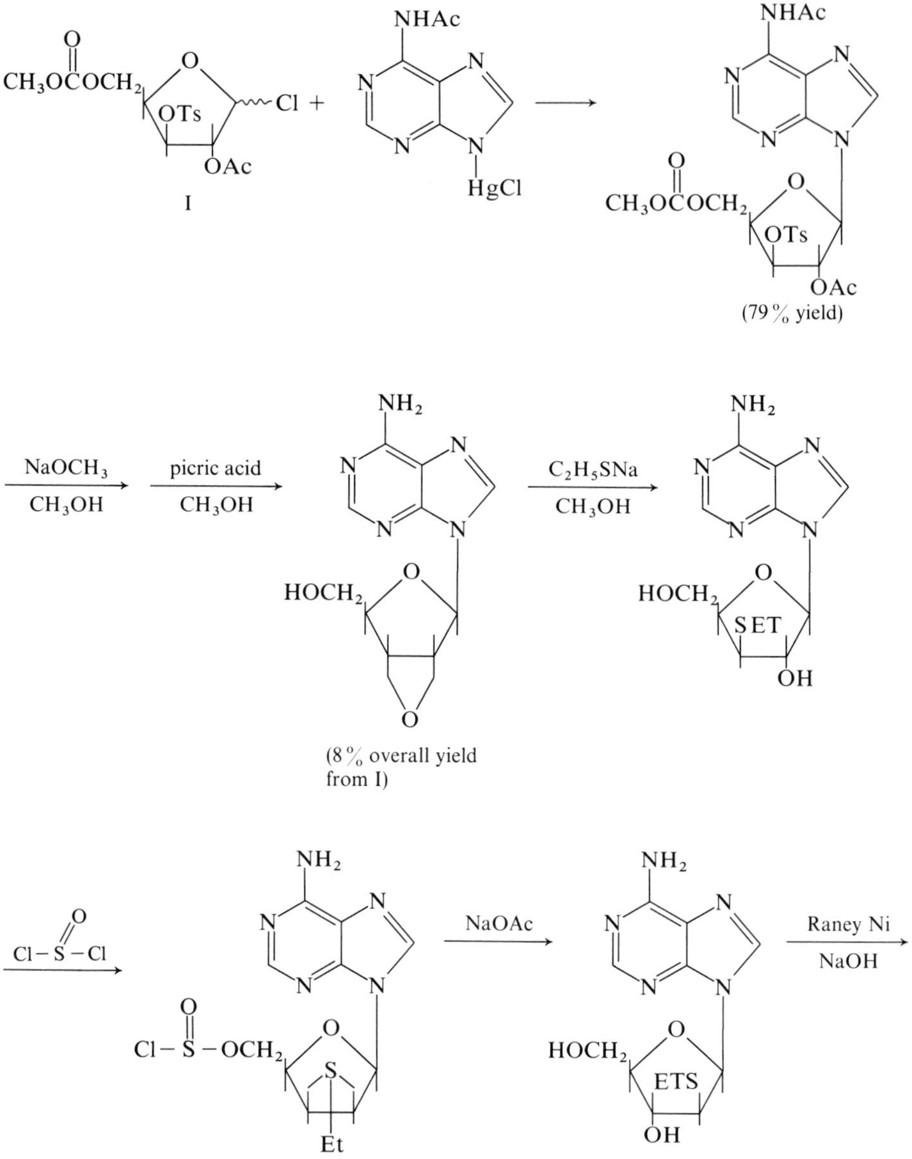

(79% yield)

(8% overall yield
from I)

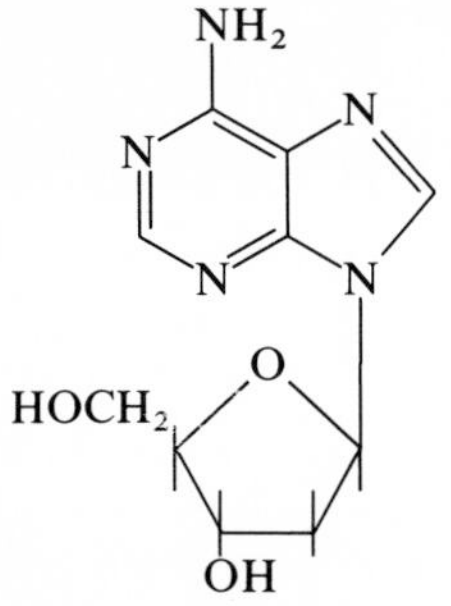

(0.5% overall yield from I)

where Ac = acetyl, Et = ethyl, and Ts = tosyl.

6. Ikehara and Tada (1968)

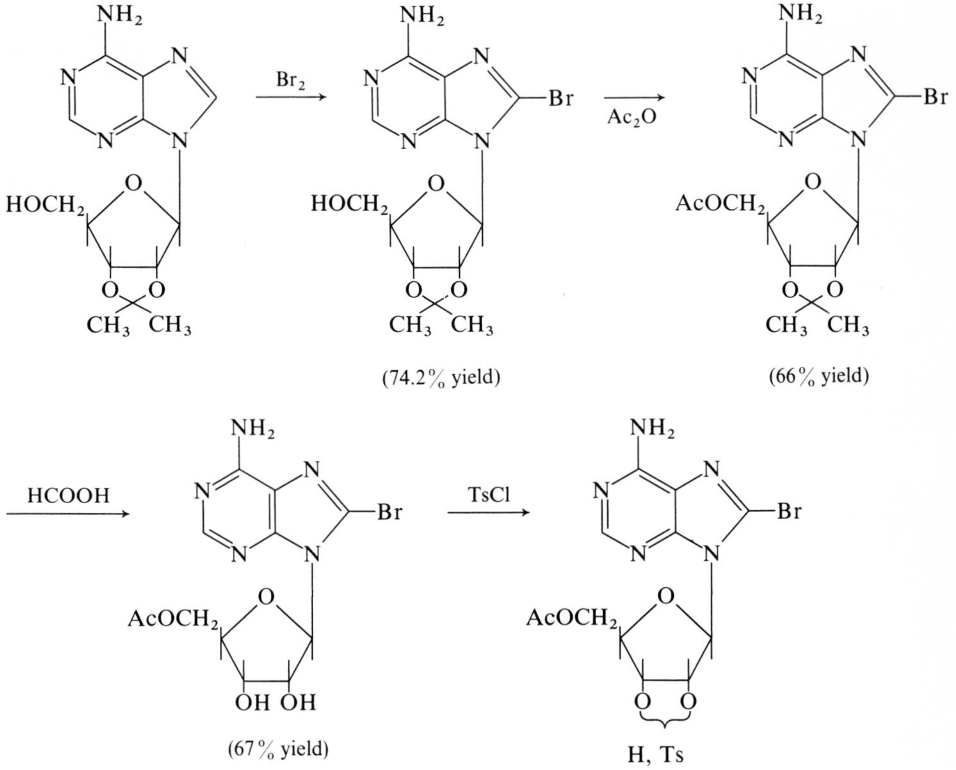

(74.2% yield)

(66% yield)

(67% yield)

H, Ts

(58% yield)   (33% yield)

where Ts = *p*-toluene sulfonyl and Ac = acetyl.

75. N⁶-METHYLDEOXYADENOSINE   6-MedAdo   m⁶dA

$C_{11}H_{15}N_5O_3$

M.W. 265.27

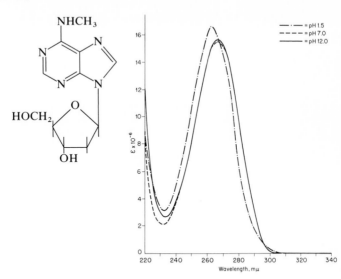

Physical Properties

M.P. 206–208°   Jones and Robins (1963)

$[\alpha]_D^{26}$ −23.5° (1.0, water)   Jones and Robins (1963)

Synthesis

1. Jones and Robins (1963)

CH₃I, N,N-dimethyl-acetamide → OH⁻ →

(40% overall yield)

76. DEOXYCYTIDINE   dCyd   dC

$C_9H_{13}N_3O_4$

M.W. 227.22

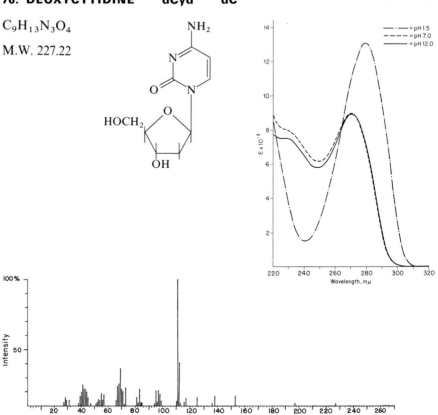

Physical Properties

M.P. 200–201°   Fox et al. (1961)

$[\alpha]_D^{19}$ +82.4° (1.31, 1 N NaOH)   Schindler (1949)

pK (basic) 4.3   Fox and Shugar (1952)

   (acidic) > 13

Synthesis

1. Fox et al. (1961)

(16% overall yield)

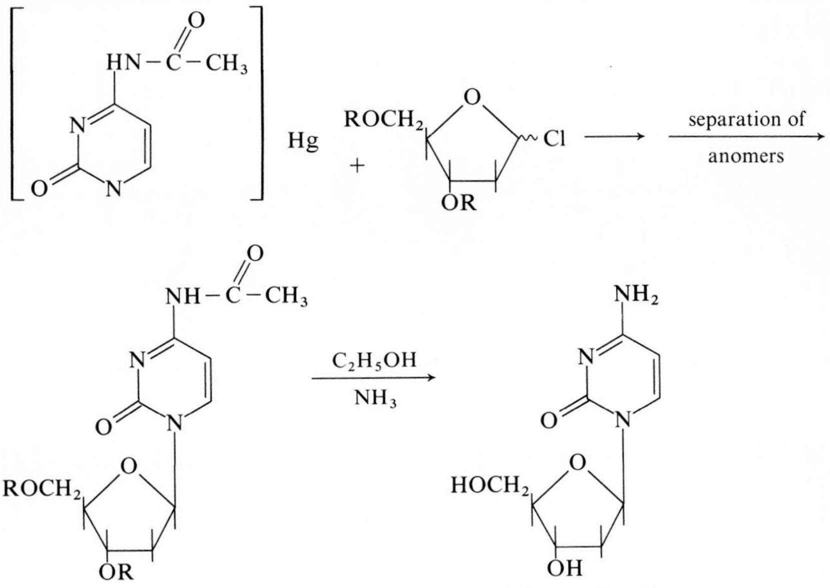

where R = *p*-chlorobenzoyl.

2. Wempen et al. (1961)

(74% yield)

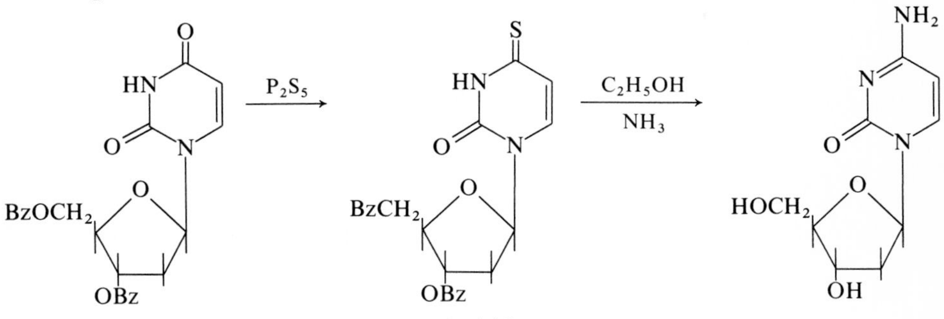

where Bz = benzoyl.

77. 5-METHYLDEOXYCYTIDINE   5-MedCyd   m⁵dC

$C_{10}H_{15}N_3O_4$

M.W. 241.24

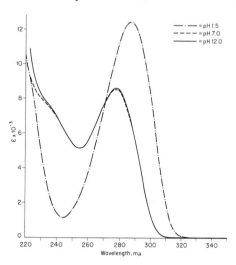

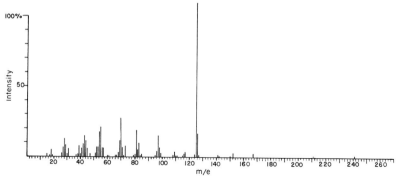

Physical Properties

M.P. 211–212°	Wempen and Fox (1967b)
154–155° (dec.) (HCl salt)	Fox et al. (1959)
$[\alpha]_D^{22}$ +43° (1.4, water)	Wempen and Fox (1967b)
pK (basic) 4.4	Fox et al. (1959)
(acidic) > 13	Fox et al. (1959)

Synthesis

Fox et al. (1959); Wempen and Fox (1967b)

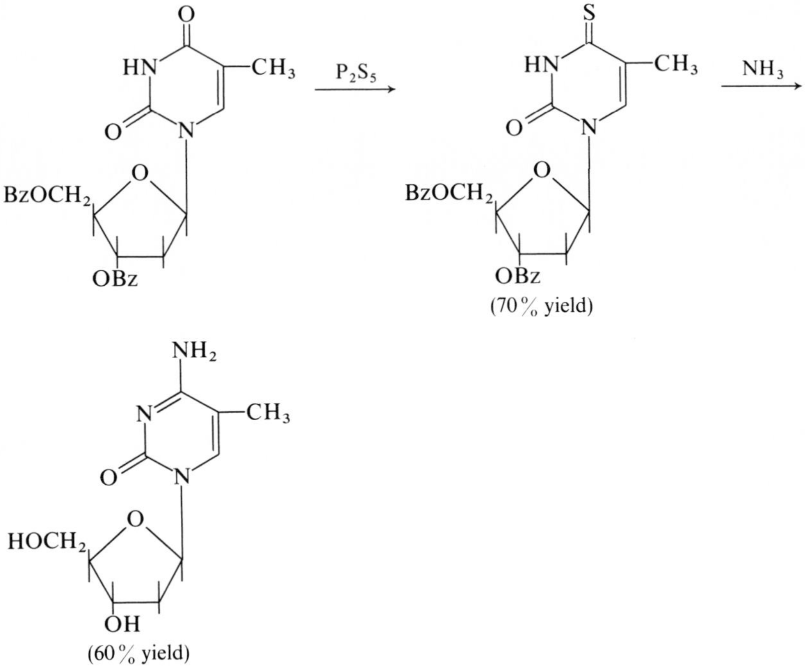

(70% yield)

(60% yield)

where Bz = benzoyl.

78. 5-HYDROXYMETHYLDEOXYCYTIDINE   5-HmdCyd   hm⁵dC

$C_{10}H_{15}N_3O_5$

M.W. 257.25

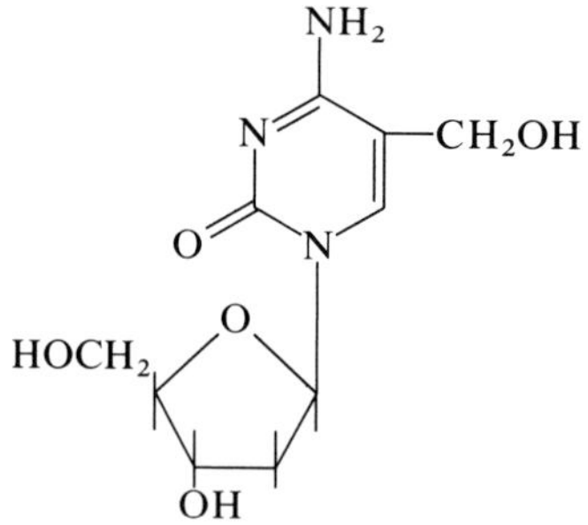

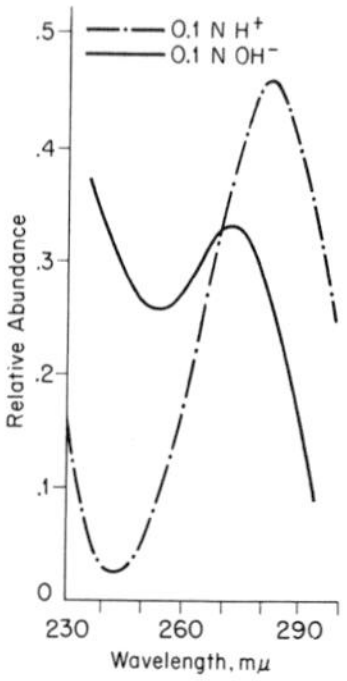

Loeb and Cohen (1959)

Physical Properties

M.P. 203° (dec.)   Loeb and Cohen (1959)
$[\alpha]_D^{20} + 51°$ (water)   Brossmer and Röhm (1963)
pK (basic) 3.5   Loeb and Cohen (1959)

Synthesis

1. Brossmer and Röhm (1963)

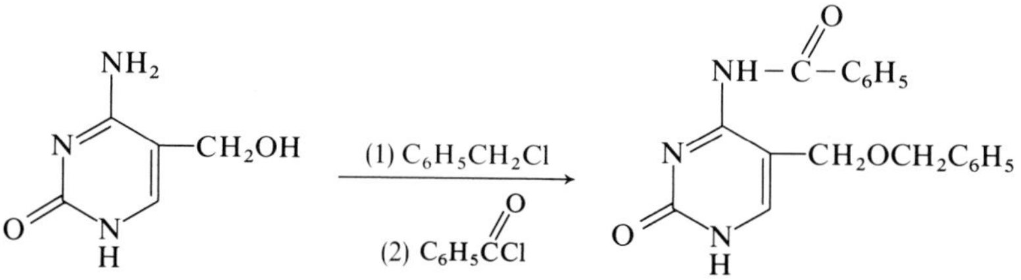

I (80% yield)

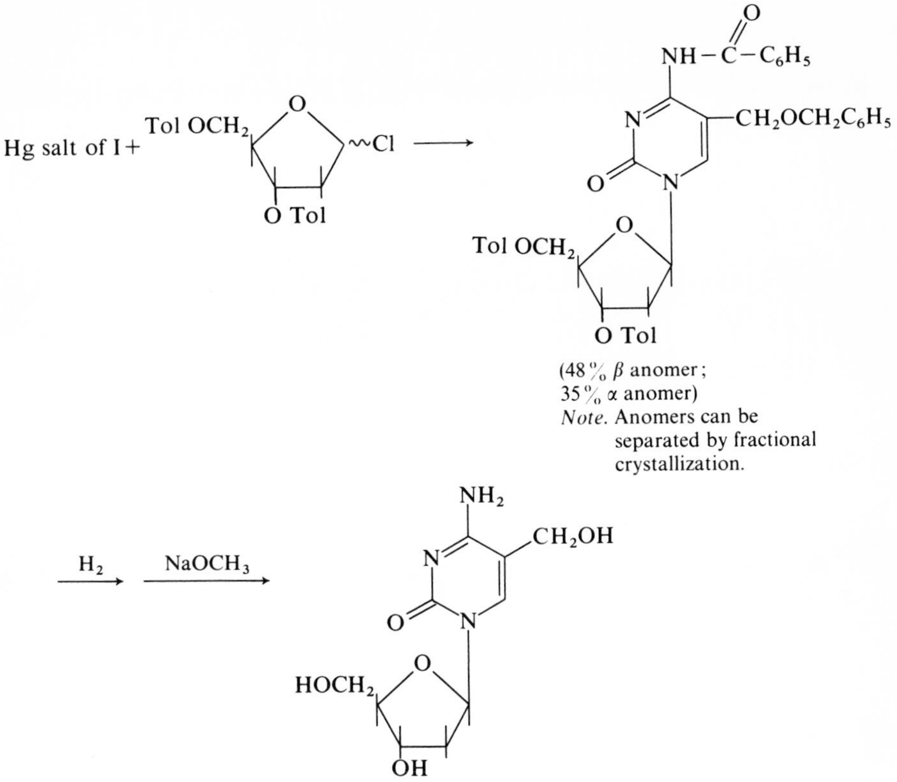

Hg salt of I+
Tol OCH2
O
~~Cl
O Tol
NH−C−C6H5
CH2OCH2C6H5
N
Tol OCH2
(48% β anomer;
35% α anomer)
Note. Anomers can be
separated by fractional
crystallization.
H2
NaOCH3
NH2
CH2OH
HOCH2
OH

(35% yield, based on Hg salt of I)

where Tol = *p*-toluyl.

2. Alegria (1967)

NH2
HCH
HO−P−OCH2
OH
CH2OH

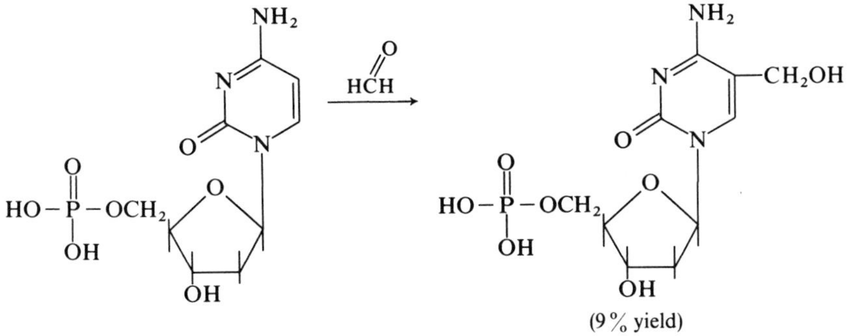

(9% yield)

79. DEOXYGUANOSINE   dGuo   dG

$C_{10}H_{13}N_5N_4$

M.W. 267.24

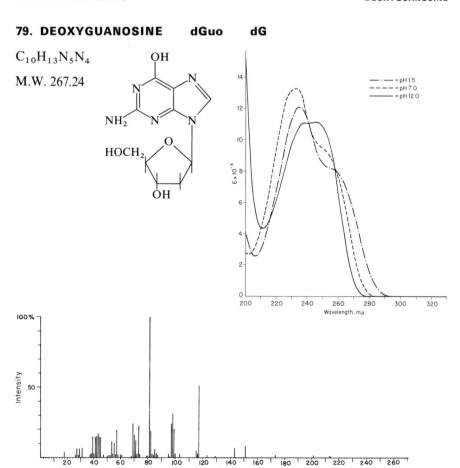

Physical Properties

M.P. 250°   Venner (1960)
$[\alpha]_D^{19} -47.7°$ (0.86, 1 N NaOH)   Schindler (1949)
pK (basic) 2.5

Additional Reference for Synthesis

Andersen et al. (1952) (isolation from DNA)

Synthesis

Venner (1960)

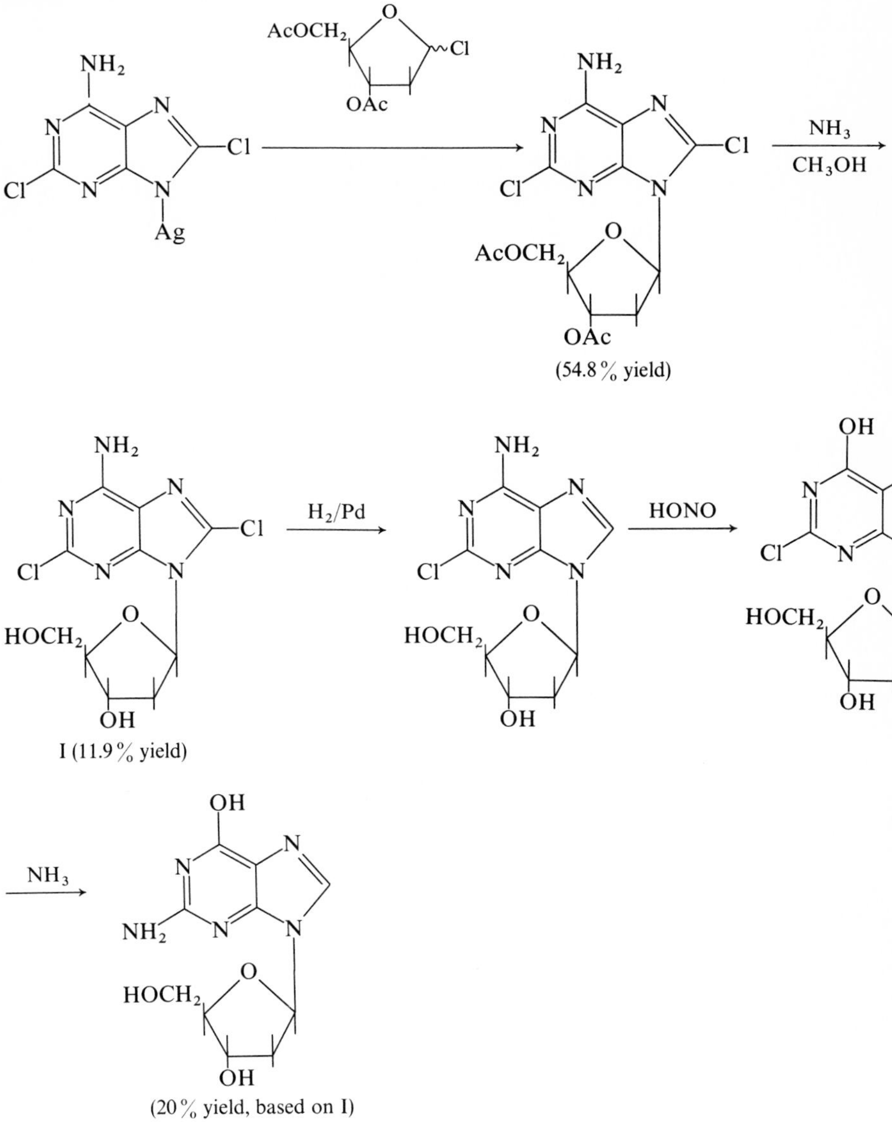

AcOCH2
OAc
Cl
NH2
N
Ag
NH3
CH3OH
(54.8% yield)
I (11.9% yield)
H2/Pd
HONO
HOCH2
OH
NH3
(20% yield, based on I)

where Ac = acetyl.

80. DEOXYURIDINE   dUrd   dU

$C_9H_{12}N_2O_5$

M.W. 228.20

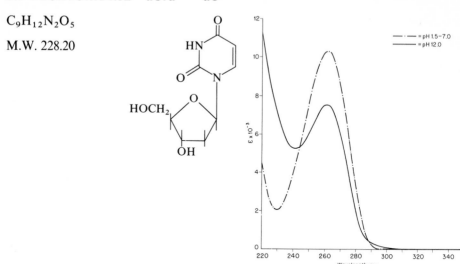

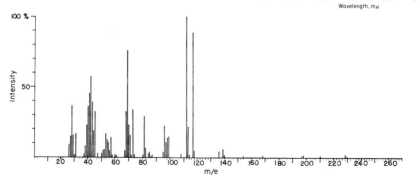

Physical Properties

M.P. 163°   Dekker and Todd (1950)
$[\alpha]_D^{22}$ +50° (1.1, 1 N NaOH)   Dekker and Todd (1950)
pK (acidic) 9.3   Fox and Shugar (1952)

Synthesis

1. Fox and Shugar (1952)

81. THYMIDINE (5-METHYLDEOXYURIDINE)   dThd   dT

$C_{10}H_{14}N_2O_5$

M.W. 242.23

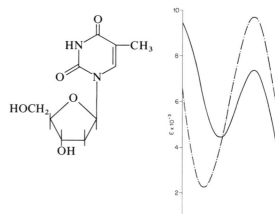

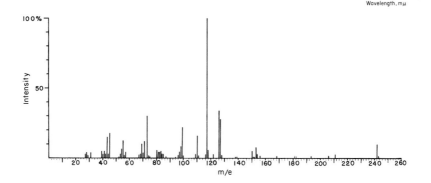

Physical Properties

M.P. 182.5–183.5°	Fox et al. (1959)
$[\alpha]_D^{16}$ +32.8° (1.04, 1 N NaOH)	Schindler (1949)
pK (acidic) 9.8, >13	Fox and Shugar (1952)

Synthesis

1. Ulbricht (1959)

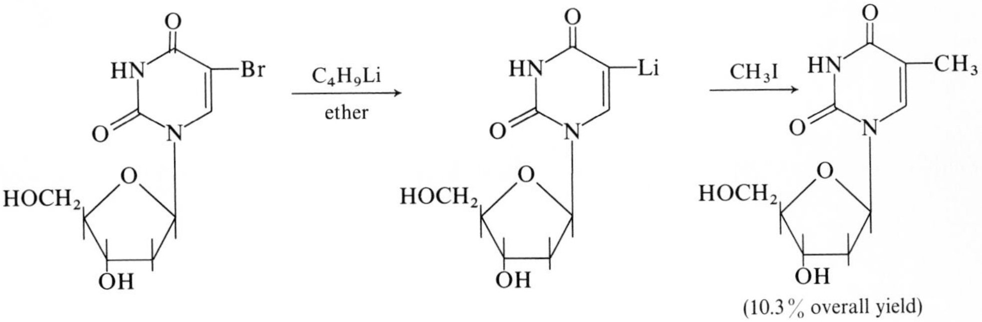

(10.3% overall yield)

2. Cline et al. (1959)

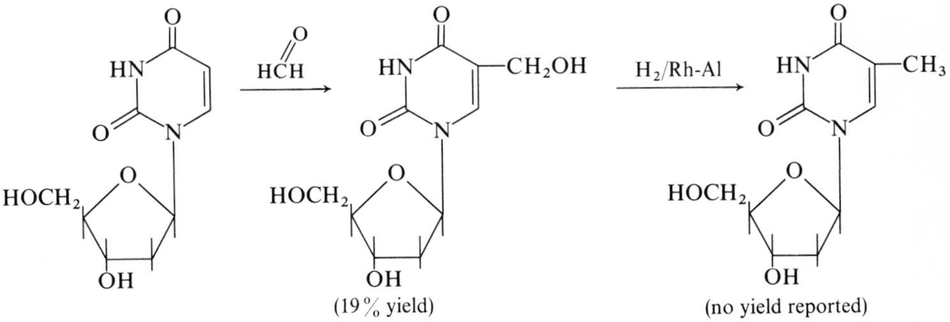

(19% yield)   (no yield reported)

3. Fox et al. (1959)

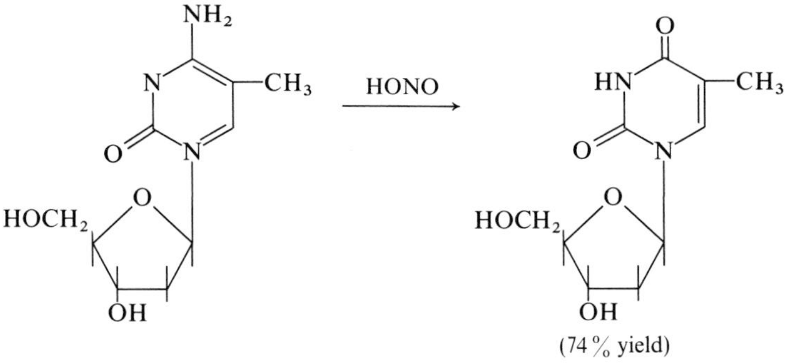

(74% yield)

See synthesis 77 for 5-methyldeoxycytidine.

82. 5-HYDROXYMETHYLDEOXYURIDINE   5-HmUrd   hm⁵dU

$C_{10}H_{14}N_2O_6$

M.W. 258.23

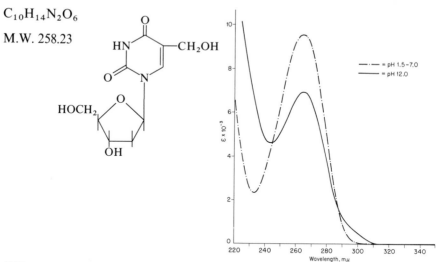

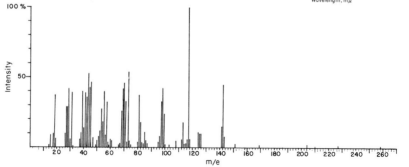

Physical Properties

> M.P. 180–182°   Brossmer and Röhm (1964)
> $[\alpha]_D^{20}$ +19° (water)   Brossmer and Röhm (1964)

Synthesis

1. Cline et al. (1959); Alegria (1967)

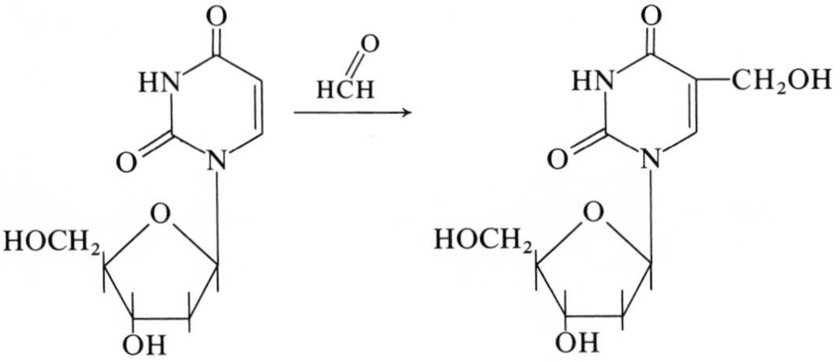

(19 % yield)

2. Brossmer and Röhm (1964)

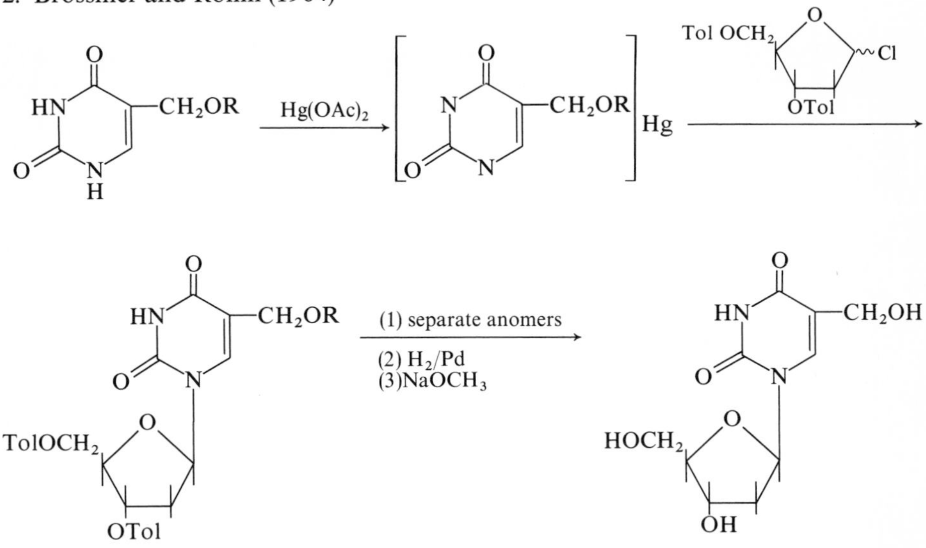

(1) separate anomers

(2) H₂/Pd
(3) NaOCH₃

(mixture of anomers: 30 % yield α;
49 % yield β)

where R = benzyl, and Tol = toluyl.

MODIFIED NUCLEOSIDE CONTENT OF ORGANISMS AND TISSUES

TABLE 1

Modified Nucleoside Content of Organisms and Tissues

The data in this table were compiled from the lists in the chapter. The values for each source are not directly comparable, since they may be the results of the work of more than one investigator.

		tRNA		
Ribonucleoside	*E. coli*	*Mouse Liver*	*Neurospora*	*Rat Liver*
1-(N^6)-Methyladenosine	0.10	0.15		1.6
2-Methyladenosine	0.25			
N^6-Methyladenosine				
N^6,N^6-Dimethyladenosine		0.11		0.02
2'-O-Methyladenosine			pres.	pres.
N^6-(Δ^2-Isopentenyl)adenosine				pres.
N^6-(Δ^2-Isopentenyl)-2-methylthioadenosine	pres.			
N-(Nebularin-6-ylcarbamoyl)threonine	0.07			pres.
3-Methylcytidine				pres.
$N^4,O^{2'}$-Dimethylcytidine				
5-Methylcytidine	1.5		0.8	2.0
2'-O-Methylcytidine				pres.
N^4-Acetylcytidine				
2-Thiocytidine	pres.			
1-Methylguanosine	0.14	0.45	0.5	0.8
N^2-Methylguanosine		1.6	0.10	0.4
N^2,N^2-Dimethylguanosine		0.28	0.65	0.65
7-Methylguanosine	0.50			pres.
2'-O-Methylguanosine				pres
Inosine				
1-Methylinosine				
5-Methyluridine	1.1		1.1	0.40
2'-O-Methyluridine			0.07	0.30
5'(2'-O-Methylribosyl)uracil				pres.
2-Thio-5-carboxymethyluridine methyl ester				
4-Thiouridine	0.9			
5-Carboxymethyluridine				
5-Ribosyluracil	2.0		4.3	4.3
5,6-Dihydrouridine	2.2–3.5			1.1
2-Thio-5-(N-methylaminomethyl)uridine	pres.			

[a] 16S RNA.   [b] 23S RNA.

| | | | rRNA | | | | |
Tobacco Leaves	Wheat Germ	Yeast	E. coli	Neurospora	Rat Liver	Wheat Germ	Yeast
	0.95	0.90			0.10		
0.04	pres.		0.02[b]				
			0.13[b]				
0.07			0.27[a]		0.02		
	pres.	0.03		pres.	pres.	0.54	pres.
		0.07					
		0.28					
		0.1					
			0.07[a]				
0.75	1.37	1.2	0.13[a]	0.02	0.02		0.02
			0.07[b]				
	0.04	0.28	0.03	pres.	pres.	0.34	pres.
		pres.					
0.50	0.73	1.00	0.07[b]		0.02		
0.40	0.29	0.35	0.27[a]		0.02		
			0.13[b]				
0.55	0.55	0.70			0.02		
0.20		0.35	0.13[a]		pres.		
			0.07[b]				
	0.47	0.25	0.02	pres.	pres.	0.38	pres.
0.10	0.09	0.33					
		0.05					
0.75		1.2	0.13[b]				
	0.30	0.06	0.07[b]	0.16	0.18	0.47	0.07
	0.12	pres.					
		pres.					
		0.34					
2.6	2.7	4.5	0.12[a]	1.2	1.6		0.5
			0.30[b]				
		3.0–3.6					

Chapter 3

METHODS FOR ISOLATION OF MODIFIED
NUCLEOSIDES; CHROMATOGRAPHIC DATA

1. INTRODUCTION

THE PROCEDURES DEVELOPED for the separation and detection of the modified components of nucleic acids are based on techniques of paper chromatography and paper electrophoresis. These basic techniques can be supplemented by column separation techniques for handling larger samples of nucleic acids. No one procedure enables the separation of all the known modified constituents, although some permit the identification of about twenty known modified constituents in a single sample of RNA.

Many of the methods were developed to enable isolation of a particular modified component and can be used only for this purpose. Nevertheless, such methods provide useful information about the separation characteristics of nucleic acid components. These methods, together with the more general procedures, provide a body of technical information that can serve as a basis for the design of a variety of procedures for the isolation of undetected nucleic acid components.

The procedures are classified in four sections. Section A describes several techniques common to many of the analytical procedures. Section B contains three general procedures that permit the separation of all the methylated components of RNA, as well as other components such as 5-ribosyluracil and inosine, starting from a single sample. These procedures use a sufficient sample (5–200 mg) which enables detection of the modified components by spectroscopy. Moreover, they have the added advantage that they can be scaled up;

one method, based on column partition chromatography, permits isolation of the modified constituents from gram quantities of RNA.

Section C describes methods suitable for the detection of modified constituents that have been selectively labeled, for example, $[^{14}C]$-methyl groups. In these procedures marker compounds can be added and the separation carried out expeditiously since there is no problem of detecting a small amount of a modified component in the presence of a large amount of the major components. Section D also contains several procedures developed for a specific analysis for a particular modified component. Section E lists analytical procedures useful for separating purine and pyrimidine derivatives, although these procedures have not been applied to the analysis of nucleic acid hydrolyzates.

A. General Procedures

1. COLUMN PARTITION CHROMATOGRAPHY (HALL, 1962, 1967): PURIFICATION OF DIATOMACEOUS EARTH

A slurry of Celite-545 (Johns Manville brand of diatomaceous earth) in 3 N HCl is poured into a column or large Büchner funnel. The Celite is washed with 3 N HCl until the effluent is clear. In the column method, a visible yellow band moves down the length of the Celite column. The material is washed with water until the effluent is neutral, then it is washed with ethanol. The Celite is spread in thin layers in trays and dried for 16 hr at 100°C. Microcel-E (Johns Manville brand of diatomaceous earth) can be purified in a similar manner, but since it compacts readily the washing should be carried out in a Büchner funnel.

Solvents. All solvents should be reagent grade and redistilled in an all-glass apparatus. The solvent systems are prepared by mixing the indicated proportions (by volume) and shaking the mixture in a reciprocating shaker for 30 min. The mixture should stand overnight in order to permit the phases to separate completely (see Section 8 for composition of solvents).

Temperature. The laboratory should be maintained at a constant temperature as precisely as possible, at least within $\pm 2°C$.

Packing of the Columns. The most simple and rapid method for the preparation of partition columns is the "dry pack" procedure, in which a quantity of Celite-545 is mixed with an amount of the aqueous (lower) phase sufficient to saturate the binding capacity of the Celite. In practice this amounts to approximately 1 part aqueous phase to 2 parts Celite-545. The mixture remains a free-flowing powder and can be packed into a glass column by means of a plunger. The column must be constructed of precision-bore tubing (heavy-wall tubing is preferable), and the plunger must be machined to fit the tube like a piston in a cylinder. Suitable glass tubing is available from the Fischer-Porter Co., Warminster, Pa. Figure 1 shows the design of a column 2.45 cm in diameter and its corresponding plunger. A column of this size will hold approximately 160 g of Celite-545. The Celite-aqueous phase mixture is compressed in small portions to form compacted layers, each equal in height to the diameter of the tubing. This procedure provides a column of even density with respect to cross-section and avoids "channeling" of the eluent.

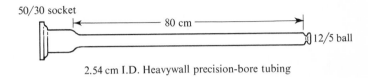

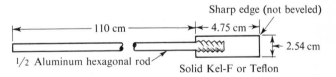

FIGURE 1. Dimensions of a glass tube and corresponding plunger suitable for partition chromatography. This column will hold approximately 160 g of Celite-545.

The most convenient method for introducing the sample onto the column is to dissolve the sample in a small amount of the aqueous (lower) phase of the solvent system and mix this solution with Celite-545. The Celite mixture is packed onto the top of the column in the manner described above.

2. ELECTROPHORESIS

Techniques for the electrophoresis of nucleotides are discussed in detail by Smith (1967). The relative mobilities of some of the modified nucleotides are presented in Table 1.

METHODS FOR ISOLATION OF MODIFIED NUCLEOSIDES

TABLE 1

Mobilities of Minor Ribonucleotides Relative to Up

Compound	pH 3.5[a]	pH 3.5[b]	pH 2.1[c]
4,5-Dihydrouridylic acid	1	...	...
Ribothymidylic acid	0.98	...	...
5-Ribosyluracil phosphate	0.98	...	...
Inosinic acid	0.9	...	...
Guanylic acid	0.73	...	0.38
7-Methylguanylic acid[d]	0.98	...	...
1-Methylguanylic acid	...	...	0.3
N^2-Methylguanylic acid	...	...	0.3
N^2,N^2-Dimethylguanylic acid	0.76	...	0.25
Adenylic acid	...	0.50	0
N^6-Methyladenylic acid	...	0.46	...
N^6,N^6-Dimethyladenylic acid	...	0.42	...
2-Methyladenylic acid	0.25	0.19	...
Cytidylic acid	...	...	0

[a] Sanger et al. (1965).
[b] Ammonium formate, 0.05 M, pH 3.5, Littlefield and Dunn (1958).
[c] Phosphate buffer, 0.05 M, pH 2.1.
[d] This compound presumably is migrating as the ring-opened derivative, 2-methyl-4-hydroxy-5-N-methylformamido-6-ribosylaminopyrimidine 2′(3′)-phosphate.

3. ENZYMIC HYDROLYSIS OF RNA TO YIELD A MIXTURE OF NUCLEOSIDES (HALL, 1965)

Five grams of yeast tRNA is dissolved in 500 ml of 0.005 M magnesium chloride solution. To this are added 500 mg of lyophilized *Crotalus adamanteus* venom (Ross Allen's Snake Farms, Silver Springs, Fla.) and 15 mg of bacterial alkaline phosphatase (Worthington Biochemicals Corp., Freehold, N.J.), grade BAP-C. A few drops of toluene are added to inhibit bacterial contamination, and the mixture is incubated at 37°. The pH of the solution is maintained at 8.6 by the addition of 0.5 N NaOH while the mixture is stirred continuously by means of a slow-moving paddle. After 7 hr, 0.5 ml of 1 M magnesium chloride, 5 mg of bacterial alkaline phosphatase, and 200 mg of snake venom are added. Incubation is continued until the uptake of NaOH has ceased (a total of 20–24 hr). The solution is heated at 60° for 30 min, then kept at 4° for 4 hr, and finally centrifuged at 15,000 × g for 1 hr. The clear supernatant is lyophilized.

4. ALKALINE HYDROLYSIS OF RNA (BOCK, 1967; SINGH AND LANE, 1964b)

Approximately 96% of the internucleotide bonds of RNA hydrolyze to give the 2'(3')-mononucleotides on treatment with 0.3 M potassium hydroxide for 16 hr at 37°. The potassium ions can be precipitated by neutralization of the solution with perchloric acid.

Under these conditions, however, a significant percentage of the phosphate bonds attached to the 3' position of the ribonucleosides remains unhydrolyzed. In particular, those bonds attached to the 3' hydroxyl of adenosine are resistant. Moreover, certain combinations of dinucleotides such as m$_2^6$Apm$_2^6$A are very resistant. This fact must be taken into consideration if one is using this hydrolytic technique to release modified nucleosides. And if the object of the experiment is to obtain the normally "alkali-resistant" dinucleotides, namely, those containing 2'-O-methyl-nucleosides, incomplete hydrolysis can yield false results. Singh and Lane (1964b) recommend that the hydrolysis be carried out in 1 M sodium hydroxide for 90 hr at 22° (see p. 219).

5. RELEASE OF BASES BY PERCHLORIC ACID DIGESTION*

Dry nucleic acid is dissolved in 72% perchloric acid to give a concentration of nucleic acid in solution of 6–8%. After the solution is heated at 100° for 1–2 hr, a sample is diluted with an equal volume of water and centrifuged briefly in order to remove the carbon residue. The solution can be applied directly to the paper, but it is important that the spot not be allowed to dry before chromatography, or the acid will attack the paper. The most satisfactory solvent system is C. Under these conditions perchloric acid does not interfere with separation of the bases. Direct chromatography of the hydrolyzate works well for DNA since this solvent system separates the five bases. Because of variability in the Rf values of this procedure, it is advisable to apply markers dissolved in perchloric acid to the chromatogram. If some other chromatographic procedure is used for the first step, it is preferable to dilute the sample with water, neutralize the solution with potassium hydroxide, and filter off the potassium perchlorate. Dr. D. B. Dunn (personal communication) points out, however, that there is a danger of loss of purines on neutralization due to their insolubility.

*Littlefield and Dunn, 1958; Marshak and Vogel, 1950; Sluyser and Bosch, 1962; Wyatt, 1951a.

6. PAPER CHROMATOGRAPHY

Most procedures described in this chapter rely on paper chromatographic techniques. According to the size of the sample to be chromatographed, either Whatman No. 1 or Whatman 3 MM paper is satisfactory. Although prewashing of the papers may be helpful, it is not essential for most routine work.

Recovery of the separated products from the developed chromatogram can be made by elution with water or, in the case of guanine derivatives, dilute ammonium hydroxide or dilute formic acid. A method for recovering trace amounts of material from paper is described by Heppel (1967). Rf values for most of the modified nucleosides are tabulated in the Appendix (p. 409).

7. DEPHOSPHORYLATION WITH PHOSPHOMONOESTERASE; DESALTING OF SOLUTIONS

Many analytical procedures require the removal of phosphate groups, for example, conversion of mononucleotides to nucleosides or dinucleotides to dinucleoside phosphates. A number of phosphomonoesterases can be used, such as prostatic phosphomonoesterase (Littlefield and Dunn, 1958; Markham and Smith, 1952b). Perhaps the most convenient enzyme commercially available (Worthington Biochemical Corp., Freehold, N.J.) is the alkaline phosphatase obtained from *Escherichia coli* (E.C. 3.1.3.1).

The following typical dephosphorylation procedure is described by Tamaoki and Lane (1968). A solution of 1 μmole of dinucleotides in 50 μl of 1 M ammonium formate buffer (pH 9.2) is mixed with 10 μl of *E. coli* alkaline phosphatase (10 mg/ml; 32 units/mg), and the solution is incubated for 3 hr at 37° in order to remove the phosphomonoester groups. The pH of the solution is adjusted to 4.0, and the nucleotide mixture is desalted by passing the solution through 200 mg of charcoal; the charcoal is mounted between two one-inch layers of Celite-545 in a chromatographic tube 2.4 cm in diameter. The charcoal is washed with three 30 ml lots of water, and the dinucleoside phosphates are eluted with approximately 100 ml of 10% pyridine in 50% aqueous alcohol.

Alternatively, the nucleotides can be desalted by chromatography of the mixture on paper in solvent O. Inorganic phosphate has an Rf similar to that of guanylic acid and will contaminate this product or any other that migrates at this or a slower rate. If the ratio of nucleotides to buffer and enzyme is increased, it may be possible to apply the digest directly to paper and achieve a satisfactory separation of the products. The possibility of incomplete digestion of the sample, as well as the adverse effects of the salt, generally precludes the use of this shortcut, except for less precise separations.

8. SOLVENT SYSTEMS (MEASUREMENT BY VOLUME)

A. 1-Butanol : water : conc. ammonium hydroxide (86 : 14 : 5)

B. 2-Propanol : 1 % aqueous ammonium sulfate (2 : 1) (Anand et al., 1952)

C. 2-Propanol : conc. hydrochloric acid : water (680 : 176 : 144) (Wyatt, 1951b)

D. 2-Propanol : water : conc. ammonium hydroxide (7 : 2 : 1)

E. Ethyl acetate : 1-propanol : water (4 : 1 : 2) (Hall, 1963d)

F. Ethyl acetate : 2-ethoxyethanol : water (4 : 1 : 2) (Hall, 1965)

G. Ethyl acetate : 1-butanol : ligroin (b.p. 66–75°) : water (1 : 2 : 1 : 1) (Hall, 1965)

H. 1-Butanol : water : conc. ammonium hydroxide (3 : 1 : 0.05)

J. Ethyl acetate : 1-butanol : water (1 : 1 : 1) (Hall, 1965)

K. 1-Butanol : conc. ammonium hydroxide : sodium borate solution (38 g of $Na_2B_4O_7 \cdot 10H_2O$ per liter) (3 : 0.05 : 1) (Hall, 1964b)

L. 1-Butanol : 95 % ethanol : water (50 : 17 : 35) (Markham and Smith, 1949)

M. 95 % Ethanol : water (80 : 20) (Paper dipped in solution of 1 part saturated $(NH_4)_2SO_4$ solution and 9 parts water; air dry before use) Lane (1963)

N. 2-Butanone : water (3 : 1)

O. 2-Propanol : water (7 : 3), NH_3 in vapor phase (Markham and Smith, 1952a)

P. 1-Butanol : water : 98 % formic acid (77 : 13 : 10) (Markham and Smith, 1949)

Q. 1-Butanol : water (86 : 14), NH_3 in vapor phase (Hotchkiss, 1948 ; Partridge, 1948)

R. 2-Propanol : conc. hydrochloric acid : water (170 : 41 : 39)

S. 2-Propanol : conc. hydrochloric acid : water (65 : 16.7 : 18.3)

T. 1-Propanol : 25 % ammonia : water (6 : 3 : 1)

U. 2-Propanol : sat'd ammonium sulfate solution : water (2 : 79 : 19) (Markham and Smith, 1951)

V. 1-Butanol : acetic acid : water (5 : 3 : 2)

W. 2-Propanol : sat'd ammonium sulfate solution : 1 M sodium acetate (2 : 80 : 18) (Markham and Smith, 1952a)

X. Isobutyric acid : 0.5 N ammonium hydroxide (5 : 3) (Magasanik et al., 1950)

Y. t-Amyl alcohol : water : 89 % formic acid (6 : 3 : 1) (Hanes and Isherwood, 1949)

Z. 2-Propanol : 0.5 N acetic acid (80 : 20) (D. B. Dunn, personal communication)

B. General Procedures for the Simultaneous Isolation of Several Modified Nucleosides

9. THE ISOLATION OF N-METHYLATED NUCLEOSIDES, 5-METHYLURIDINE, 5-RIBOSYLURACIL, AND INOSINE BY PAPER CHROMATOGRAPHY AND PAPER ELECTROPHORESIS*

The RNA sample (2–50 mg, depending on the level of the modified nucleosides in the RNA; 2–5 mg is sufficient to estimate the above compounds by spectroscopy if they are present in amounts $> 0.12\%$ of the constituent nucleosides) is hydrolyzed in $1.0\,M$ potassium hydroxide for 18 hr at $37°$, and the potassium ions are precipitated by neutralization with perchloric acid. The hydrolyzate is chromatographed twice for 18–24 hr in solvent O (the chromatogram is dried between runs). Under these conditions three bands are obtained. The slowest moving band (1) contains principally guanylic acid, the next band (2) contains adenylic, cytidylic, and uridylic acids, and the fastest moving band (3) contains the methylated adenylic acid derivatives and 5-methyluridylic acid. The material in the three bands is analyzed as follows:

*Band 1. Guanylic Acid, Inosinic Acid, 5-Ribosyluridylic Acid, and the Degradation Product of 7-Methylguanylic Acid.*** The band is eluted in water and the material is subjected to electrophoresis in $0.05\,M$ phosphate (pH 2.2). Several bands containing dinucleotides, including the 2′-O-methyl products (NmpNp) are obtained. 5-Ribosyluracil phosphate migrates as a distinct band (7.1 cm/hr at 20 V/cm), and inosinic acid and the product of 7-methylguanylic acid migrate together with the dinucleotide GpUp (6.4 cm/hr at 20 V/cm). These three compounds are eluted and hydrolyzed with 0.1 N hydrochloric acid for 18 hr at $37°$. Under these conditions the 7-methylguanylic acid degradation product is converted to 2,6-diamino-4-hydroxy-5-N-methylformamidopyrimidine. The three products are separated by chromatography in solvent O. The 7-methylguanylic acid product migrates at about the rate of guanosine. Inosinic acid moves

---

* Dunn, 1960; Littlefield and Dunn, 1958; Smith and Dunn, 1959a,b. I am indebted to Dr. Dunn for providing this composite procedure which is based on his methods described in the cited publications.

** 2-Amino-4-hydroxy-5-N-methylformamido-6-ribosylaminopyrimidine 2′(3′)-phosphate.

slightly faster than GpUp, and it can be dephosphorylated and identified as the nucleoside.

The alkali-stable dinucleotide bands containing 2'-O-methylribonucleosides (NmpNp) can be eluted together and the products separated by procedures described in Section 10.

Band 2. Adenylic, Cytidylic, and Uridylic Acids; Methylated Guanylic Acid Derivatives. The material is eluted and the products are separated by paper electrophoresis in 0.05 M-phosphate buffer, pH 2.5. At this pH adenylic and cytidylic acid have almost a net charge of zero, whereas uridylic acid has a negative charge of one. When electrophoresis is carried out at 20 V/cm, N^2,N^2-dimethylguanylic acid moves towards the anode at 2.6 cm per hr, and 1-methylguanylic acid and N^2-methylguanylic acid migrate together at a rate of 2.9 cm per hr. Uridylic acid migrates at 7.4 cm per hr.

The three methylated guanylic acids are eluted together and the material is chromatographed on paper in solvent O to remove phosphate ions. The nucleotides are eluted and dephosphorylated with phosphomonoesterase, and the nucleoside mixture is chromatographed on paper in solvent P. N^2,N^2-dimethylguanosine separates from the two monomethyl derivatives, which in turn can be separated from each other either by paper chromatography in solvent Q (m^1G migrates at a slower rate than m^2G) or by electrophoresis in 0.5 M sodium tetraborate (pH 9.2). The mobilities in cm/hr (20 V/cm) for 1-methylguanosine and N^2-methylguanosine are $+5.0$ and $+5.9$, respectively.

5-Methylcytidylic acid can be separated from cytidylic acid in this mixture if electrophoresis is carried out at pH 3.5. The separated product is dephosphorylated and isolated as described in Section 21.

Band 3. Methylated Adenylic Acid Derivatives and 5-Methyluridylic Acid. The chromatogram section between the leading edge of band 2 and close to the solvent front is eluted with water, and the nucleotides are dephosphorylated with phosphomonoesterase. The hydrolyzate is rechromatographed on paper in solvent O to remove ions. Adenosine could be run on this sheet as a marker. The chromatogram area from and including the leading third of the marker to close to the solvent front is eluted and the nucleoside mixture is resolved by two-dimensional paper chromatography, as shown in Figure 2.

Under the conditions of the isolation procedure, 1-methyladenosine rearranges to N^6-methyladenosine. The methylated adenosines and 5-methyluridine can also be isolated by the general procedures in Sections 10 and 11.

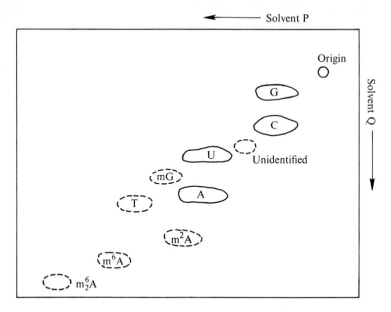

FIGURE 2. Diagram of an autoradiograph of [^{14}C-methyl]-methylated nucleosides derived from the rRNA of *E. coli* by digesting RNA with potassium hydroxide followed by enzymic removal of the phosphate by a phosphatase. The chromatogram is developed for 70 hr in solvent P (the ratio of water and formic acid is changed to 11 and 12, respectively), and in the second dimension for 40 hr in solvent Q. Any 1-methyladenosine in the RNA will be isolated as N^6-methyladenosine. Data taken from Starr and Fefferman (1964), who used the separation procedure described by Littlefield and Dunn (1958).

10. THE ISOLATION OF *N*-METHYLATED NUCLEOSIDES, 2'-*O*-METHYLATED NUCLEOSIDES, 5-METHYLURIDINE, AND INOSINE BY ION-EXCHANGE CHROMATOGRAPHY AND PAPER CHROMATOGRAPHY*

In this procedure the RNA is digested exhaustively with alkali to give a mixture of the nucleoside 2'(3')-phosphate. In addition, the hydrolyzate contains a small fraction of nucleosides derived from the 3' end of the RNA chains and alkali-stable dinucleotides. The latter fraction contains the 2'-*O*-methylribonucleosides

* Hudson et al., 1965; Lane, 1965; Nichols and Lane, 1966a, 1968c; Singh and Lane, 1964a,b. I wish to thank Dr. Lane for his assistance in preparing this account of the overall procedure.

(NmpNp); to avoid misleading results, therefore, the normal NpN phosphate bonds must be cleaved (see p. 213 for further comment).

This method is excellent for processing relatively large samples of RNA, and the following procedure is described in terms of a 200-mg sample of tRNA. To a solution of RNA (200 mg) in 5 ml of water is added 0.56 ml of 10 M sodium hydroxide solution. The solution is allowed to stand for 90 hr at 22° and is then neutralized with formic acid. The neutralized hydrolyzate is diluted to 700 ml with water and applied to a column [2.5 cm (diam.) × 20 cm] of DEAE-cellulose (formate). The column is washed with 100 ml of 0.025 M TRIS-formate, pH 7.8 buffer, to remove the nucleosides.

The nucleoside monophosphates, the alkali-stable dinucleotides, the alkali-stable trinucleotides, and the nucleoside diphosphates (from the 5′ end of the RNA) have net charges of -2, -3, -4, and -4, respectively, at pH 7.8. Removal of these materials from DEAE-cellulose is accomplished by stepwise elution with TRIS-formate buffers, pH 7.8 in 7 M urea, according to Tomlinson and Tener (1963); pH values are measured in the absence of urea.

The nucleoside monophosphates are eluted from the column with 0.085 M TRIS-formate, pH 7.8 (7 M urea); approximately 750–1000 ml is required. The alkali-stable dinucleotides are eluted with 0.17 M TRIS-formate, pH 7.8 buffer (7 M urea); approximately 600–800 ml is required. Nucleoside diphosphate and alkali-stable trinucleotides (NmpNmpNp) are eluted with 1 M pyridinium-formate (pH 4.5) solution (spectral grade pyridine should be used).

Nucleoside Fraction. The effluent from the charging solution and the 0.025 M TRIS-formate wash contains the nucleosides derived from the chain terminal of the RNA. The nucleosides can be desalted by charcoal adsorption (Section 7) and separated by paper chromatography for 15–24 hr in solvent M (Lane, 1963).

Mononucleotide Fraction. This fraction is desalted by diluting it threefold and adsorbing the nucleotides to a column [2.5 cm (diam.) × 20 cm] of DEAE-formate (Tomlinson and Tener, 1963). The column is washed with water, and the nucleotides are desorbed with 1 M pyridinium-formate. This buffer is volatile and can be removed *in vacuo*. The level of methylated components in the tRNA of yeast and higher organisms is sufficient to enable their detection in this mixture by direct two-dimensional chromatography (i.e., Figure 3a). For the tRNA of bacteria and rRNA which have a relatively low content of modified nucleosides, prior removal of the bulk of the major ribonucleotide fraction may be necessary (see procedures 9 and 11).

METHODS FOR ISOLATION OF MODIFIED NUCLEOSIDES

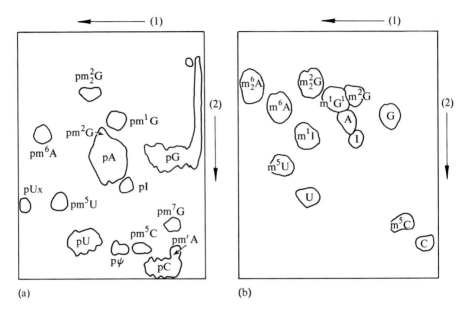

(a)   (b)

FIGURE 3. (a) Two-dimensional separation of *N*-methylated ribonucleoside 5'-phosphates using the solvent systems listed in Figure 4. The chromatogram is developed for 24 hr in each direction. In this example a venom phosphodiesterase digest of wheat germ tRNA is applied directly to the paper. The paper is slightly overloaded in order to enable detection of the methylated nucleotides. Data taken from Hudson et al. (1965). (b) The identical procedure is used to separate a mixture of the methylated ribonucleosides (Nichols and Lane, 1968c).

Dinucleotide Fraction. This fraction is desalted by absorbtion to DEAE-cellulose (formate) as described for the mononucleotide fraction using a corresponaingly smaller column, and the dinucleotides are treated with *E. coli* phosphomonoesterase to remove the phosphomonoester groups. The resulting dinucleoside phosphates are resolved by two-dimensional paper chromatography (Figure 4).

The alkali-stable trinucleoside diphosphates can be partially resolved by two-dimensional paper chromatography as described by Lane (1965). One version of this overall procedure useful for the detection of radioactive labeled *N*- and $O^{2'}$-methylated components of RNA is described in Section 14.

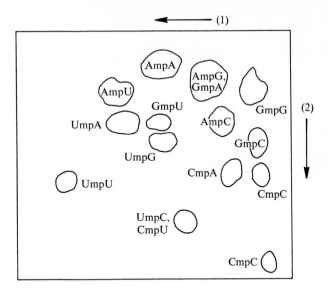

FIGURE 4. Two-dimensional chromatogram showing the separation of the alkali-stable dinucleoside phosphates (NmpN). Whatman No. 1 paper is dipped into a solution consisting of 1 volume of saturated ammonium sulfate solution and 9 volumes of water and air-dried before applying the sample. The chromatogram is developed for 24 hr in the first dimension in solvent M and for 24 hr in the second dimension in solvent W (modified to consist of 80 volumes of saturated ammonium sulfate solution and 2 volumes of isopropanol). Data taken from Singh and Lane (1964a, b).

11. THE ISOLATION OF *N*-METHYLATED NUCLEOSIDES, 2'-*O*-METHYLATED NUCLEOSIDES, 5-METHYLURIDINE, 5-METHYLCYTIDINE, 5-RIBOSYLURACIL, *N*-(NEBULARIN-6-YLCARBAMOYL)-L-THREONINE, AND INOSINE BY COLUMN PARTITION CHROMATOGRAPHY (HALL, 1965)

The following procedure is particularly suitable for large-scale fractionation of mixtures of nucleosides. The procedure is described in terms of a 5-g sample of yeast tRNA.

Step 1. Primary Column. The nucleosides in the digest are first separated into six primary fractions, and the mixture in each of these fractions is resolved by using a second system. The column for the primary separation consists of a precision-bore glass pipe, 5.08 (diam.) × 105 cm. The column is charged with 690 g of Celite-545 : Microcel-E (9 : 1) which has been mixed thoroughly with 308 ml of the lower phase (aqueous) of solvent F. A lyophilized enzymic digest (p. 212), starting from 5.0 g of tRNA, is dissolved in 35 ml of the lower phase of solvent F, and the pH is adjusted to 7.0. The solution is filtered to remove insolubles, which include a portion of the guanosine, and the clear solution is mixed with 80 g of the Celite : Microcel-E (9 : 1) mixture, which is packed into the column by means of the procedure described in Section 1. The column is developed with the upper phase of solvent F at a flow rate of 600 ml/hr. The optical density of the effluent is monitored continuously at 270 mμ, using a flow cell with a path length of 1 mm.

The elution pattern obtained is shown in Figure 5. When the uridine fraction is completely eluted (point A, Figure 5), the eluent is changed from the upper phase of solvent F to the upper phase of solvent G. The change in solvents, on

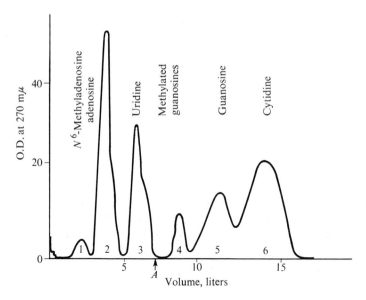

FIGURE 5. Fractionation of a nucleoside mixture obtained by enzymic hydrolysis of 5.4 g of yeast tRNA on a column [5.08 cm (diam.) × 80 cm] containing 690 g of Celite-545, Microcel-E (9 : 1) mixed with 308 ml of aqueous phase of solvent F. The hydrolyzate, dissolved in 35 ml of aqueous phase, is mixed with 80 g of Celite-Microcel mixture. The column is developed with the upper phase of solvent F until the uridine fraction is eluted (point A). Then the solvent is changed to the upper phase of solvent G. The flow rate is 600 ml/hr. Data taken from Hall (1965, 1967).

occasion, causes the separation of a small amount of water from the next few hundred milliliters of the effluent. This condition appears to be due to the mixing of the two solvent systems, as well as to the fact that solutes with low solubility (methylated guanosines and guanosine) start to appear. There is no noticeable impairment in the resolving power of the column. About 30 hr is required to complete the separation, and for convenience in laboratory scheduling the column may be operated on three consecutive days. The interruptions do not appear to diminish the resolving power or to cause the appearance of specious fractions.

Each of the six fractions corresponding to the six peaks shown in Figure 5 is evaporated *in vacuo* to near dryness in a rotating glass evaporator (bath temperature 30°). Water (250 ml) is added, and the mixture is re-evaporated. This procedure should be repeated at least three times in order to remove organic solvent. The residue is finally redissolved in water, and the solution is filtered and lyophilized.

Step 2. Secondary Columns. Four of the primary fractions, excluding fractions 2 (adenosine) and 3 (uridine), are refractionated on smaller partition columns

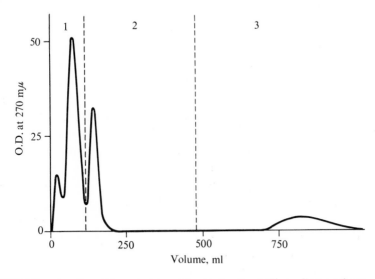

FIGURE 6. Fractionation of the material contained in fraction 1 (Figure 5) on a column [2.54 cm (diam.) × 80 cm] containing 150 g of Celite-545 mixed with 72 ml of lower phase of solvent H. The sample, dissolved in 9 ml of lower phase, is mixed with 18 g of Celite-545. Flow rate is 150 ml/hr.

METHODS FOR ISOLATION OF MODIFIED NUCLEOSIDES

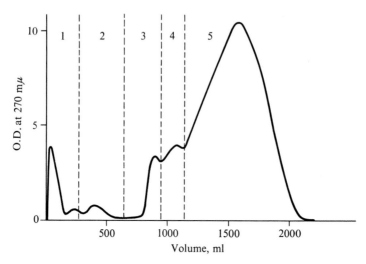

FIGURE 7. Fractionation of the material contained in fraction 4 (Figure 5) on a column [2.54 cm (diam.) × 80 cm] containing 150 g of Celite-545 mixed with 66 ml of the lower phase of solvent E. The sample, dissolved in 11 ml of lower phase, is mixed with 22.5 g of Celite-545. The column is developed with the upper phase of solvent E at a flow rate of 150 ml/hr.

[2.54 cm (diam.) × 86 cm]. Each column is packed with 150 g of Celite-545 mixed with the specified amount of the aqueous phase of the solvent system used, and the columns are developed at a flow rate of 50 ml/hr. The elution profiles of the four columns are shown in Figures 6–9.

Step 3. Isolation by Paper Chromatography. Final isolation of the modified nucleosides is accomplished by means of paper chromatography. The effluent from each secondary column is divided according to the subfractions shown in the elution profiles. In some cases divisions are arbitrary, with the result that certain modified nucleosides appear in more than one fraction. The subfractions are concentrated *in vacuo* so that the concentration of nucleosides in the final solution is approximately 5 mg/ml. The entire concentrate (for larger fractions an aliquot can be taken) is streaked on Whatman No. 3 MM paper, which is developed in the indicated solvent system for a specified period of time (see Figure 10). The separated nucleosides are identified by their ultraviolet absorption spectra and by their chromatographic mobilities in several solvent systems (see Appendix). The entire isolation procedure is presented schematically in Figure 10. The modified nucleosides detected in a hydrolyzate of yeast tRNA (Hall, 1965) are listed in Table 2.

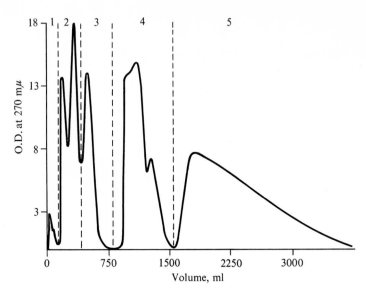

FIGURE 8. Fractionation of material contained in fraction 5 (Figure 5) on a column [2.54 cm (diam.) × 80 cm] containing 150 g of Celite-545, mixed with 72 ml of lower phase. The sample is dissolved in 10 ml of hot water, filtered, and then kept at 4° for several days. The precipitate, consisting mostly of guanosine, is filtered off and the filtrate is lyophylized. The residue is dissolved in 10 ml of lower phase and mixed with 22 g of Celite-545. The column is developed with the upper phase of solvent H at a flow rate of 150 ml/hr.

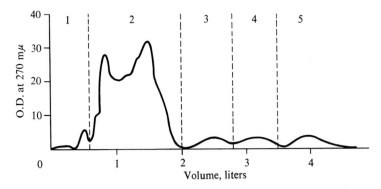

FIGURE 9. Fractionation of material contained in fraction 6 (Figure 5) on a column [2.54 cm (diam.) × 80 cm] containing 150 g of Celite-545 mixed with 65 ml of lower phase. The solution of the sample in 14 ml of lower phase is mixed with 32 ml of Celite-545. The column is developed with the upper phase of solvent J at a flow rate of 150 ml/hr.

METHODS FOR ISOLATION OF MODIFIED NUCLEOSIDES

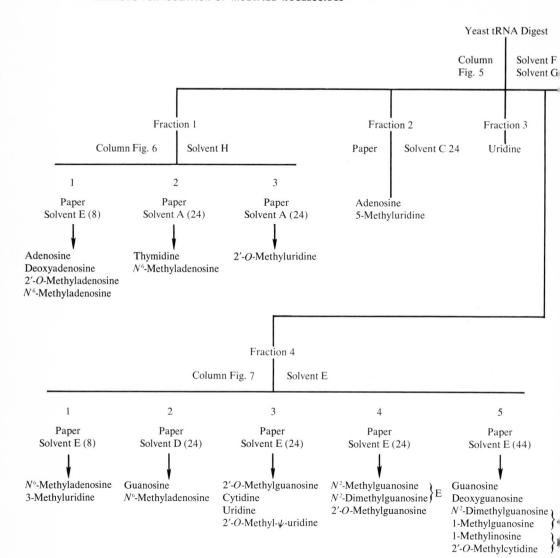

FIGURE 10. Flow diagram for separation of the modified nucleosides of enzymic digest of tRNA. (The numbers in parentheses after each solvent letter indicate the number of hours the chromatogram is developed at 25°.)

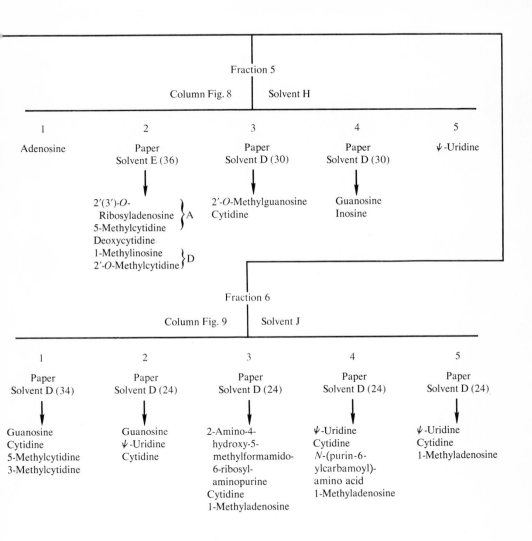

METHODS FOR ISOLATION OF MODIFIED NUCLEOSIDES

TABLE 2

Summary of the Modified Nucleosides Isolated by the Procedure Shown in Figure 10 (Data taken from Hall, 1965)

The amounts listed are those isolated. For a more accurate representation of the amount of each modified nucleoside in yeast tRNA see Table 1, Chapter 2.

Nucleoside	Moles/100 Moles of Total nucleosides
2'-O-Methyladenosine	0.03
1-Methyladenosine	0.13
N^6-Methyladenosine	0.4
N-(Nebularin-6-ylcarbamoyl)-threonine	0.04
2'(3')-O-Ribosyladenosine	0.01
2'-O-Methylcytidine	0.1
3-Methylcytidine	0.01
5-Methylcytidine	0.7
2'-O-Methylguanosine	0.3
1-Methylguanosine	0.1
N^2-Methylguanosine	0.02
N^2,N^2-Dimethylguanosine	0.2
7-Methylguanosine	0.05
2-Amino-4-hydroxy-5-N-methylform-amido-6-ribosylaminopyrimidine	
Inosine	0.2
1-Methylinosine	0.04
2'-O-Methyluridine	0.03
3-Methyluridine	0.01
5-Methyluridine	0.7
5-Ribosyluracil	4.5
5-(2'-O-Methylribosyl)uracil	0.001

Comments. This separation procedure, which is designed for approximately 4–6 g of RNA, can be readily scaled down and smaller-sized columns used. In our laboratory, in addition to the two sizes of columns mentioned, 1.9 cm (diam.) × 60 cm (60 g of Celite) and 1.2 cm (diam.) × 42 cm (30 g of Celite) are used routinely.

12. SEPARATION OF THE 2'(3')-MONONUCLEOTIDES ON AN ANION-EXCHANGE COLUMN (BANK ET AL., 1964; DUBIN AND GUNALP, 1967)

A sample of unfractionated tRNA (approximately 100 μmoles) is hydrolyzed in potassium hydroxide, and the hydrolyzate is neutralized with perchloric acid. The solution is run onto a column (0.25 cm^2 × 8 cm) of Dowex-1 × 8 [formate] (<400 mesh), which is eluted consecutively with 25 ml of water, 90 ml of a linear gradient 0.1→0.2 M formic acid, 120 ml of a linear gradient 0.2→1.0 M formic acid, and 70 ml of 1.0 M formic acid. Two milliliter fractions are collected, and the radioactivity of each fraction and the optical density at 260 mμ are determined. Figure 11 shows the elution profile of this column. The resolution of the column depends greatly on the fact that its loading is very low.

13. SEPARATION OF METHYL-LABELED COMPONENTS BY TWO DIMENSIONAL CHROMATOGRAPHY (BJORK AND SVENSSON, 1967)

The tRNA is hydrolyzed in 1 M hydrochloric acid for 30 minutes at 100°, which gives a mixture of purine bases and pyrimidine nucleotides (Markham and Smith, 1952b). A sample of the hydrolyzate not exceeding 0.15 mg of tRNA is diluted with 10 μg of authentic N^6,N^6-dimethyladenine and 5 μg of each of the other methylated compounds as markers. The mixture is applied to a thin-layer plate prepared from superfine grade of Avicel microcrystalline cellulose (American Viscose Corp., Marcushook, Pa.). The plates are prepared by suspending 40 g of the cellulose powder in 160 ml of distilled water. The suspension is allowed to stand for 15 min and then homogenized in a Waring Blender for 90 sec at 6000 rpm. The suspension is degassed under vacuum and poured into a spreading device. The plates are made 0.2 mm thick and dried overnight at room temperature and then for 30 min at 70°.

The thin-layer plate is developed in the first direction for about 7 hr in solvent Q (15 cm) and for about 17 hr in the second direction in solvent R (15 cm). After drying, the chromatogram is exposed to ammonia vapor for about 15 min in order to convert the guanine-fluorescent spots into ultraviolet-absorbing spots; this facilitates the detection of the guanine derivatives. A developed plate is shown in Figure 12.

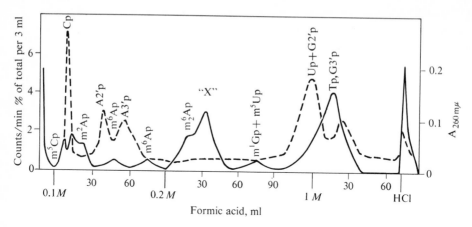

FIGURE 11. Chromatographic separation of the 2'(3')-mononucleotides obtained from an alkaline digest of [14C-methyl]-labeled tRNA on a column of Dowex-1 × 8. The peak labeled "X" has been identified as the ring-opened 7-methylguanylic acid derivative, 2-methyl-4-hydroxy-5-N-methylformamido-6-ribosyl-amino-pyrimidino 2'(3')-phosphate (Dunn and Spahr, private communication). An elution profile of the major nucleotides has been superimposed on top of the [14C] profile. See Section 12 for details of the procedure. Data taken from Dubin and Gunalp (1967).

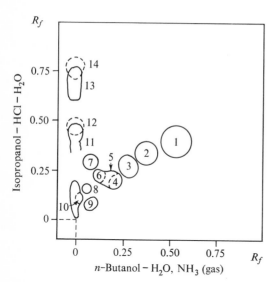

FIGURE 12. Chromatography of a mixture of purine bases and pyrimidine nucleotides obtained from a 1 M hydrochloric acid digest of tRNA on thin-layer plate containing microcrystalline cellulose. The plate is developed in the first direction for 7 hr in solvent Q and in the second direction for 17 hr in solvent R. The plate is 15 cm square. See Section 13 for details of the procedure. Markers are added to the hydrolyzate in order that the radioactive compounds can be readily identified. The marker compounds are as follows: 1, N^6,N^6-dimethyladenine; 2, N^6-methyladenine; 3, 2-methyladenine; 4, 1-methyladenine; 5, adenine; 6, N^2,N^2-dimethylguanine; 7, N^2-methylguanine; 8, 7-methylguanine; 9, 1-methylguanine; 10, guanine; 11, cytidylic acid; 12, methylated cytidylic acid; 13, uridylic acid; 14, methylated uridylic acid. Data taken from Bjork and Svensson (1967).

14. RESOLUTION BY PAPER CHROMATOGRAPHY; SEPARATION OF METHYLATED COMPONENTS INTO TWO GROUPS; N-METHYL AND $O^{2'}$-METHYL DERIVATIVES (TAMAOKI AND LANE, 1968)

$[^{14}C$-Methyl]-RNA (1–5 mg) is dissolved in 225 μl of water and mixed with 25 μl of 10 M sodium hydroxide, and the solution is incubated for 90 hr at room temperature. The hydrolyzate is neutralized by the addition of 100 μl of 10% acetic acid and evaporated to dryness. The dry residue is mixed with 200 μl of carrier NmpNp compounds (1 μmole total), and the mixture is treated with *E. coli* alkaline phosphatase (Section 7).

The digest is desalted by the charcoal procedure (Section 7) and is chromatographed on Whatman No. 1 paper in solvent L. The alkali-stable dinucleoside phosphate compounds separate clearly from the nucleosides as shown in Figure 13. The developed chromatogram is sectioned into two parts: section A contains

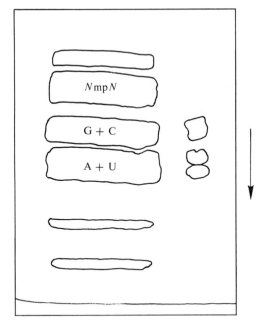

FIGURE 13. Paper chromatography of a mixture of ribonucleosides and alkali-stable dinucleoside phosphates, in solvent L. Data taken from Tamaoki and Lane (1968).

material between the origin and the leading edge of the NmpNp band; section B, material between the rear edge of the G+C band and the solvent front (Figure 13). The dinucleoside phosphates contained in section A are separated by two-dimensional paper chromatography as shown in Figure 4. The N-methylated nucleosides contained in section B are resolved in the same two-dimensional paper chromatographic system; the results are shown in Figure 3b.

C. Specific Procedures for Isolation of Modified Nucleosides

15. 1-METHYLADENINE, 3-METHYLCYTOSINE, AND 7-METHYLGUANINE*

Under conditions of alkaline cleavage of the internucleotide bond, 1-methyladenosine and 7-methylguanosine components of the RNA sample undergo the following changes:

*Dunn, 1963; Dunn and Flack, 1967a; Dunn, personal communication. Dr. Dunn has kindly provided the details of the procedure used to obtain the data in the cited publications.

Both these RNA components can be detected as their alkali-degradation products (see Section 9); however, it is sometimes desirable to isolate them in the naturally occurring form. This is particularly true for 1-methyladenosine, since both 1-methyl- and N^6-methyladenosine occur in RNA (Chapter 2). In the following procedure the RNA is hydrolyzed in cold acid, and the components are identified as the mononucleotides. Cleavage of purine glycosylic bonds under these conditions is less than 10%. At no point in the procedure does the pH of the working solutions rise above 7.0.

The RNA (5–20 mg) is dissolved in 1 N perchloric acid (2–5 ml) and the solution is kept at 28° for 18 hr. The solution is carefully neutralized to pH 6.0 with potassium hydroxide (prior to neutralization sufficient ammonium acetate is added to make the solution 0.1 M in order to prevent any local elevation of the pH). The potassium perchlorate is removed at 0°; the solution is diluted to 25–50 ml, run onto a column [0.9 cm (diam.) × 2 cm] of Dowex-1 (acetate) equilibrated with 0.01 M acetate (pH 5.5). The immediate eluate and one wash of 50 ml of 0.02 M ammonium acetate (pH 5.0) are combined and evaporated to dryness. The ammonium acetate can be removed by placing the sample in a vacuum desiccator over potassium hydroxide pellets and concentrated sulfuric acid.

The products are chromatographed on paper in solvent Z. The nucleotides separate from the faster-moving free bases and terminal nucleosides. 1-Methyladenylic acid and 7-methylguanylic acid migrate together, while 3-methylcytidylic acid moves slightly faster. The 3-methylcytidylic acid band is eluted and rechromatographed in solvent Z to remove impurities. It can be further characterized by electrophoresis at pH 7.0, at which pH it has a mobility similar to that of 1-methyladenylic acid.

1-Methyladenylic acid and 7-methylguanylic acid have 50% and 70% of the mobility of adenylic acid on electrophoresis in phosphate buffer at pH 7.0.

The nucleotides retained on the Dowex-1 column can be eluted with $2 N$ formic acid and subjected to the same separation procedure used for the products of alkaline hydrolysis of RNA (Section 10).

1-Methyladenine can be detected in an acid hydrolyzate of RNA (Dunn 1961a). The RNA is hydrolyzed in $1 N$ hydrochloric acid for one hour at $100°$, and the hydrolyzate is chromatographed on a Whatman No. 1 paper in solvent C. In this system, 1-methyladenine migrates with adenine. The adenine band is eluted with 0.1 N formic acid, and the material is chromatographed successively in solvent O (to remove hydrochloric acid) and solvent P. The Rf values of 1-methyladenine, N^6-methyladenine, and adenine are shown in Table 3.

TABLE 3

Rf Values of 1-Methyladenine, N^6-Methyladenine, and Adenine in Four Solvent Systems

	C	O	P	Q
1-Methyladenine	0.39	0.53	0.20	0.25
N^6-Methyladenine	0.49	0.75	0.40	0.65
Adenine	0.38	0.60	0.26	0.35

If any N^6-methyladenine is present in the acid hydrolyzate, it migrates with cytidine in the first chromatographic system. The band containing cytidine can be eluted with 0.1 N formic acid, and the two compounds can be separated by chromatography in solvent O.

16. N^6-METHYLDEOXYADENOSINE; ISOLATION FROM DNA (DUNN AND SMITH, 1958)

A sample of DNA (20 mg per ml in aqueous 0.002 M MgSO$_4$) is adjusted to pH 7.4 and treated with 10 μg per ml of DNAse for 6 hr at $37°$. The pH is maintained at 7.4 by periodic addition of dilute ammonia. The pH of the solution is adjusted to 9.6 by the addition of glycine buffer to give a final concentration of 0.02 M glycine, and 100 μg per ml of *Crotalus adamanteus* venom is added. Incubation is continued for 5 hr with periodic adjustment of the pH to 9–9.6. The hydrolyzate containing the nucleoside mixture is applied as a band on a sheet of Whatman 3 MM paper, and the chromatogram is developed in solvent O. N^6-Methyldeoxyadenosine migrates slightly ahead of thymidine. The band containing thymidine and N^6-methyldeoxyadenosine is eluted in water, and the two compounds are separated by paper electrophoresis at pH 2.5. Thymidine has no charge and remains near the origin, while N^6-methyldeoxyadenosine migrates toward the cathode.

17. N^6-(Δ^2-ISOPENTENYL)ADENOSINE; ASSAY METHOD (ROBINS ET AL., 1967; UNPUBLISHED DATA)

N^6-(Δ^2-Isopentenyl)adenosine can be detected as its acid degradation product. Mild acid hydrolysis of N^6-(Δ^2-isopentenyl)adenosine does not produce the free base, N^6-(Δ^2-isopentenyl)adenine, but gives the hydrated product, N^6-(3-methyl-3-hydroxybutyl)adenine, *17*, and the cyclic product, 3-H-7,7-dimethyl-7, 8, 9-trihydropyrimido-[2-1-*i*]purine, *18* (see p. 324, Chapter 7). The ratio of *17* to *18* obtained under these conditions is 60 to 40; therefore, isolation of compound *17* represents a method for the analysis of N^6-(Δ^2-isopentenyl)-adenosine.

Column Partition Chromatography. The following procedure is described in terms of a 5 g sample of yeast tRNA; and, in addition to compound *17*, small amounts of N^6-methyladenine and N^6,N^6-dimethyladenine are detected. Five grams of tRNA is dissolved in 100 ml of water, and the pH of the solution is adjusted to 4.0 with sulfuric acid. The flask containing the solution is placed in a boiling water bath; when the temperature of the solution reaches 100°, 25 ml of 5 N sulfuric acid is added. The solution is kept at 100° for 15 min, then rapidly cooled to room temperature. Barium hydroxide [Ba(OH)$_2$·8H$_2$O, 19.7 g] is dissolved in 50 ml of hot water, and this solution is added slowly to the vigorously stirred solution of the digest. The precipitate is filtered off (or centrifuged) and washed with warm water. The combined aqueous solution is lyophilized to yield a product that contains 40% of the original A_{260} units of the starting sample.

This lyophilized sample is fractionated as shown in Figure 14. Fraction 1 of the column contains a small amount of the free base, N^6-(Δ^2-isopentenyl)-adenine, 0.5 mg. Fraction 2 contains 0.3 mg of N^6,N^6-dimethyladenine, and fraction 3 contains a mixture of compound *17* and N^6-methyladenine. These two compounds are separated by paper chromatography in solvent D; the yield is 5.5 mg of compound *17* and 2.8 mg of N^6-methyladenine. This entire procedure can be easily scaled down.

Ion-Exchange Chromatography. The two acid degradation products, compounds *17* and *18* (see Chapter 7, p. 324), can be isolated by means of ion-exchange chromatography as follows. Yeast tRNA (870 mg) is hydrolyzed with

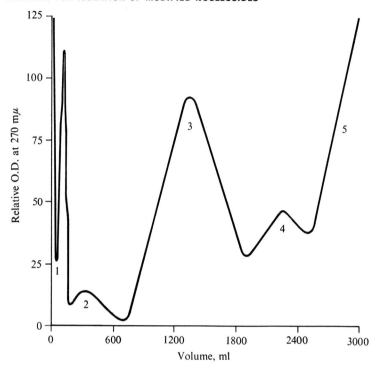

FIGURE 14. Isolation of N^6-(3-hydroxy-3-methylbutyl)adenine from an acid hydro-lyzate of tRNA. Fractionation of hydrolyzate obtained from 5 g of yeast tRNA on a partition column (5.08 × 80 cm) in solvent system E at a flow rate of 600 ml/hr.

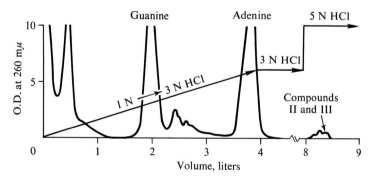

FIGURE 15. Elution profile of the resolution of purines from a mild acid hydrolyzate of 870 mg of yeast tRNA on an ion-exchange column [1 cm (diam.) × 30 cm] of Dowex-50 W-X8 [H$^+$] (200–400 mesh). The column was developed with a gradient of hydro-chloric acid. Data taken from Robins et al. (1967). Compounds II and III equal **17** and **18**.

sulfuric acid as described above, and the digest in neutral solution is run onto a column of Dowex-50 W × 8 (Cohn, 1949). The column is developed with a gradient of increasing concentration of hydrochloric acid. The resulting elution profile is shown in Figure 15. The fraction containing compounds *17* and *18* is evaporated to dryness *in vacuo*, and the residue is chromatographed on Whatman No. 1 paper in solvent D. Six bands of ultraviolet-absorbing material are obtained. Band 6, counting from the origin, consists of compound *17* (230–260 μg); and band 5, of compound *18* (25–35 μg).

18. N^6-(Δ^2-ISOPENTENYL)ADENOSINE AND N^6-(*CIS*-4-HYDROXY-3-METHYLBUT-2-ENYL)ADENOSINE; ISOLATION FROM tRNA (HALL ET AL., 1967b)

The procedure is described in terms of corn tRNA. The tRNA (10.0 g) is hydrolyzed enzymically (Section 3), and the lyophilized digest is stirred with 75 ml lower phase of solvent E for 1 hr. The mixture is centrifuged at 3000 × g for 5 min; then the supernatant is mixed with 140 g of Celite-545, and the mixture is packed on top of a previously prepared column (5.08 cm diam.) containing 500 g of Celite-545 mixed with 230 ml of lower phase of solvent E. The residue in the centrifuge tube is stirred with 100 ml of upper phase of solvent E for 30 min, and the solution is clarified by centrifugation. This solvent is introduced on top of the column, followed by fresh solvent in the standard manner. The elution profile of this column is shown in Figure 16.

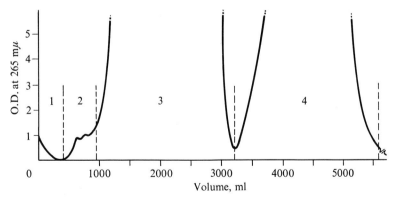

FIGURE 16. Partition chromatography of an enzymic digest of 10 g of corn tRNA on a column [5.08 cm (diam.) × 80 cm] of Celite-545; solvent system E. Data taken from Hall et al. (1967a).

Corn tRNA does not contain N^6-(Δ^2-isopentenyl)adenosine, but if it were present it would be eluted in fraction 1. (It can be recovered readily by concentrating the solution and chromatographing the residue for 5 hr on Whatman No. 3 MM paper in solvent E.) The eluate corresponding to fraction 2 is concentrated, and the residue is chromatographed on Whatman No. 3 MM paper in solvent D. This fraction contains N^6-(cis-4-hydroxy-3-methylbut-2-enyl)-adenosine and N^6,N^6-dimethyladenosine, which migrate together (Rf 0.73). This solvent system is used, however, because it separates other components of the mixture. N^6,N^6-Dimethyladenosine, Rf 0.73, and N^6-(cis-4-hydroxy-3-methylbut-2-enyl)adenosine, Rf 0.79, are separated by rechromatography in solvent V. The yield of the latter compound is 109 OD_{270} mμ units.

19. N^6-(Δ^2-ISOPENTENYL)-2-METHYLTHIOADENOSINE (BURROWS ET AL., 1968)

Escherichia coli tRNA is hydrolyzed enzymically to its constituent nucleosides, and the nucleoside mixture is fractionated by column partition chromatography, using solvent system F (2-methoxyethanol is substituted for 2-ethoxyethanol). The procedure used is similar to that described in Section 18. The nucleoside emerges with the solvent front and is purified by thin-layer chromatography in cellulose MN 300 F_{254} (Machery, Nagel and Co., Düren, West Germany) plates, with distilled water as the solvent. The material in the band at R_f 0.06–0.12 is rechromatographed, using 10% ethanol in water. The nucleoside migrates with an Rf of 0.31.

20. N-(PURIN-6-YLCARBAMOYL)-L-THREONINE; ISOLATION FROM AN ACID HYDROLYZATE OF tRNA (CHHEDA ET AL., 1969a)

A solution of tRNA (100 mg) in 1.3 ml of water is heated to 100° in a 10 ml Erlenmeyer flask; hot 2 N hydrochloric acid (1.3 ml) is added, and the solution is maintained at 100° for another 10 min. The flask is rapidly cooled in ice water. The solution is evaporated in vacuo (bath temperature < 30°), and the gummy residue is re-evaporated twice with small portions of water and triturated with several changes of anhydrous ether. The solid residue is placed in a vacuum desiccator over potassium hydroxide pellets for 4 hr; then the residue is dissolved in 3 ml of water and the solution clarified by centrifugation. The pH of the solution is adjusted to 7.0 by the addition of 2 N sodium hydroxide,

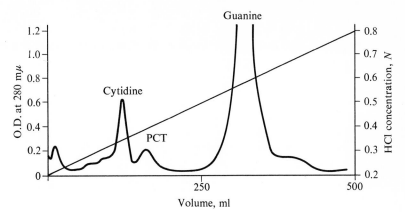

FIGURE 17. Elution profile of an acid hydrolyzate of calf liver tRNA fractionated on a column 0.9 × 5 cm) of Dowex-50 W-X8 [H$^+$] (200–400 mesh). The column is developed with a linear gradient of hydrochloric acid (0.2 N→0.8 N); total volume 500 ml, at a flow rate of 1.2 ml/min. PCT = N-(purin-6-ylcarbamoyl)-L-threonine. Data taken from Chheda et al. (1969a).

and the mixture is cooled in an ice bath for 1 hr. Precipitated guanine is removed by centrifugation, and the precipitate is washed once with water. The combined supernatant and wash is applied to a column [0.9 cm (diam.) × 5 cm] of Dowex-50W × 8[H$^+$]; (200–400 mesh). The column is washed with water (approximately 20 ml) until the pH of the effluent reaches approximately 5. Washing is continued with 0.001 N hydrochloric acid (75 ml), and the column is then developed by a linear gradient (0.2 N→0.8 N HCl; total volume, 500 ml) at a flow rate of 1.2 ml/1 min. The elution profile is shown in Figure 17. The amount of N^6-(purin-6-ylcarbamoyl)-L-threonine obtained from various samples of tRNA is shown on p. 109, Chapter 2.

21. 5-METHYLCYTIDINE, 5-METHYLDEOXYCYTIDINE

Isolation of Free Base from an Acid Hydrolysis of DNA (Wyatt, 1951a). The DNA is hydrolyzed in perchloric acid to yield the free bases (Section 5), and the hydrolyzate is resolved by paper chromatography in solvent C. The 5-methyl-cytosine migrates just ahead of cytosine; however, if the quantity of cytosine is much greater than that of 5-methylcytosine, the separation is unsatisfactory.

5-Methylcytosine can also be isolated from an acid hydrolyzate of DNA or

METHODS FOR ISOLATION OF MODIFIED NUCLEOSIDES

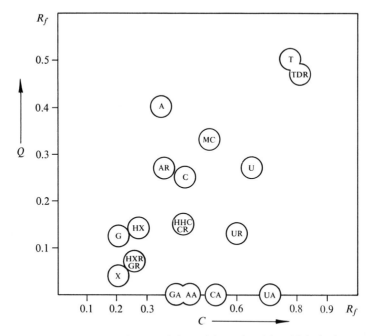

FIGURE 18. Diagram of the positions of nucleic acid derivatives on two-dimensional chromatogram run on Whatman No. 1 paper by the descending technique, first in solvent C, then in solvent Q. A, adenine; AA, adenylic acid; AR, adenosine; C, cytosine; CA, cytidylic acid; CR, cytidine; G, guanine; GA, guanylic acid; GR, guanosine; HMC, 5-hydroxymethylcytosine; HX, hypoxanthine; U, uracil; UA, uridylic acid; UR, uridine; X, xanthine. Data taken from Wyatt (1955).

RNA by two-dimensional paper chromatography in solvents C and Q, as shown in Figure 18.

Isolation from RNA (Dunn, 1960). A sample of RNA is hydrolyzed in 1 N potassium hydroxide for 18 hr at 30°, the alkali is neutralized with perchloric acid, and the nucleotides are desalted by paper chromatography in solvent O before being subjected to electrophoresis in 0.05 M phosphate buffer, pH 2.6. The adenylic and cytidylic acids have almost a zero net charge and are separated from uridylic, guanylic, and methylated guanylic acids. The separated nucleotides are chromatographed in solvent O to remove phosphate. Then the nucleotides are treated with a phosphomonoesterase, and the resulting nucleosides are separated by chromatography in solvent C. 5-Methylcytidine migrates slightly ahead of cytidine. The leading edge of the cytidine band is eluted, and the material is rechromatographed in solvent Q for 36–72 hr. Alternatively before treating the nucleotides with the phosphomonoesterase, they are subjected to electrophoresis at pH 3.5 (formate buffer). The eluted products are dephosphorylated and chromatographed successively in solvents O and Q.

This method is rather lengthy; if the objective is to demonstrate the presence of 5-methylcytosine, the paper chromatographic separation of the bases described above may be preferred. However, the risk of obtaining misleading results exists if the RNA sample is contaminated with DNA. The utility of the above procedure lies in the fact that the compound is identified as the ribonucleoside.

5-Methylcytidine can also be isolated by the general methods outlined in Sections 9–11, and 5-methylcytidylic acid and cytidylic acid can also be isolated together by ion-exchange chromatography (Section 26, p. 247). 5-Methyldeoxycytidine can be separated as the 5'-mononucleotide from an enzymic digest of DNA (see Section 22).

22. 5-METHYLDEOXYCYTIDYLIC ACID; 5-HYDROXYMETHYLDEOXYCYTIDYLIC ACID AND ITS GLUCOSYLATED DERIVATIVES (SINSHEIMER AND KOERNER, 1951)

The DNA is hydrolyzed to its constituent mononucleotides in the following manner. The sample (200 mg) is dissolved in 20 ml of 0.5 M acetate buffer (pH 6.5); to this solution is added 10 ml of 0.1 M magnesium sulfate solution and 10 ml of an aqueous solution of pancreatic DNAse (400 μg/ml). The solution is incubated for 72 hr at 37° under a layer of hexane with constant stirring. The second stage of the digestion can be carried out with a preparation from intestinal mucosa, according to the method of Klein (1945). The pH of the digest resulting from the action of DNAse is adjusted to 8.5 with 2 N sodium hydroxide, and to this solution is added 10 ml of 0.36 M sodium arsenate, 10 ml of 1 M ammonium sulfate buffer (pH 8.5), and 10 ml of a solution of the intestinal enzyme (10 mg/ml). The mixture is stirred gently for 20 hr at 37° under a layer of hexane.

The second stage of the enzymic hydrolysis may be more conveniently carried out with snake venom (*Crotalus adamanteus*) phosphodiesterase, which is commercially available (Worthington Biochemical Corp., Freehold, N.J.).

If the intestinal preparation is used, the digest is adjusted to pH 4.7 with glacial acetic acid and centrifuged. The precipitate is washed with 0.01 M

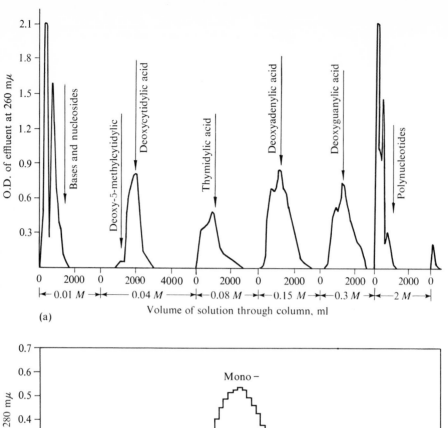

(a)

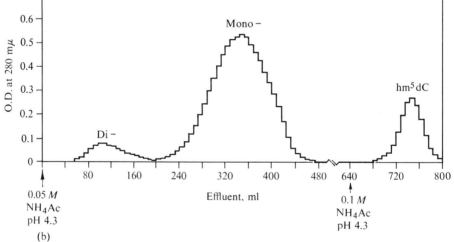

(b)

FIGURE 19. (a) Separation of 5-methyldeoxycytidine from an enzymic hydrolyzate of DNA. Elution profile obtained on a column (3.4 cm² × 11 cm) of Dowex-1 [acetate] (200–500 mesh) Column developed with acetate buffers (pH 4.7). Data taken from Sinsheimer and Koerner (1951). (b) Elution profile of the resolution of 5-hydroxymethyldeoxycytidylic acid and its mono- and diglucosylated derivatives obtained from T-even phage DNA. Procedure same as in 19a. Data taken from Lehman and Pratt (1960).

acetate buffer (pH 4.7), the combined wash and supernatant is adjusted to pH 9.0, and the volume of the solution is increased to 250 ml. This solution is run onto a column (3.4 cm² × 11 cm) of Dowex-1 (acetate) (200–500) mesh. The column is developed with acetate buffers (pH 4.7) according to the chart in Figure 19a.

Lehman and Pratt (1960) have adapted this method to separate 5-hydroxy-methyldeoxycytidylic acid and its mono- and diglucosylated derivatives from digests of T-even phage DNA (Figure 19b).

23. 7-METHYLGUANINE; ISOLATION FROM RNA (CRADDOCK ET AL., 1968)

A sample of RNA (32 mg) in which the methyl groups are labeled with [^{14}C] is hydrolyzed with perchloric acid, and the hydrolyzate is adjusted to 3 ml with water and centrifuged. The hydrolyzate is fractionated on a column [1 cm (diam.) × 10 cm] of Dowex-50[H$^+$] as shown in Figure 20. In the elution profile shown in the figure, a small amount of carrier 7-methylguanine was added to the sample in order to make it visible (ultraviolet absorption).

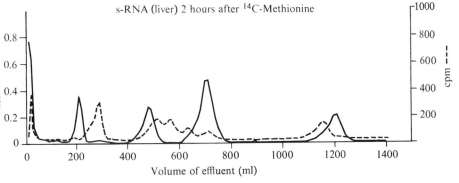

FIGURE 20. Ion-exchange chromatography of a perchloric acid hydrolyzate of the total RNA (32 mg) of rat liver labeled with [^{14}C-methyl], on a column [1 cm (diam.) × 10 cm] of Dowex-50 [H$^+$]. After the sample is applied, the column is washed with 300 ml of N hydrochloric acid at a flow rate of 12.5 ml/hr. The column is developed with a linear gradient of hydrochloric acid (1 $N\rightarrow4\,N$). In this elution profile, a small amount of 7-methylguanine was added to the sample. The probable identities of the UV peaks are: uracil, cytosine, guanine, 7-methylguanine, and adenine. The probable identities of the radioactive peaks are: 5-methyluracil, 5-methylcytosine, N^2-dimethylguanine, N^2-methylguanine, 1-methylguanine, 7-methylguanine, and 1-methyladenine. Data taken from Craddock et al. (1968).

METHODS FOR ISOLATION OF MODIFIED NUCLEOSIDES

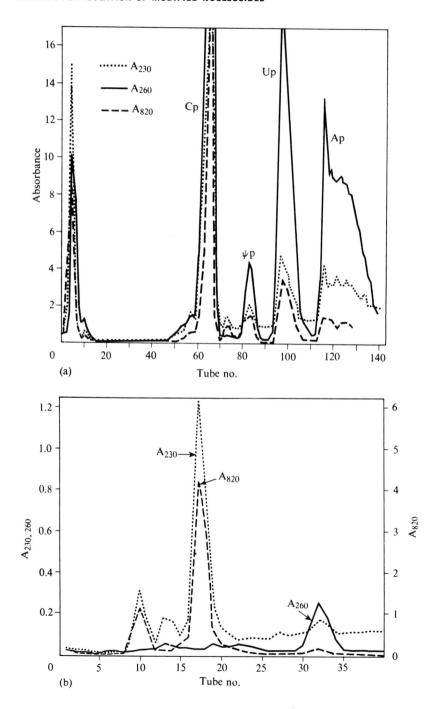

FIGURE 21 (*on facing page*). (a) Chromatography of a ribonuclease T_1 digest of 100 mg of yeast tRNA on a column [0.25 cm (diam.) × 125 cm] of Dowex-1 × 8. Elution is carried out with a gradient of ammonium formate, pH 3.5. The gradient is produced from 240 ml each of water and of 0.1 M, 0.3 M, and 10.0 M ammonium formate in four chambers of a Varigrad (Peterson and Sober, 1959). The peak of guanylic acid is eluted considerably after adenylic acid and is not shown on the chart. Fraction size is about 4 ml. (b) Rechromatography of material in tubes 71–76 (Figure 21a) on DEAE-Sephadex. Elution is carried out with a linear gradient of 0→0.17 M ammonium carbonate. Data taken from Madison and Holley (1965) and Madison (1967).

24. 5,6-DIHYDROURIDYLIC ACID; ISOLATION FROM tRNA (MADISON AND HOLLEY, 1965)

Unfractionated tRNA (100 mg) is hydrolyzed with RNAse T_1, and the digest is chromatographed on a column of Dowex-1 × 8. The elution profile is shown in Figure 21. The 5,6-dihydrouridylic acid is dephosphorylated with bacterial alkaline phosphatase, and the nucleoside is purified by paper chromatography. 5,6-Dihydrouridine does not absorb in the 260 mμ region; it can be visualized with the *p*-dimethylaminobenzaldehyde reagent described by Fink et al. (1956). Paper chromatograms are sprayed with 0.5 N sodium hydroxide, allowed to dry, and then sprayed with a solution of 1 g of *p*-dimethylaminobenzaldehyde (crystallized from ethanol) in 10 ml of concentrated hydrochloric acid and 100 ml of ethanol. Control chromatograms are sprayed only with the *p*-dimethyl-aminobenzaldehyde reagent. Madison (1967) lists five solvent systems suitable for chromatographing 5,6-dihydrouridine.

5,6-Dihydrouridine can also be detected in tRNA by hydrolysis of the sample in 0.2 N sodium hydroxide for 3 hr at 100°. Under these conditions about 78 % of the dihydrouridylic acid residues is converted to β-alanine, which can be detected by quantitative ion-exchange analysis for amino acids or by paper chromatography (Magrath and Shaw, 1967).

25. 4-THIOURIDYLIC ACID, ISOLATION FROM tRNA OF *E. COLI* (LIPSETT, 1965)

The tRNA (1 g) is hydrolyzed in 0.3 N potassium hydroxide for 18 hr at 37°. The solution is neutralized with Dowex-50 [H^+] to pH 9–10 and centrifuged, and the pH of the solution is adjusted to 8.6 by addition of hydrochloric acid.

METHODS FOR ISOLATION OF MODIFIED NUCLEOSIDES

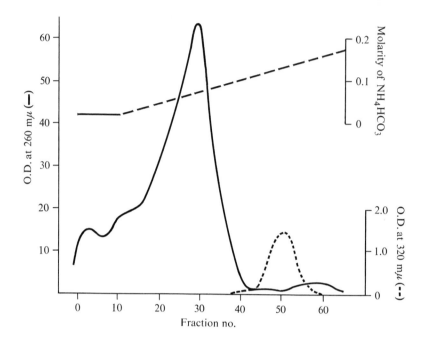

FIGURE 22. Isolation of 4-thiouridylic acid from an alkaline hydrolyzate of *E. coli* tRNA. The alkaline hydrolyzate of 1 g of tRNA is fractionated on a column [4 cm (diam.) × 24 cm] of DEAE-cellulose, previously equilibrated with 0.01 M ammonium bicarbonate, pH 8.6. The column is washed with 200 ml of 0.025 M ammonium bicarbonate, pH 8.6, and then eluted with a linear gradient of ammonium bicarbonate in 7 M urea (0.05 M→0.25 M); total volume is 1500 ml. Fraction size 20 ml each. Data taken from Lipsett (1965).

The solution (approximately 250 ml) is made 0.02 M with respect to ammonium bicarbonate and is then run onto a column [4 cm (diam.) × 24 cm] of DEAE-cellulose, which has been previously equilibrated with 0.01 M ammonium bicarbonate, pH 8.6. The sample is fractionated as shown in Figure 22. Urea is removed from the fraction (tubes 45–55) containing 4-thiouridylic acid according to the method of Rushizky and Sober (1962), and the 4-thiouridylic acid is purified by electrophoresis in 0.05 M ammonium bicarbonate, pH 8.6. The thio compound has a mobility 1.17 times that of uridylic acid. Trace amounts of impurities can be removed by chromatography of the sample in solvent W.

26. 5-RIBOSYLURACIL

Isolation from an Alkaline Digest of tRNA (Cohn, 1967; Cohn and Uziel, 1961).
An alkaline hydrolyzate (25–50 g) of RNA is neutralized, and 1 volume of
ethanol is added. After standing overnight at 4°, the solution is filtered to remove
guanylates and some of the cytidylates and adenylates. The hydrolyzate is
adjusted to pH 8–10 with ammonium hydroxide and run onto a column
(15 cm × 33 cm²) of Dowex-1 × 8 [formate] (200–400 mesh). The column is
washed with several volumes of water (flow rate approximately 3 ml/cm²) and
with 0.02 M ammonium formate until the pH falls to 7 (10–20 column volumes);
then the column is developed with 0.15 M formic acid (about 60 column
volumes) until all the adenylic acid is removed. The column is freed of formic
acid by washing with 5–10 column volumes of water and then eluted with
0.05 M ammonium formate in 0.001 M formic acid. This buffer elutes 5-ribosyl-
uracil 2'(3')-phosphate and uridylic acid. The former compound is eluted first.
It is an easy matter to test the eluate for the presence of 5-ribosyluracil by
measuring the maximum ultraviolet absorption at alkaline pH. The ratio of
$A_{290} : A_{260}$ at pH 12 is 1.6–2.1 and 0.03 for 5-ribosyluracil 5'-phosphate and
uridylic acid, respectively.

The fractions containing 5-ribosyluracil 2'(3')-phosphate are pooled, and the
pH is adjusted to 10 by the addition of ammonium hydroxide. The solution is
absorbed onto a column of Dowex-1 × 8 [formate] (8 cm × 5 cm²). Fifteen
column volumes of 0.1 M sodium carbonate are passed through the column in
order to convert it to the carbonate form and at the same time eliminate any
cytidylic or adenylic acid that may have remained. The column is washed with
5 column volumes of water followed by 10–30 column volumes of 0.4 M
ammonium bicarbonate. The 5-ribosyluracil 2'(3')-phosphate and uridylic acid,
as well as any guanylic acid, are eluted together. The solution is boiled to
eliminate the ammonium bicarbonate.

The natural form of 5-ribosyluracil in the tRNA is the β-ribofuranosyl
configuration (Michelson and Cohn, 1962). This nucleoside (or its nucleotides)
in alkali or acid isomerizes to the four possible sugar configurations: α- and
β-furanose and α- and β-pyranose (Cohn, 1960; Tomasz et al., 1965). Consequently, this isolation procedure may lead to production of the other isomers,
although under the conditions of this procedure, little isomerization occurs.
The material recovered in the second column procedure is recycled through a
column (15 cm × 8 cm²) of Dowex-1 × 8 [formate] (200–400 mesh). The

METHODS FOR ISOLATION OF MODIFIED NUCLEOSIDES

elution sequence described for the first column is repeated, and the last buffer first elutes 5-(α-ribosyl)uracil 2'-phosphate contaminated by uridine 5'-phosphate (if any is present), then the 2'- and 3'-phosphate isomers together, and finally 5-(β-ribosyl)uracil 3'-phosphate. The desired peaks are recovered by recycling through the second column procedure.

The nucleotides can be hydrolyzed to the nucleosides by use of a suitable monophosphotase, and the isomers are separated on a column (10 cm $\times$ 1 cm^2) of Dowex-1 $\times$ 8 [formate] (200–400 mesh). The nucleosides are absorbed to the resin at a pH above 10 (ammonium hydroxide). The column is developed by means of a linear gradient increasing in chloride concentration and made by using 1 liter of 0.005 M ammonium chloride + 0.01 M ammonium hydroxide + 0.0025 M potassium borate and 1 liter of 0.02 M ammonium chloride. The elution pattern is shown in the table.

Isomer	Column Volumes at			pH 12 $A_{280} : A_{260}$
	Start	Peak	End	
Pyranose	40	46	52	1.4
Pyranose	50	56	62	2.5
α-furanose	76	86	96	1.5
β-furanose	110	125	140	2.0
Uridine	136	150	170	0.3

The nucleosides can be recovered from the eluate of this column by reabsorption on a small anion-exchange column (3 cm $\times$ 1 cm^2) (pH greater than 10, ammonium hydroxide). The column is washed with water and eluted with 0.05 M acetic acid.

For the isolation of smaller quantities of 5-ribosyluracil, the paper chromatographic procedures described in Section 27 would be preferable.

Isolation from an RNAse Digest of Yeast tRNA. The pancreatic RNAse digestion of yeast tRNA releases a considerable amount of 5-ribosyluracil as the mononucleotide. Staehelin (1964) has reported a procedure for resolving the RNAse digest of unfractionated yeast tRNA on a column of DEAE-cellulose, using triethylammonium bicarbonate as the eluting buffer. Cytidylic acid, together with 5-methylcytidylic acid and uridylic acid, is eluted early in the elution pattern. 5-Ribosyluracil phosphate appears either as a separate peak or as a shoulder on the trailing edge of the uridylic acid (see Figure 23).

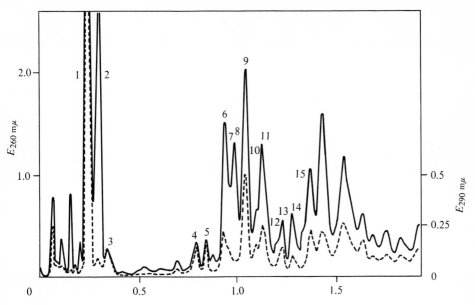

FIGURE 23. Fractionation of a pancreatic RNAse digest of tRNA (1360 O.D.$_{260}$ units) on a column [1.25 cm (diam.) × 50 cm] of DEAE-cullulose. The column is developed with a linear gradient of triethylammonium bicarbonate, pH 8.6 (0.01 M→0.25 M). The total volume is 2.4 litres. ———, O.D. 260 mμ; . . . , O.D. 290 mμ. Key 1, m^{5}C + C; 2, U; 3, ψ; 4, m$_2^2$G-C; 5, m$_2^2$G-ψ; 6, A-C; 7, A-ψ; 8, A-U + GT; 9, G-C; 10, G-ψ; 11, G-U; 12, A-A-ψ; 13, m^{1}G-m^{2}G-C; 14, A-A-C; 15, A-A-U. Data taken from Staehelin (1964).

27. 5-METHYLURIDINE, 5-RIBOSYLURACIL, N^6-METHYLADENO-SINE, 1- METHYLGUANOSINE AND N^2,N^2-DIMETHYLGUANOSINE AS THE 2′(3′)-PHOSPHATES*

These compounds are separated as the 2′(3′)-nucleotides. The RNA (0.3 mg) is hydrolyzed from an alkaline digest with alkali (Section 4), and the RNA is resolved by two-dimensional paper chromatography on Whatman No. 1 paper as shown in Figure 24. Brawerman et al. (1962) achieve a similar separation using two-dimensional paper chromatography. The paper is washed before use by autoclaving for 1 hr in 1 M acetic acid (Crestfield and Allen, 1957). The chromatogram is eluted in the first direction with solvent X for 2–3 days and in the second direction in solvent Y for 18 hr.

*Davis et al., 1959; Price et al., 1963.

METHODS FOR ISOLATION OF MODIFIED NUCLEOSIDES

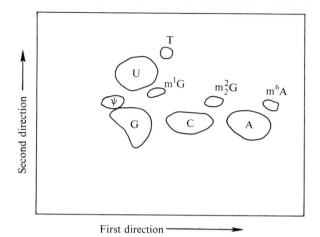

FIGURE 24. Chromatographic separation of an alkaline hydrolyzate as yeast tRNA on a sheet of Whatman No. 1 filter paper (23 × 29 cm). The chromatogram is developed in the first direction for 12 hr in solvent system X and in the second direction for 6–8 hr in solvent system Y. (It is sometimes necessary to redevelop the chromatogram in the second direction 2–3 times in order to separate the major nucleotides from 5-ribosyluracil phosphate.) Data taken from Davis et al. (1959). N^2,N^2-Dimethylguanylic acid was identified using this procedure by D. B. Dunn (private communication).

In another variation of this procedure, Hayashi et al. (1966) separate from an alkaline hydrolyzate of *E. coli* rRNA the 2'(3')-phosphates of 5-methyluridine, 5-ribosyluracil, N^2-methylguanosine, 3-methylcytidine, and N^6-methyladenosine. (Some of these identifications are tentative. For example, 3-methylcytidine is alkali labile and might not survive the extensive hydrolysis.) The chromatogram (40 × 40 cm) is developed in the first direction in solvent X and in the second direction in solvent C.

28. ISOLATION OF 1-METHYLGUANINE AND 7-METHYLGUANINE BY ION-EXCHANGE CHROMATOGRAPHY (KRIEK AND EMMELOT, 1963)

This procedure is based upon the cation exchange procedure devised by Cohn (1949) for separation of the four major bases. An RNA sample (200 μmole) is hydrolyzed in 1 N hydrochloric acid for 15 min, the hydrolyzate is diluted with water and applied to a column of Dowex-50 resin (1.0 cm (diam.) × 15 cm) previously equilibrated with 0.5 N hydrochloric acid (Figure 25).

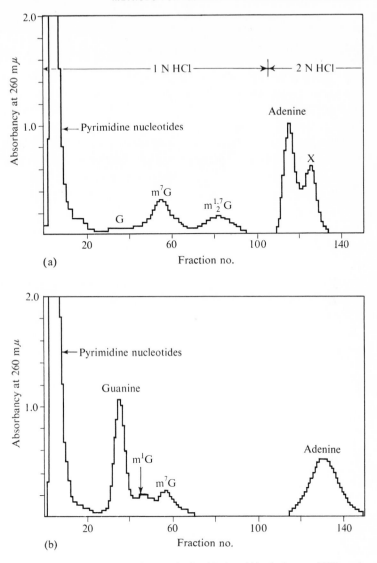

FIGURE 25. Separation of the purines present in a 1 N hydrochloric acid hydrolyzate of 200 μmole of tRNA. The amount of 1-methylguanine and 7-methylguanine shown in these elution profiles is exaggerated, since the tRNA samples were methylated with diazomethane before hydrolysis. The sample was applied to a column [1.0 cm (diam.) × 15 cm] of Dowex-50 × 8 [H^+] (200–400 mesh) equilibrated with 0.5 N hydrochloric acid. The column was eluted at a flow rate of 60 ml/hr. Fraction size is 5 ml. This figure shows two separate runs. Data taken from Kriek and Emmelot (1963). Adler et al. (1958), using a similar procedure, were able to detect N^2-methylguanine in the fraction between guanine and 1-methylguanine.

29. 2'-O-METHYLRIBONUCLEOSIDES BY COLUMN PARTITION CHROMATOGRAPHY (HALL, 1964b)

The 2'-O-methylribonucleosides can be isolated as the alkali-stable dinucleotides NmpNp (Section 10). In this method these nucleosides separate readily from ribonucleosides using a solvent system containing borate ions. In principle, this procedure works well, except that any deoxyribonucleosides contaminating the RNA digest are eluted concomitantly with the 2'-O-methylribonucleosides. The following procedure, designed to separate all eight nucleosides, is described in terms of a sample of yeast tRNA contaminated with about 2% DNA.

The solvent system is prepared by dissolving 38 g of $Na_2B_4O_7 \cdot 10H_2O$ in 1 liter of warm water and allowing the solution to stand at room temperature overnight. The solution is filtered and 5 ml of concentrated ammonium hydroxide and 3 liters of 1-butanol are added. The mixture is shaken vigorously for 30 min, and the layers are allowed to separate. Sometimes the lower phase requires clarification, which can be accomplished by means of gravity filtration through Whatman No. 1 filter paper.

The sample of tRNA (635 mg) is digested enzymically as described in Section 3 (p. 212), and the lyophilized digest is resolved on a partition column as shown in Figure 26. Only the first part of the elution pattern is shown. The ribonucleo-

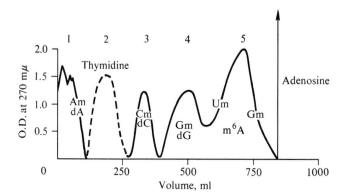

FIGURE 26. Isolation of the 2'-O-methylribonucleosides from an enzymic hydrolyzate of yeast tRNA (450 mg of mixed nucleosides). The column [2.54 cm (diam.) × 80 cm] contains 146 g of Celite-545 mixed with 65.3 ml of the lower phase of solvent K. A solution of the lyophilized hydrolyzate in 10 ml of the aqueous phase is mixed with 21.5 g of Celite-545. The column is developed with the upper phase of solvent K at a flow rate of 120 ml/hr. Data taken from Hall (1964b).

sides are eluted eventually, but their recovery from the borate-containing solvent is not very practical.

The 2'-O-methylribonucleosides are isolated from the column fractions by means of paper chromatography. The fractions corresponding to peaks 1, 3, 4, and 5 of Figure 26 are concentrated *in vacuo*, and the concentrate is applied to strips of Whatman No. 3 MM paper (16.5 cm wide). The paper strips are developed according to the protocol of Table 4, and the separated nucleosides are eluted from the paper with water. If the resolution between peaks 3 and 4 is not as sharp as that shown in Figure 26, separation of the material can be achieved by chromatography in solvent E for 16 hr. This procedure separates the deoxynucleosides from the 2'-O-methylribonucleosides; 2'-O-methyl-guanosine and 2'-O-methylcytidine, obtained in the same band, are separated by rechromatography in system D.

TABLE 4

Isolation of 2'-O-Methylribonucleosides from Column Fractions (Figure 26)

Fraction (Fig. 26)	Solvent	Development Time, hr	Nucleoside	Distance from Origin, cm
1	E	8	Deoxyadenosine	20.0
			2'-O-Methyladenosine	27.3
3	E	30	Deoxycytidine	13.0
			2'-O-Methylcytidine	22.5
4	E	16	Deoxyguanosine	5.9
			2'-O-Methylguanosine	10.5
5	A	40	2'-O-Methylguanosine	7.9
			2'-O-Methyluridine	13.0
			N^6-Methyladenosine	27.7

D. General Analytical Techniques That Might Be Applied to the Separation of Modified Nucleosides Obtained from Nucleic Acid Hydrolyzates

The following techniques have been used to separate synthetic mixtures of modified nucleosides and derivatives. Some may be of value in developing new analytical procedures for the detection of modified nucleosides in nucleic acid samples.

Hori (1967) describes a method for the separation of the free bases, nucleosides, and mononucleosides on a column of Dowex-1 × 8 (acetate form), based on the procedures of Cohn (1955), Anderson and Ladd (1962), Anderson et al. (1963), and Anderson (1962). The method can be used to separate mixtures of either nucleosides or nucleotides.

A mixture of the methylated guanines, methylated adenines, 5-methylcytosine, and thymine can be resolved by a combination of two-dimensional electrophoresis and chromatography on a thin-layer plate made from microcrystalline cellulose (Avicel) (see Section 13) (Augusti-Tocco et al., 1968). The plate (30 × 20 cm) is first electrophoresed (30 cm) in 0.1 M sodium acetate buffer at pH 3.5 (1000 V, 20 mamp for 3 hr), and then chromatographed in the second direction (20 cm) in solvent S.

Goldstein (1967) has applied ligand exchange to the resolution of modified nucleosides. In this procedure a cation-exchange resin is loaded with copper ions; even when attached to the resin, the metal ion retains its ability to coordinate ligands, so that complexing agents can be absorbed onto the column and then eluted by displacement with another ligand. This method has the advantage of speed; for example, a mixture of uridine, 5-methyluridine, and dihydrouridine can be completely resolved in 45 min. This technique has also been applied to the separation of a mixture of 1-methyladenosine, N^6-methyladenosine, and N^6,N^6-dimethyladenosine and a mixture of 7-methylguanosine, 1-methylguanosine, and guanosine.

Pataki (1967) has used two-dimensional thin-layer chromatography on cellulose to separate a synthetic mixture of modified bases and nucleosides. In his method, the plate is developed first in solvent T, and in the second dimension, in solvent U.

An alternative method for the purification of 2'-O-methylribonucleosides

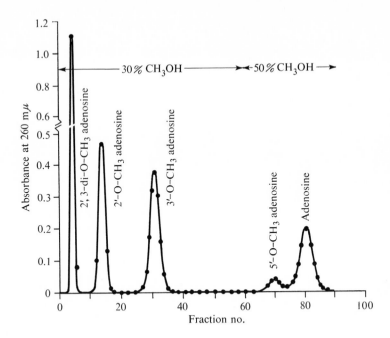

FIGURE 27. Separation of a synthetic mixture (600 mg total) of adenosine and *O*-methylated adenosine derivatives on a Dowex (bio-RAD AG)-1-2 × [OH⁻] column (200–400 mesh) [1.6 cm (diam.) × 22 cm]. The column is developed with 30 % methanol for fractions 1–56 at 12 ml/tube and with 50 % methanol for fractions 57–100 at 18 ml/tube. The flow rate is equivalent to 1 tube per 12 min. Data taken from Gin and Dekker (1968).

has been developed by Gin and Dekker (1968). This method has not been applied to RNA digests but has been used for the purification of 2'-*O*-methyladenosine from a reaction mixture consisting of the 3', 5' and 2',3'-*O*-methylated derivatives. The mixture is dissolved in a small volume of 30 % aqueous methanol and is applied to a column [1.6 cm (diam.) × 22 cm] of the anion exchange resin Dowex-1-2 × [OH⁻] (200–400 mesh). The column is developed with aqueous methanol; the elution profile is shown in Figure 27.

Uziel et al. (1968) have developed a rapid ion-exchange method for the separation of nmole amounts of modified nucleosides.

Hedgcoth and Jacobsen (1968) report a method, using thin-layer chromatography, for the determination of the nucleoside composition of a 600 μg enzymic digest of tRNA. The method detects the presence of 5-ribosyluracil and 5-methyluridine in the mixture.

Chapter 4

MODIFIED NUCLEOSIDES IN THE PRIMARY

STRUCTURE OF SPECIFIC tRNA MOLECULES

1. INTRODUCTION

THE DETERMINATION of the primary sequences of seven molecular species of yeast tRNA, four of *E. coli* tRNA, and one of mammalian tissue tRNA has revealed much about tRNA structure. The tRNA molecule consists of about eighty nucleosides; although this would theoretically permit a vast number of permutations, the known primary sequences conform to a highly consistent pattern. It is particularly instructive to examine these sequences with respect to the modified nucleosides, since the types of structural modification and the location of the individual modified nucleosides in the primary sequence are not random but highly specific.

The nucleoside components of the known tRNA structures are listed in Table 1. The average yeast tRNA molecule, for example, contains fourteen modified nucleosides (about 17–18 % of the total nucleosides), which consist of about ten structural types. The two *E. coli* sequences, on the other hand, contain relatively few modified nucleosides. The primary structure of *E. coli* tRNA^{Phe} contains two methylated nucleosides, 7-methylguanosine and 5-methyluridine (Uziel and Gassen, 1968). This fact is in striking contrast to yeast tRNA^{Phe}, which contains nine methylated nucleosides. These facts are in accord with the general analytical results reported in Chapter 2, which indicate that the tRNA of *E. coli* contains significantly fewer methyl groups than that of yeast.

2. COMMON SECONDARY STRUCTURE; CLOVERLEAF PATTERN

The primary sequence of each tRNA molecule can be arranged in a common pattern for the secondary structure. Although the exact form of this structure

MODIFIED NUCLEOSIDES IN THE PRIMARY STRUCTURE OF SPECIFIC tRNA MOLECULES

TABLE 1

Composition of Specific tRNA Molecules (Data taken from Figure 1)

Molecule	Abbrev.	Yeast							E. coli				Rat Liver
		Ala	Ile	Phe	Ser I	Ser II	Tyr	Val	Su_{III}^{+Tyr}	m Met	f Met	Val	Ser
Adenosine	A	8	14	17	17	15	15	14	18	18	14	14	14
Cytidine	C	23	20	15	18	17	20	20	28	19	25	23	19
Guanosine	G	25	21	18	23	25	20	19	21	18	24	23	25
Uridine	U	11	9	12	13	14	7	11	12	10	8	9	10
1-Methyladenosine	m^1A		1	1			1	1					1
N^6,N^6-Dimethyl-adenosine	m_2^6A												
2'-O-Methyladenosine	Am												
N^6-(Δ^2-Isopentenyl)-adenosine	iA			1	1	1							1
N^6-(Δ^2-Isopentenyl)-2-methylthio-adenosine	m^2siA								1				
N-(Nebularin-6-yl-carbamoyl)threonine			1										
3-Methylcytidine	m^3C												2
5-Methylcytidine	m^5C		1	2	1	1	1	1					1
2'-O-Methylcytidine	Cm			1							1		
N^4-Acetylcytidine	acC			1	1								1
1-Methylguanosine	m^1G	1						1					
N^2-Methylguanosine	m^2G		1			1							
N^2,N^2-Dimethyl-guanosine	m_2^2G	1	1	1	1	1	1						1
7-Methylguanosine	m^7G			1						1	1	1	
2'-O-Methylguanosine	Gm			1	1	1	1		1	1			1
Inosine	I	1	1		1	1		1					1
1-Methylinosine	m^1I	1											
5-Methyluridine	T	1	1	1	1	1	1	1	1	1	1	1	1
2'-O-Methyluridine	Um				1	1							1
Dihydrouridine	hU	3	5	2	3	3	6	4		3	1	1	3
5-Ribosyluracil	ψ	2	2	2	3	3	3	4	2	2	1	1	2
2'-O-Methyl-5-ribosyluracil	ψm												1
4-Thiouridine	s^4U								1	1	1	1	
Nucleoside, unidentified	N			1						3		2	

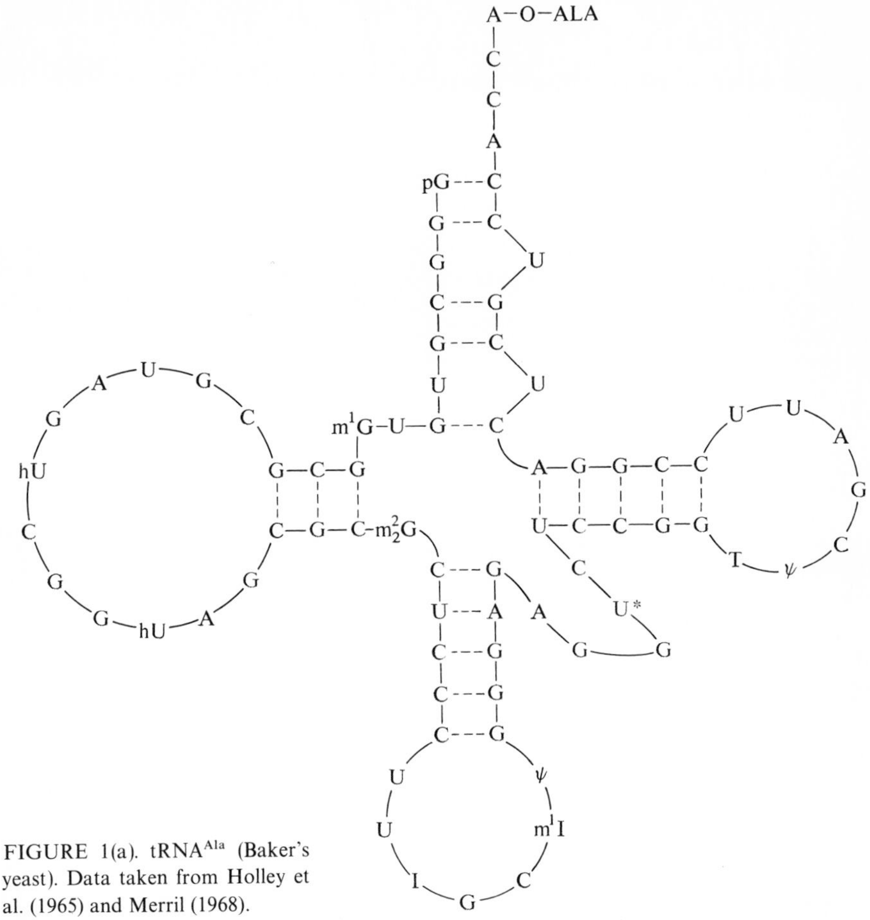

FIGURE 1(a). tRNA^{Ala} (Baker's yeast). Data taken from Holley et al. (1965) and Merril (1968).

has not been determined, it is instructive to arrange the sequence in the so-called cloverleaf pattern, first suggested by Holley et al. (1965) (Figure 1a). A certain amount of support for a structure containing a mixture of double helical and single-stranded regions has been obtained from a measurement of the physical properties of tRNA. Optical rotary dispersion studies on yeast tRNA^{Tyr} and tRNA^{Ala} by Vournakis and Scheraga (1966) indicate that the molecules contain regions of intramolecular hydrogen bonding and regions of neighbor-neighbor base stacking. Data obtained by measurement of the intensity of X-ray scattering and measurement of the radius of gyration on un-fractionated yeast tRNA (Lake and Beeman, 1967) are consistent with a folded

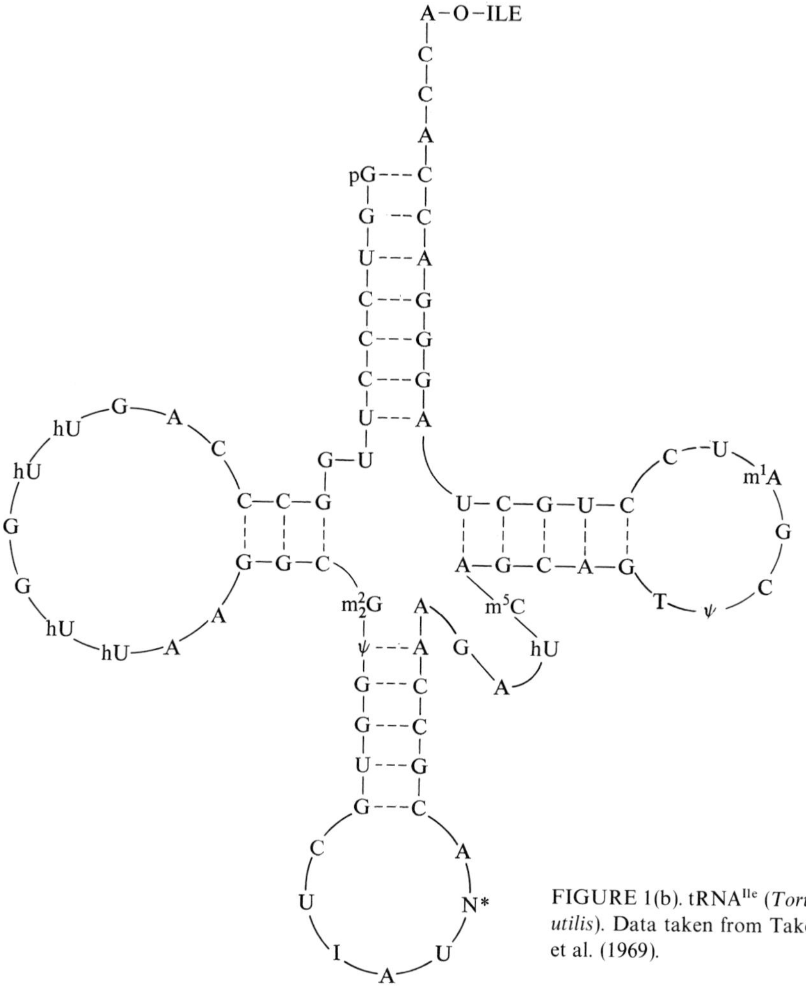

FIGURE 1(b). tRNA^Ile (*Torulopsis utilis*). Data taken from Takemura et al. (1969).

cloverleaf conformation and eliminate from consideration models consisting of only a single-strand or a hairpin configuration. These data, in fact, suggest that the tRNA molecule exists either with three arms folded tightly together and the fourth arm extended in the opposite direction, or possibly with two arms up and two down. A number of more general physical studies using the techniques of optical rotary dispersion (Fasman et al., 1965), thermal denatura-

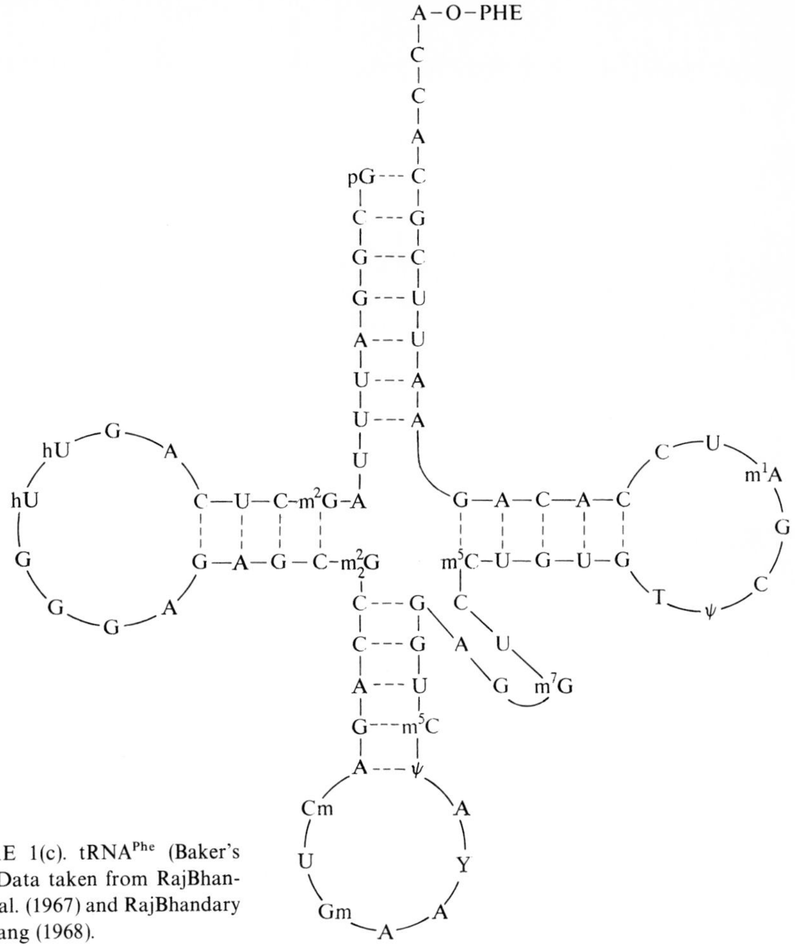

FIGURE 1(c). tRNA^Phe (Baker's yeast). Data taken from RajBhandary et al. (1967) and RajBhandary and Chang (1968).

tion (Fresco, 1963), and measurement of hydrodynamic properties (Adams et al., 1967) further support the view that the tRNA molecule is endowed with a unique asymmetric conformation that contains helical regions. Direct experimental confirmation of the cloverleaf structure, however, is still lacking. Nevertheless, regardless of the exact nature of the secondary and tertiary structure, it is evident that tRNA exists in highly ordered secondary and tertiary forms.

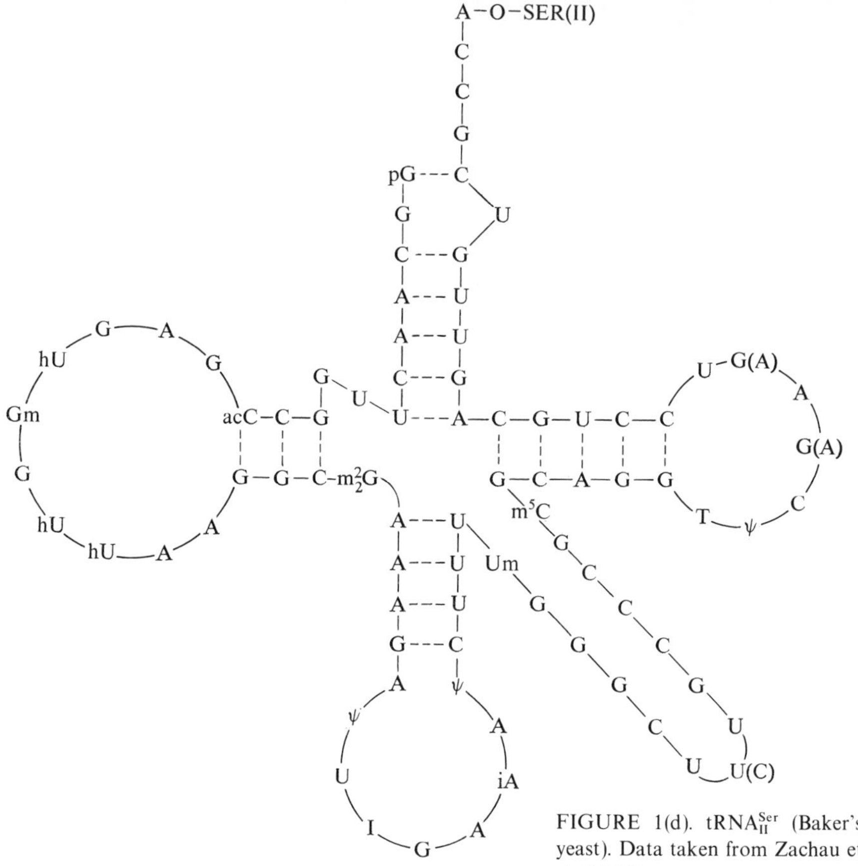

FIGURE 1(d). tRNA$_{II}^{Ser}$ (Baker's yeast). Data taken from Zachau et al. (1966b).

3. FEATURES COMMON TO ALL THE PRIMARY SEQUENCES

A schematic diagram using the basic cloverleaf pattern is shown in Figure 2. The tRNA molecule is postulated as consisting of three loops of unpaired nucleosides supported by short runs of paired bases; the left-hand or dihydrouridine loop contains eight to twelve unpaired nucleosides supported by an arm of three or four base pairs. The center or anticodon loop consists of seven unpaired nucleosides supported by an arm of five base pairs. The right-hand loop also consists of seven unpaired nucleosides supported by an arm of four

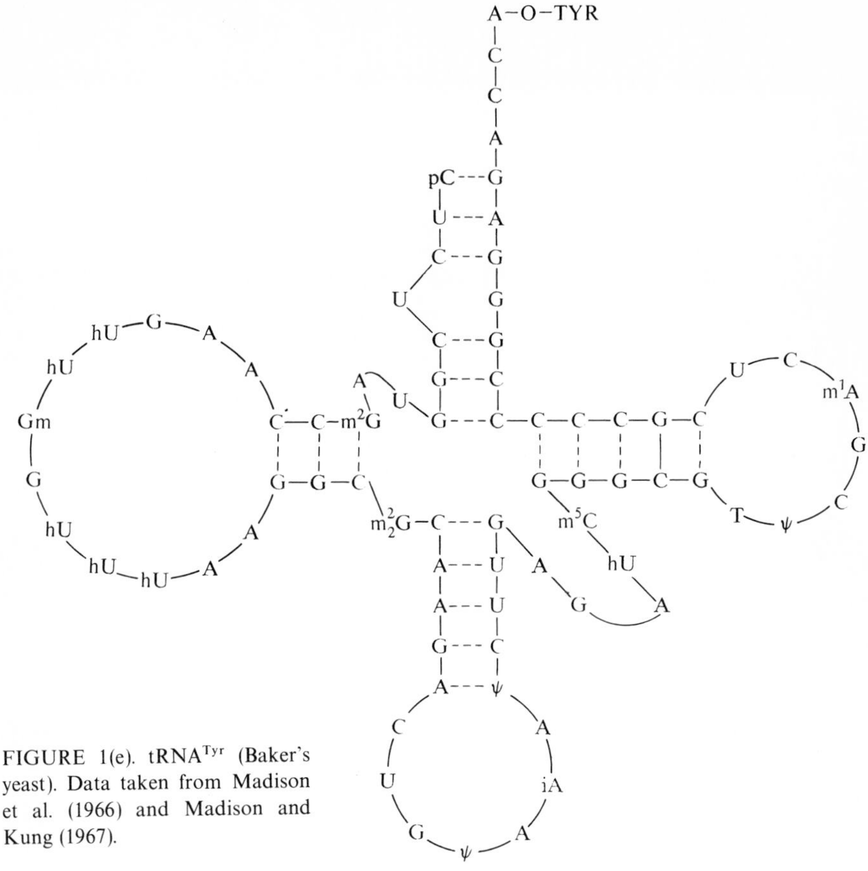

FIGURE 1(e). tRNA^{Tyr} (Baker's yeast). Data taken from Madison et al. (1966) and Madison and Kung (1967).

base pairs. The amino acid acceptor end of the molecule is at the end of an arm consisting of seven base pairs. In addition, an extra arm or loop varying in size extends from the molecule between the center and the right-hand loops.

Several additional structural features common to the yeast tRNA molecules can be summarized as follows:

1. The presumed anticodon is located in the middle loop and is bracketed by a uridine residue on the 5′ side and a modified adenosine residue on the 3′ side (except in the case of yeast tRNA^{Val} and *E. coli* tRNA$_F^{Met}$). This feature of the anticodon loop is discussed in more detail below.

2. The modified nucleosides, in general, are located in unpaired regions. A few of these compounds, however, such as 5-methylcytidine, N^2-methyl-

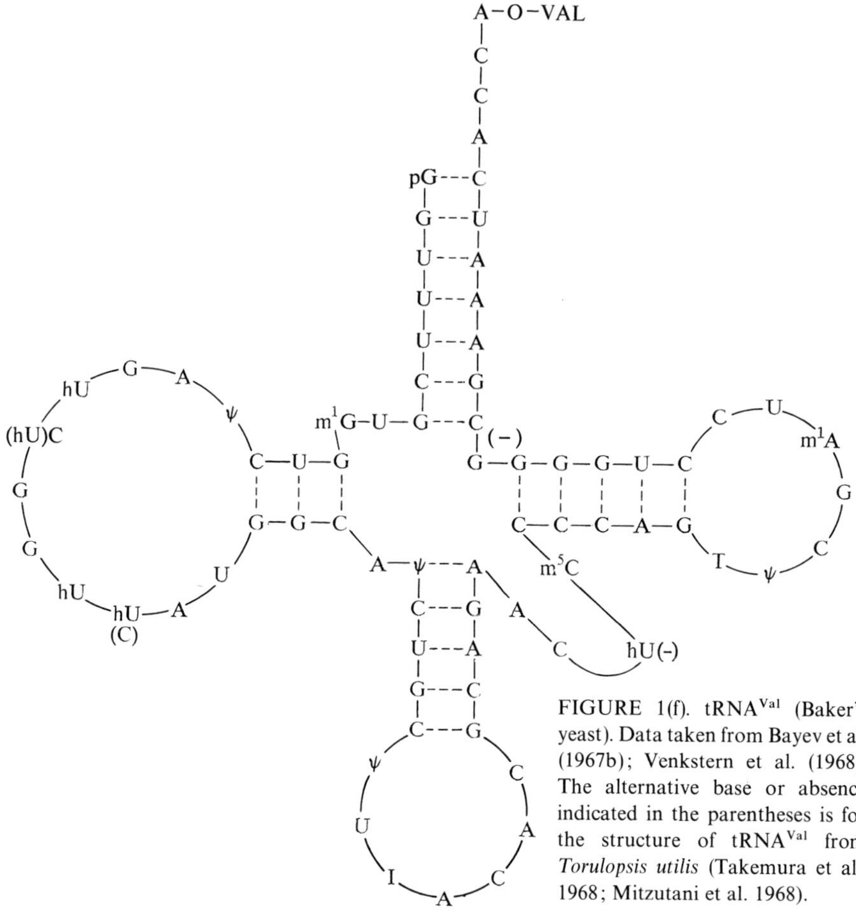

FIGURE 1(f). tRNA^{Val} (Baker's yeast). Data taken from Bayev et al. (1967b); Venkstern et al. (1968). The alternative base or absence indicated in the parentheses is for the structure of tRNA^{Val} from *Torulopsis utilis* (Takemura et al., 1968; Mitzutani et al. 1968).

guanosine, and 5-ribosyluracil, capable of hydrogen bonding in the same manner as their parent nucleosides, occur in paired regions.

3. The dihydrouridine residues occur in the left-hand loop, sometimes referred to as the dihydrouridine loop. In addition, a dihydrouridine residue is found in the extra loop of yeast alanine, tyrosine, and valine tRNA (Chapter 2). It is not found in the extra arm of either the serine of phenylalanine tRNAs. The extra arms of the phenylalanine and serine molecules, especially that of the latter, are much longer than the extra arms of the other tRNA molecules mentioned. Interestingly, in the case of tRNA^{Ala}, about one-half of the isolated

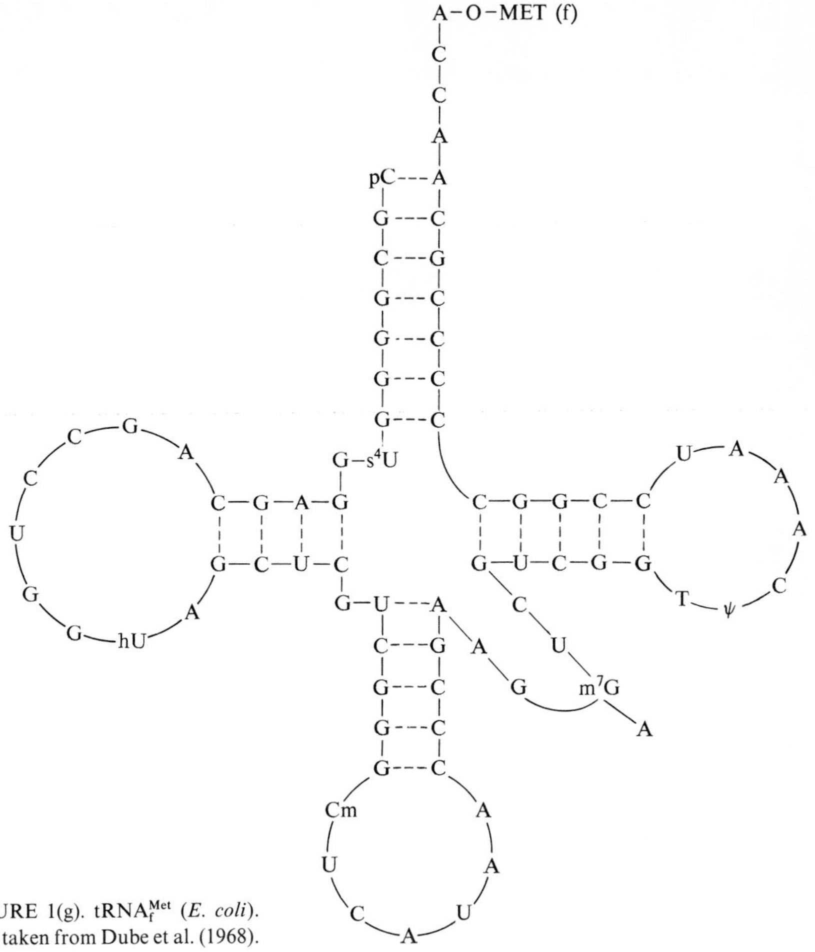
A–O–MET (f)
C
C
A
pC---A
G---C
C---G
G---C
G---C
G---C
G---C
G–s^{4}U
C–G–A–G
G–C–U–C
C–G–G–C–C
G–U–C–G–G
G–U---A
C---G
G---C
G---C
G---C
C G A C U G G hU A
U A A A C ψ T
C U m^{7}G A G A
Cm U C A U A A

FIGURE 1(g). tRNA$^{Met}_f$ (*E. coli*).
Data taken from Dube et al. (1968).

sample contains a uridine residue at this position. This fact suggests that at this site in the tRNA^{Ala} molecule a facile and reversible hydrogenation occurs and that at the stage in which the yeast is harvested a substantial portion of the tRNA^{Ala} is not reduced at this residue.

4. None of the modified nucleosides occurs in the amino acid acceptor arm of the molecule.

5. N^2,N^2-Dimethylguanosine or 1-methylguanosine occurs in the transition point between the amino acid acceptor arm and the left-hand loop. The presence

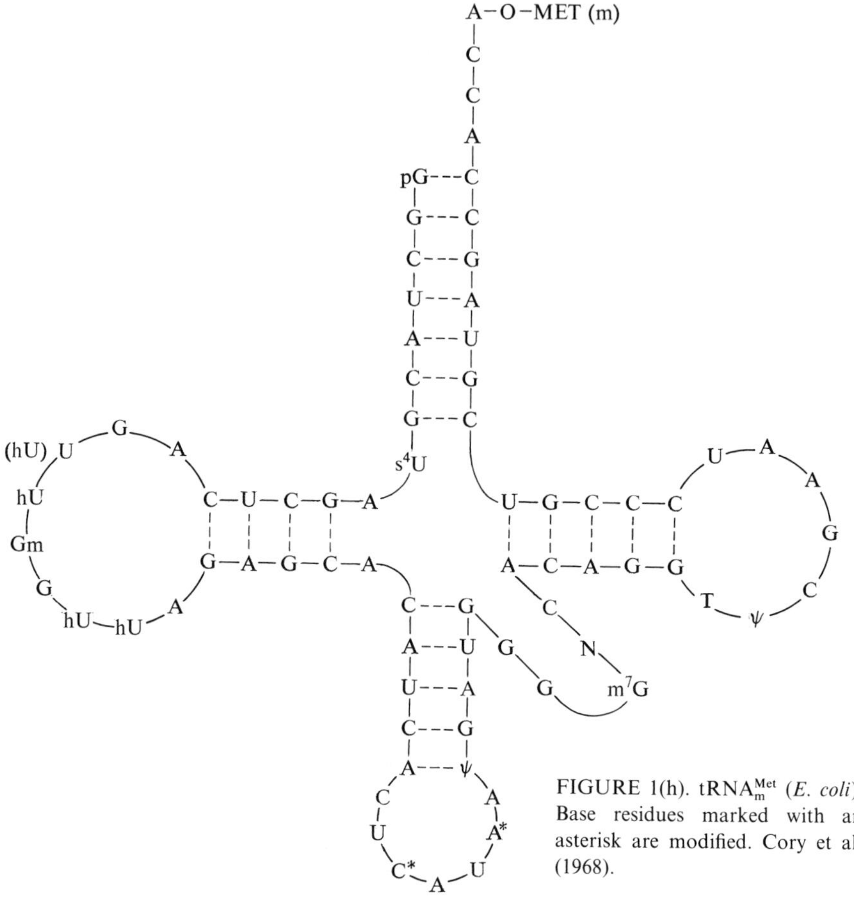

MODIFIED NUCLEOSIDES IN THE PRIMARY STRUCTURE OF SPECIFIC tRNA MOLECULES

A—O—MET (m)

FIGURE 1(h). tRNA$_m^{Met}$ (*E. coli*). Base residues marked with an asterisk are modified. Cory et al. (1968).

of these components may facilitate the formation of the three-dimensional structure at the transition point.

6. N^2,N^2-Dimethylguanosine residue occurs in the transition point between the left-hand loop and the middle (anticodon) loop (except in the case of tRNA^{Val}).

7. Cytosine or 5-methylcytosine occurs in the transition point between the extra arm and the right-hand loop. In the case of yeast alanine, isoleucine, tyrosine, phenylalanine, and valine tRNAs, the dinucleotide hU-C(m^{5}C) occurs at this point.

FIGURE 1(i) tRNA$_{Su}^{Tyr}$ (*E. coli*). Data taken from Goodman et al. (1968). A* = N^6-(Δ^2-isopentenyl)-2-methylthioadenosine. In Su$_{III}^-$ tRNA^{Tyr}, the C of the antocodon is replaced by a modified G. In regular tRNA$_{II}^{Tyr}$, the two base changes in the extra arm occur.

8. Figure 2 indicates sequences common to most or all of the tRNA molecules. The most notable of these is G-T-ψ-C-G, first described by Zamir et al. (1965), which occurs in the right-hand loop of each of the tRNA molecules. Other common sequences are A-C-C-A at the acceptor end of the molecule, A-G-hU in the left-hand loop, and G-C-m$_2^2$G (A in tRNA^{Val}) in the left-hand supporting arm of the yeast tRNAs.

Aside from the short oligonucleotide sequences common to most or all tRNA

MODIFIED NUCLEOSIDES IN THE PRIMARY STRUCTURE OF SPECIFIC tRNA MOLECULES

```
                          A–O–VAL
                           C
                           C
                           A
                     pG---C
                      G---C
                      G---C
                      U---A
                      G---C
                      A---U
          C  G   A    U---A                        C  U
       /           \                            /        \
     hU    C–U–C–G–A–s⁴U      C–U–G–C–C        A
                  | | | | |     | | | | |                 G
      G    G–A–G–C–A        G–G–C–G–G         T  ψ  C
       \         /              \                \     /
         G  G  A      C---G      C                  
                      C---G G      U
                      U---A   \
                      C---G    G   m⁷G
                      C---G
                   C          A
                    \        /
                     U     m⁶A
                      \    /
                       N  C
                        A
```

FIGURE 1(j). tRNA$_I^{Val}$ (*E. coli*). Data taken from Yaniv and Barrell (1969).

molecules, certain oligonucleotide sequences are common to a group of aminoacyl-accepting species. In such cases the homology may extend over a longer section of the molecule. A rather striking example is the sequence A-iA-A-ψ-C-U-U, which occurs in the anticodon loop and supporting arm of tRNA^{Tyr} and tRNA$_{I,II}^{Ser}$ of yeast. The complementary sequence of the opposite side of the arm is consequently identical (see Figure 3). *E. coli* Su$_{III}^+$ tRNA^{Tyr} and rat liver tRNA^{Tyr} also contain the first six and five nucleosides, respectively, of this sequence [in the case of *E. coli*, iA is N^6-(Δ^2-isopentenyl)-2-methyl-thioadenosine]. These five tRNA molecules have the hypermodified nucleoside,

A—O—SER
|
C
|
C
|
G
|
pG---C
| |
U---A
| |
A---U
| |
G---C
| |
U---A
| |
C---G
| |
G---C
|
U

G—A
hU   G
Gm   AcC—C—G—G   C—G—U—C—C   U—A
| | | | | | | | m¹A
G   G—G—C—m²₂G   G—C—A—G—G   G
hU   A   m⁵C   T—ψ   C
hU—A   A---U
ψ---A Um   G
G---C   G   C
G---C   G   C
A---ψm   G   C
m³C   A   G   C
U   iA   U   U
I—G—A   m³C

FIGURE 1(k). tRNA^Ser (rat liver).
Data taken from Staehelin et al.
(1968).

N^6-(Δ^2-isopentenyl)adenosine (iA), in common. In addition, the A-A-A triplet (the Δ^2-isopentenyl group is attached to the middle adenosine) appears to be almost unique in the published sequences, whereas runs of three or more guanosine residues are not uncommon. This sequence peculiar to the Δ^2-isopentenyl-containing tRNA molecules may have some function with respect to the mechanism of attachment of the Δ-isopentenyl group or with respect to a particular biological activity of these molecules.

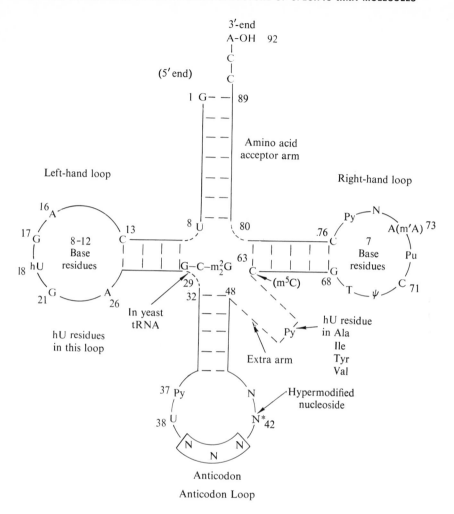

FIGURE 2. Schematic diagram of the cloverleaf pattern with features common to the known tRNA sequences.

4. ANTICODON LOOPS: STRUCTURAL CHARACTERISTICS

The loop containing the presumed anticodon consists of seven unpaired nucleosides. The presumed anticodon triplet is bracketed by a uridine residue on one side and a modified nucleoside on the other. A pyrimidine nucleoside

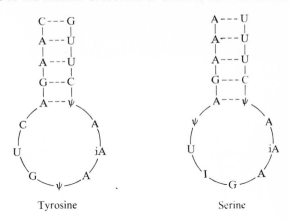

Tyrosine   Serine

FIGURE 3. Sequence of anticodon loop and arm for tRNA^{Tyr} and tRNA^{Ser}.

occupies the top left position adjacent to the uridine residue. In terms of the known amino acid codons the center triplet of this loop in the known primary sequence would seem to be the actual anticodon. Supporting evidence for this view has now been obtained. Clark et al. (1968) isolated a fragment of *E. coli* tRNA$_f^{Met}$ containing the anticodon loop sequence and tested its binding to 70S ribosomes as stimulated by various oligonucleotide triplets. The fragment is bound by AUG (the correct codon) but not by any trinucleotides complementary to the other four triplet sequences in the loop. Yoshida et al. (1968b), using acrylonitrile, selectively cyanoethylated the inosine residues of yeast tRNA^{Ala} (see p. 366 for details of the reaction). The treated tRNA was inactive in the binding ability to ribosomes in the presence of its triplet codons. These workers conclude, therefore, that the anticodon may be the middle three nucleotides of the loop.

The seven-membered anticodon loop probably exists in a well-defined configuration. Fuller and Hodgson (1967) have suggested that the loop adopts a helical conformation that maximizes single-stranded base stacking. They further suggest that this polynucleotide loop has surprisingly little conformational freedom. This rigidity may be necessary to provide the proper configuration for infallible binding with messenger codons. Many questions arise, however, concerning the actual mechanism of interaction of the anticodon loop with the codon, enzymes, and ribosomes; and superimposed on these mechanistic details is the question of modulation of tRNA activity. Perhaps the hypermodified nucleoside adjacent to the 3' end (first letter read) is the key.

MODIFIED NUCLEOSIDES IN THE PRIMARY STRUCTURE OF SPECIFIC tRNA MOLECULES

TABLE 2

Nucleosides Adjacent to the 3' Side of the Anticodon (Adjacent to the first letter of the anticodon)

Species of tRNA	Anticodon 5'→3'	Adjacent Nucleoside
Yeast alanine	I-G-C	1-Methylinosine
Yeast isoleucine	I-A-U	N-(Nebularin-6-ylcarbamoyl)-L-threonine
Yeast phenylalanine	Gm-A-A	Unidentified
Yeast serine I	I-G-A	N^6-(Δ^2-Isopentenyl)adenosine
Yeast serine II	I-G-A	N^6-(Δ^2-Isopentenyl)adenosine
Rat liver serine	I-G-A	N^6-(Δ^2-Isopentenyl)adenosine
Yeast tyrosine	G-ψ-A	N^6-(Δ^2-Isopentenyl)adenosine
Yeast valine	I-A-C	Adenosine
E. coli Su$_{\text{III}}^+$ tyrosine	C-U-A	N^6-(Δ^2-Isopentenyl)-2-methylthioadenosine
E. coli f-methionine	C-A-U	Adenosine
E. coli m-methionine	C-A-U	Modified adenosine
E. coli valine	N-A-C	Methylated adenosine

The structure of this particular modified nucleoside is considerably more complex than that of other modified nucleosides. Table 2 lists the anticodons of the tRNA molecules with their adjacent modified nucleosides. Two of the identified nucleosides in this position, N^6-(Δ^2-isopentenyl)adenosine (and its 2-methylthio derivative) and N-(nebularin-6-ylcarbamoyl)-L-threonine, are distinguished by the bulk of the modifying side chain and perhaps more importantly by the presence of a reactive chemical group (i.e., allylic double bond, carboxylic, hydroxyl) in the side chain. These reactive groups have three important characteristics. First, they are essentially unique in their chemical reactivity (i.e., the allylic double bond of the Δ^2-isopentenyl group exhibits a reactivity unlike that of any other known component of tRNA). Second, the group occurs in a single position of the given tRNA molecule. Third, each of the known modifications in this category occurs in only a few of the tRNA molecular species of an organism. Thus this nucleoside, called a hypermodified nucleoside, endows this critical region of the tRNA molecule with unique chemical reactivity. This reactivity may be essential for the tRNA to carry out its function (see p. 13, Chapter 1, for a more detailed discussion). Gefter and Russell (1969) have shown, for example, that the Δ^2-isopentenyl group of tRNA$_{\text{SuIII}}^{\text{Tyr}}$ is essential for the binding of this tRNA molecule to the ribosome.

The basic analytical data (see Chapter 2) seem to indicate that the tRNA of higher organisms quantitatively contains a greater number of modified nucleosides than bacterial tRNA. The tRNA of the lowest organism capable of self-replication, mycoplasma, contains few modified nucleosides (Hall et al., 1967b; Hayashi et al., 1969). This relationship, discussed in Chapter 1, p. 15, suggests that the more sophisticated tRNA structure, as represented by a greater degree of structural modification, may correlate with the more sophisticated metabolic control processes occurring in the cells of higher organisms.

5. HOMOLOGIES IN THE PRIMARY SEQUENCES

TABLE 3

Homologous Alignment of RNA Sequences

	Yeast						E. coli				Rat Liver
	Ala	Ile	Phe	Ser II[a]	Tyr	Val	f Met	m Met	Su$^+$Tyr	Val	Ser
1	pG	pG	pG	pG	pC	pG	pC	pG	pG	pG	pG
2	G	G	C	G	U	G	G	G	G	G	U
3	G	U	G	C	C	U	C	C	U	G	A
4	C	C	G	A	U	U	G	U	G	U	G
5	G	C	A	A	C	U	G	A	G	G	U
6	U	C	U	C	G	C	G	C	G	A	C
7	G	U	U	U	G	G	G	G	G	U	G
8	U	U	U	U	U	U	s^{4}U	s^{4}U	s^{4}U	s^{4}U	U
9	m^{1}G		A	G	A		G	A	s^{4}U	A	
10	G	G	m^{2}G	G	m^{2}G	m^{1}G	G	G		G	G
11	C	G	C	C	C	G	A	C	C		G
12	G	C	U			U	G	U	C		C
13	C	C	C	AcC	C	C	C	C	C	C	AcC
14	G			G	A				G	U	G
15	U	C				ψ				C	
16	A	A	A	A	A	A	A	A	A	A	A
17	G	G	G	G	G	G	G	G	G	G	G
18	hU	hU	hU	hU	hU	hU	C	U (hU)		C	hU
19	C	hU	hU		hU	C(hU)	C	hU	C	hU	
20	G	G	G	Gm	Gm	G	U	Gm	Gm	G	Gm
21	G	G	G	G	G	G	G	G	G	G	G

MODIFIED NUCLEOSIDES IN THE PRIMARY STRUCTURE OF SPECIFIC tRNA MOLECULES

TABLE 3 continued

	Yeast						E. coli				Rat Liver
	Ala	Ile	Phe	Ser II[a]	Tyr	Val	f Met	m Met	Su[+]Tyr	Val	Ser
22			G						C	G	
23		hU		hU	hU	hU	G	hU	C		hU
24	hU	hU		hU	hU	hU(C)	hU	hU	A		hU
25		A			hU				A		A
26	A	A	A	A	A	A	A	A	A	A	A
27	G		G	A	A	U	G	G	G	G	G
28	C	G	A	G	G	G	C	A	G	A	G
29	G	G	G	G	G	G	U	G	G	G	C
30	C	C	C	C	C	C	C	C	A	C	
31	m_2^2G	m_2^2G	m_2^2G	m_2^2G	m_2^2G	A	G	A	G	A	m_2^2G
32	C	ψ	C	A	C	ψ	U	C		C	A
33	U	G	C	A	A	C	C	A	C	C	ψ
34	C	G	A	A	A	U	G	U	A	U	G
35	C	U	G	G	G	G	G	C	G	C	G
36	C	G	A	A	A	C	G	A	A	C	A
37	U	C	Cm	ψ	C	ψ	Cm	C	C	C	m^3C
38	U	U	U	U	U	U	U	U	U	U	U
39	I	I	Gm	I	G	I	C	C^d	C (G)	N	I
40	G	A	A	G	ψ	A	A	A	U	A	G
41	C	U	A	A	A	C	U	U	A	C	A
42	m^1I	N^b	Y^c	iA	iA	A	A	A^d	m^2siA	mA	iA
43	ψ	A	A	A	A	C	A	A	A	A	A
44	G	C	ψ	ψ	ψ	G	C	ψ	ψ	G	ψm
45	G	G	m^5C	C	C	C	C	G	C	G	C
46	G	C	U	U	U	A	C	A	U	A	C
47	A	C	G	U	U	G	G	U	G	G	A
48	G	A		U		A	A	G	C	G	U
49	A	A		Um		A	A	G	C	G	Um
50	G	G	G	G	G	C	G	G	G	G	G
51	G	A	A	G	A				U		G
52		G		G	G						G
53				C	A				C		G
54				U					A		U
55	hU (U)	hU	m^7G	U (C)	hU	hU(−)	m^7G(A)	m^7G	U (C)	m^7G	m^3C
56		U	U				U	N	C (A)	U	U
57			G						G		C
58			C						A		C

MODIFIED NUCLEOSIDES IN THE PRIMARY STRUCTURE OF SPECIFIC tRNA MOLECULES

TABLE 3 continued

	Yeast						E. coli				Rat Liver
	Ala	Ile	Phe	Ser II[a]	Tyr	Val	f Met	m Met	Su$^+$Tyr	Val	Ser
59		C	C						C		C
60			C						U		C
61			G						U		G
62	C	m^{5}C	m^{5}C	m^{5}C	m^{5}C	m^{5}C	C	C	C	C	m^{5}C
63	U	A		G	G	C	G	A	G	G	G
64	C	G	U	C	G	C	U	C	A	G	C
65	C	C	G	A	G	C	C	A	A	C	A
66	G	A	U	G	C	A	G	G	G	G	G
67	G	G	G	G	G	G	G	G	G	G	G
68	T	T	T	T	T	T	T	T	T	T	T
69	ψ	ψ	ψ	ψ	ψ	ψ	ψ	ψ	ψ	ψ	ψ
70	C	C	C	C	C	C	C	C	C	C	C
71	G	G	G	G (A)	G	G	A	G	G	G	G
72	A	m^{1}A	m^{1}A	A	m^{1}A	m^{1}A	A	A	A	A	m^{1}A
73	U	U	U	G (A)	C	U	A	A	A	U	A
74	U	C	C	U	U	C	U	U	U	C	U
75	C	C	C	C	C	C	C	C	C	C	C
76	C	U	A	C	G	U	C	C	C	C	C
77	G	G	C	U	C	G	G	C	U	G	U
78	G	C	A	G	C	G	G	G	U	U	G
79	A	U	G	C	C	G	C	U	C	C	C
80	C	A	A	A	C	C	C	C	C	A	C
81	U	G	A	G	C	G	C	G	C	U	G
82	C	G	U	U	G	A	C	U	C	C	A
83	G	G	U	U	G	A	C	A	C	A	C
84	U	A	C	G	G	A	G	G	A	C	U
85	C	C	G	U	A	U	C	C	C	C	A
86	C	C	C	C	G	C	A	C	C	C	C
87	A	A	A	G	A	A	A	A	A	A	G
88	C	C	C	C	C	C	C	C	C	C	C
89	C	C	C	C	C	C	C	C	C	C	C
90	A-OH	A-OH	A-OH	A-OH	A-OH	A-OH	A-OH	A-OH	A-OH	A-OH	A-OH

[a]The three base changes for Ser I are shown in parentheses.
[b]N-(Nebularin-6-ylcarbamoyl)-L-threonine.
[c]Unidentified.
[d]Modification of major nucleoside.

In addition to a correlation between the degree of complexity of tRNA structure and the degree of complexity of cellular activity, a basic evolutionary relationship with respect to tRNA primary structure has become apparent. Jukes (1966) and Zachau et al. (1966b,c) suggest that tRNA molecules are derived from a common archetype and that through evolution the observed divergence in tRNA structure, subject to the constraints imposed in order to conserve a secondary and tertiary structure essential to the biological function, has resulted.

The known tRNA primary sequences are arranged according to the scheme of Jukes in Table 3. The essential concept of this arrangement is the insertion of gaps to permit alignment of common sequences designated by the boxes. These gaps could result from the deletion or the addition of genetic material during the evolutionary process. Alignment of the tRNA sequences in this fashion reveals the presence of several homologies common to all the primary sequences. In addition, sequential homologies common to a few tRNA molecules are apparent [i.e., A-iA-A-ψ-C-U-U(G), residues 41–47 in yeast phenylalanine, serine, and tyrosine, and E. coli Su$_{III}^+$ tyrosine tRNAs], suggesting a closer structural relationship for groups of tRNA molecules.

The homologies common to all molecules may bear some relationship to functions common to all tRNA molecules, for example, common but nonspecific binding sites for the ribosome or aminoacyl synthetases. The homologies within a group of tRNA molecules may result from common evolutionary divergence, but they may represent some more significant facet of tRNA function. One striking feature, for example, is the occurrence of a Δ^2-isopentenyl group at the adjacent position of those tRNA molecules for which the first letter of the codon is U. Nishimura et al. (1969) report that in addition to E. coli, tRNA^{Tyr}, tRNA$_I^{Ser}$, tRNA$_{II}^{Ser}$, and tRNA^{Phe} contain N^6-(Δ^2-isopentenyl)-2-methylthioadenosine but not tRNA^{Leu} (codons CU series) or tRNA$_{III}^{Ser}$ (codons AG series). Armstrong et al. (1969) also have evidence for the presence of an N^6-(Δ^2-isopentenyl)adenosine derivative in tRNA^{Tyr}, tRNA^{Ser}, tRNA^{Cys}, tRNA^{Phe}, and tRNA^{Leu} (UUG) of E. coli. The apparent requirement of the Δ^2-isopentenyl group for binding to the ribosome as discussed on p. 327 suggests that there may be a specific aspect of the binding process not shared by those tRNA molecules that do not have a Δ^2-isopentenyl group in this position within a cell.

The possibility of a general correlation between the first letter of the codon and the modified nucleoside adjacent to the 3' end of the anticodon is enhanced by the report of Ishikura et al. (1969) that tRNA^{Lys}, tRNA$_I^{Met}$, and tRNA$_{III}^{Ser}$ of

E. coli contain *N*-(nebularin-6-ylcarbamoyl)-L-threonine (*24*, Chapter 7). These species of tRNA respond to codons with first letter A. The sequences have not been reported, but Takemura et al. (1969) have shown that this nucleoside occurs adjacent to the 3′ end of the anticodon of tRNA^{IIe} (yeast). Species of tRNA (*E. coli*) that respond to codons with a first letter other than A, such as serine (I and II), phenylalanine, histidine, aspartic acid, and leucine, do not contain *N*-(nebularin-6-ylcarbamoyl)-L-threonine (Ishikura et al., 1969).

The physiological reason for the similarities and diversities in the primary sequence of tRNA remains an intriguing question. In this regard, the fact that there are many more tRNA molecular species (called iso-accepting tRNAs) within a cell than amino acids must be considered.

The reason for this is poorly understood. Four glycine tRNAs from yeast, for example, although structurally different, recognize the same code word (Bergquist et al., 1968). The tRNA of *E. coli* contains several iso-accepting species—five for leucine (Kelmers et al., 1965; Muench and Berg, 1966a,b), three for proline (Goldstein et al., 1964) and two for most of the other amino acids. In extending these studies Muench and Safille (1968) have separated by gradient partition chromatography of *E. coli* B tRNA fifty-six separate tRNA fractions. Separation and identification of the iso-accepting tRNA species have become of great interest because it appears that structural variation of the tRNA molecules may play an important role with respect to cellular regulatory mechanisms. (See the discussion of p. 6, Chapter 1.)

6. FRACTIONATION OF ISO-ACCEPTING tRNAs

The physical separation of these iso-accepting tRNAs is clearly based on structural differences and in most cases probably represents native sequential differences. Such a difference is demonstrated in the known structures of serine I and serine II of yeast. One fact should be kept in mind concerning the separability of tRNA molecules. Yang and Novelli (1968) have shown that a tRNA molecular species elutes differently from a column system depending on whether or not it is charged with its specific amino acid. The addition of a group of small molecular weight to a macromolecule of molecular weight 25,000 might seem to be insufficient to cause a major shift in the chromatographic behavior of the molecule. It appears likely, therefore, that conformational changes in the tRNA molecule may be responsible for this difference in elution behavior. Thus other minor structural changes that might lead to a conformational change could be expected to affect the chromatographic behavior of the molecule. The loss of a labile substituent of a RNA molecule during isolation, for example, could create a structural artifact that is separable. Such an event is particularly well illustrated by the fact that deletion of the base of the nucleoside Y^+ from yeast tRNA^{Phe} causes a dramatic change in its elution characteristics from a column of BD-cellulose (Thiebe and Zachau, 1968).

7. RECOGNITION SITE FOR THE AMINOACYL SYNTHETASE

The general function of the tRNA molecule with respect to the acceptance and transfer of amino acid is understood, but little is known about the actual molecular mechanisms. The anticodon sequences of the tRNA molecule seem to be well established, but the corresponding sequence involved in recognition of the aminoacyl synthetase has not been defined. The anticodon loop certainly provides an amino acid-specific sequence and therefore might be considered as an appropriate region to express enzyme specificity; the results of several experiments in which the anticodon loop is structurally altered, however, suggest otherwise.

Selective modification of the N^6-(Δ^2-isopentenyl)adenosine residue yeast tRNA^{Ser} with iodine diminishes the capacity of the tRNA to bind with the messenger-ribosome complexes but does not affect the ability to accept serine (Fittler and Hall, 1966). Likewise, specific removal of the base of the unidentified nucleotide Y by mild acid treatment from the anticodon loop of yeast tRNA^{Phe} causes complete loss of codon recognition but does not alter the capacity of the tRNA to accept phenylalanine (Thiebe and Zachau, 1968). The absence of the Δ^2-isopentenyl group from tRNA$^{Tyr}_{Su_{III}^+}$ prevents binding with the messenger-ribosome complex but does not interfere with the acceptance of tyrosine residues (Gefter and Russell, 1969). Such structural alterations undoubtedly perturb the normal three-dimensional structure of the loop, which may account for the loss of codon-anticodon binding; on the other hand, the possibility exists that the actual binding sequence for the codon may include the modified nucleoside at the 3' end of the anticodon. These results demonstrate, at least, that the anticodon is not involved in recognition of the aminoacyl synthetase.

Therefore, some other region of the molecule must be responsible for the specificity of enzyme recognition. This region need not be restricted to a sequence of three nucleotides; it may involve a much larger sequence, and also the tertiary structure could play an important role. It is possible, though, to experimentally damage or delete large portions of the tRNA molecule without destroying its capacity to recognize the specific aminoacyl synthetase. Schulman and Chambers (1968) found that yeast tRNA^{Ala} undergoes extensive photochemical modification without loss of acceptor activity; on the other hand, destruction of a single pyrimidine residue by ultraviolet light in the amino acid acceptor arm of the molecule is sufficient to eliminate alanine-accepting activity. On the basis of their results, they postulate that the specific recognition site of the aminoacyl synthetase involves the first three base pairs in the amino acid acceptor arm. It is difficult, however, to reconcile this postulate with the results of Stulberg and Isham (1967), who observed that sequential deletion of up to 56% of the *E. coli* tRNA^{Phe} molecule by limited digestion with snake venom diesterase does not eliminate the phenylalanine synthetase recognition site. The data of both these groups of workers indicate that, in terms of overall binding with the synthetase enzyme, the C-C-A end of the tRNA molecule contributes significantly to the binding with the synthetase but with no specificity involved except to reinforce the binding of the recognition site.

It is interesting that with respect to the structural characteristics of the tRNA molecule required for amino acid acceptance, it is possible to cleave the molecule at a residue in the anticodon loop into two large fragments without disrupting amino acid acceptance. Bayev et al. (1967a), for example, found that yeast tRNA^{Val} can be specifically cleaved between the inosinic acid and cytidylic acid residues by an enzyme from actinomycetes. Recombination of the two fragments (the two fragments do not become co-valently linked) restores the amino acid acceptance. In an analogous experiment, Philippsen et al. (1968) cleaved yeast tRNA^{Phe} by chemical treatment at the location of residue Y (see p. 371, Chapter 8, for further discussion of this reaction). The phenylalanine acceptance activity can be restored on recombination of the two fragments. These experimental results indicate that an approximation of the original tertiary structure is probably achieved.

The several homologous segments of the known tRNA sequences suggest that these segments may represent common binding sites for the aminoacyl synthetase and/or the ribosome. It is of some interest that many of these homologous nucleosides are modified. For example, residue 31 in several of the sequences is N^2,N^2-dimethylguanosine, residue 39 is the wobble position of the anticodon, residue 42 is the hypermodified nucleoside, residue 62 is 5-methylcytidine, residues 68 and 69 are 5-methyluridine and 5-ribosyluracil, respectively, and residue 72 in several molecules is 1-methyladenosine (Table 3). These common and subtle changes in structure may be critical to the efficiency and specificity of binding of the tRNA molecule with the other components of the protein-synthesizing system.

In summary, the tRNA constitutes the only class of nucleic acids for which we possess considerable information about the mechanism of its molecular interactions. The fact that the primary sequence of several molecular species is also known permits a number of significant correlations between structure and function. The importance of the modified nucleosides to the function of the tRNA molecule is now becoming evident.

Chapter 5

MODIFIED NUCLEOSIDES IN DNA

1. INTRODUCTION; GENERAL PATTERN OF DISTRIBUTION OF MODIFIED DEOXYRIBONUCLEOSIDES IN NATURE

IN CONTRAST to the number and variety of structural modifications that occur in RNA, relatively few such modifications take place in DNA. The modifications that do exist take the form of methylation of a small percentage of one of the nucleoside components. More than twenty different methylated nucleosides have been isolated from RNA, but only two have been detected in DNA, and only one of these has been detected in mammalian DNA.

The methylation pattern of individual DNA molecules has not been worked out, but evidence suggests that it is definitely not random. In fact, the DNA molecules of each organism may have a definitive pattern of distribution of methylated components; this possibility has led to speculation that one function of the methyl groups is to create molecular "fingerprinting." Such a mechanism provides cell-specific DNA molecules without interfering in any way with the encoded genetic information. (This topic, with the relevant evidence, is discussed in Chapter 6.)

Two to 8% of the deoxycytidine residues of mammalian DNA consists of 5-methyldeoxycytidine. This cytidine derivative also occurs in the DNA of cereal germ (wheat and rye), in which it constitutes about 25% of the deoxycytidine residues. In microorganisms the pattern of methylation differs sharply, and the methyl groups are attached to the N-6 position of deoxyadenosine residues. N^6-Methyldeoxyadenosine has been found in most samples of bacterial DNA; it accounts for about 2–4% of the total deoxyadenosine residues. Until recently this was the only modified nucleoside detected, but now it appears that the DNA of some bacterial strains contains small amounts of 5-methyldeoxycytidine (Doskočil and Šormová, 1965; Vanyushin et al., 1968).

MODIFIED NUCLEOSIDES IN DNA

TABLE 1

Modified Nucleosides in DNA

Only N^6-methyldeoxyadenosine has been detected in animal DNA, and only 5-methyldeoxy-cytidine in plant DNA. Both these nucleosides occur as trace components of bacterial DNA. In some bacteriophages, complete replacement of one of the major nucleoside components with a modified constituent occurs. All the reported analyses are included in this table, and the values for the most part are isolated amounts. In the case of the analyses of bacterial DNA samples, the results may vary from laboratory to laboratory, probably because of strain differences.

Part A: DNA Samples in Which a Small Percentage of One of the Major Components Is Modified

Animal	5-Methyldeoxycytidine, % of C[a]	References
Actinia ecquina	4.6	Antonov et al. (1962)
Artemia salina	0.5	Antonov et al. (1962)
Asterias rubens	4.8	Antonov et al. (1962)
Bull sperm	5.9	Wyatt (1951b)
Calf thymus	5.8–8.1	Wyatt (1951b); Sinsheimer and Koerner (1952); Laland et al. (1952); Hurst et al. (1953); Antonov et al. (1962)
Echinus esculentus sperm	8.6	Wyatt (1951b)
Embryo, chick	4.3	Smith and Stoker (1951)
Euglena gracilis, strain Z	8.9	Brawerman et al. (1962)
Frog	8.4	Dawid (1965)
HeLa cells	2	Littlefield and Gould (1960)
Herring sperm	8.3	Wyatt (1951b); Felix et al. (1956); Zahn (1958)
Herring testes	10.8	Laland et al. (1952)
Homoeodyctia palmata	6.5	Antonov et al. (1962)
Locasta migratoria	0.9	Wyatt (1951b)
Mouse liver	4.9	Antonov et al. (1962)
Nephthys ciliata	8.5	Antonov et al. (1962)
Ox spleen	5.9–6.3	Wyatt and Cohen (1953); Wyatt (1951b); Laland et al. (1952)
Paracentrotus lividus sperm	6	Chargaff et al. (1952)
Patella sperm	4.9	Antonov et al. (1962)
Pectin islandicus	2.5	Antonov et al. (1962)
Rabbit liver	4.3	Antonov et al. (1962)
Ram sperm	4.5	Wyatt (1951b)
Rat bone marrow	5.1	Wyatt (1951b)
Sheep sperm	4.5	Wyatt (1951b)
Xenopus laevis	6.9	Dawid (1965)

[a]The values are calculated as, e.g. $\dfrac{m^5C}{C+m^5C} \times 100\%$.

TABLE 1 continued

Plant	5-Methyldeoxycytidine, % of C[a]	References
Aleurites fordii	29.5	Vanyushin and Belozerskii (1959)
Allium cepa	29.7	Uryson and Belozerskii (1959)
Arachis hypagaea	29.8	Uryson and Belozerskii (1959)
Beet (*Beta vulgaris*)	26.3	Thomas and Sherratt (1956)
Bletia hiacinthiana	24.9	Vanyushin and Belozerskii (1959)
Bracken fern	23.5	Thomas and Sherratt (1956)
Brassica oleifera	9.7	Vanyushin and Belozerskii (1959)
Carrot	25.5–25.8	Thomas (1959)
Cephalotaxus fortunei	19.8	Vanyushin and Belozerskii (1959)
Ceratapteris cornuta	18.1	Vanyushin and Belozerskii (1959)
Clover (*Trifolium pratense*)	23.6	Thomas and Sherratt (1956)
Cotton		
Diploid	22.9–25.5	Ergle et al. (1964)
Tetraploid	26.9–26.4	Ergle and Katterman (1961)
Cucurbita pepo	18.7	Uryson and Belozerskii (1959)
Cudrania tricuspidata	27.8	Vanyushin and Belozerskii (1959)
Equisetum arvense	13.0	Vanyushin and Belozerskii (1959)
Helianthus annuus	28.6	Vanyushin and Belozerskii (1959)
Kale (*Brassica oteracea*)	17.8	Thomas and Sherratt (1956)
Linum usitatissimum	19.2	Vanyushin and Belozerskii (1959)
Lycopodim clavatum	18.8	Vanyushin and Belozerskii (1959)
Nephrolepsis sp.	11.6	Vanyushin and Belozerskii (1959)
Papaver somniferum	26.4	Uryson and Belozerskii (1959)
Passiflora coerula	27.3	Vanyushin and Belozerskii (1959)
Peanut (*Arachis hypogaea*)	31.7	Ergle and Katterman (1961)
Phaseolus vulgaris	25.9	Uryson and Belozerskii (1959)
Pinus sibirica	25.1	Uryson and Belozerskii (1959)
Pinus silvestris	17.9	Vanyushin and Belozerskii (1959)
Rye germ	25	Shapiro and Chargaff (1960)
Styrax obassia	23.8	Vanyushin and Belozerskii (1959)
Sunflower (*Helianthus annuus*)	37.2	Ergle and Katterman (1961)
Sweet corn	14.6–22.4	Van Schaik and Pitout (1966)
Tobacco	20.1–30.5	Tewari and Wildman (1966); Lyttleton and Petersen (1964)
Triticum vulgare	26.0–23.6	Uryson and Belozerskii (1959); Vanyushin and Belozerskii (1959)

MODIFIED NUCLEOSIDES IN DNA

TABLE 1 continued

Plant	5-Methyldeoxycytidine, % of C[a]	References
Wheat germ	25.4	Wyatt (1951b); Laland et al. (1952); Brawerman and Chargaff (1951)
Zea mays	29.3–26.7	Vanyushin and Belozerskii (1959); Ergle and Katterman (1961)

Bacteria	5-Methyl-deoxycytidine, % of C[a]	N^6-Methyl-deoxyadenosine, % of A[a]	References
Aerobacter aerogenes		2.7	Wyatt (1951b)
Agrobacterium tumefaciens	0.52	0.47	Vanyushin et al. (1968)
Alcaligenes faecalis	1.89	3.27	Vanyushin et al. (1968)
Bacillus brevis			
Strain R	0.26	1.10	Vanyushin et al. (1968)
Strain S	1.94	1.10	Vanyushin et al. (1968)
Strain P[+]	0.61	0.95	Vanyushin et al. (1968)
Bacillus cereus	pres.	pres.	Wainfan et al. (1965)
Bacillus niger	pres.	pres.	Doskočil and Šormová (1965)
Bacillus subtilis var. aterrimus	pres.	pres.	Doskočil and Šormová (1965)
Bacillus subtilis 168 wt.	pres.	pres.	Doskočil and Šormová (1965)
Bacterium morganii		1.48	Vanyushin et al. (1968)
Brucella abortus		0.81	Vanyushin et al. (1968)
Chloropseudomonas ethylicum		0.61	Vanyushin et al. (1968)
Chromatium minutissimum		0.27	Vanyushin et al. (1968)
Corynebacterium diphtheriae		0.84	Vanyushin et al. (1968)
Corynebacterium vadosum Kras.		0.58	Vanyushin et al. (1968)
Clostridium butylicum P		0.15	Vanyushin et al. (1968)
Escherichia coli C	1.0	2.0	Nikolskaya et al. (1968)
Escherichia coli CK	1.0	2.0	Nikolskaya et al. (1968)
Escherichia coli B	0.0	2.0	Nikolskaya et al. (1968)
Escherichia coli 15T	pres. +	2.4	Wyatt (1951b); Wainfan et al. (1965)
Escherichia coli K12		1.7	Wyatt (1951b); Wainfan et al. (1965)

Bacteria	5-Methyl-deoxycytidine, % of C[a]	N[6]-Methyl-deoxyadenosine, % of A[a]	References
Mycobacterium tuberculosis		0.5	Wyatt (1951b)
Mycobacterium luteum	0.30	0.30	Vanyushin et al. (1968)
Plectonema boryanum (blue-green alga)	pres.	pres.	Kaye et al. (1967)
Pneumonococcus		0.5	Wyatt (1951b)
Propionibacterium shermanii	1.53	0.26	Vanyushin et al. (1968)
Proteus vulgaris		pres.	Wainfan et al (1965)
Pseudomonas aerugenosa		pres.	Wainfan et al. (1965)
Pseudomonas fluorescens		pres.	Wainfan et al. (1965)
Pseudomonas syringae 19P		0.15	Vanyushin et al. (1968)
Rhizobium meliloti 441		0.60	Vanyushin et al. (1968)
Rhodospirillum rubrum		1.03	Vanyushin et al. (1968)
Salmonella typhosa			
Strain Ty-2	0.91	2.0	Vanyushin et al. (1968)
Strain T-5501	1.02	1.83	Vanyushin et al. (1968)
Shigella dysenteriae	pres.	pres.	Wainfan et al. (1965)
Staphylococcus aureus 909		0.15	Vanyushin et al. (1968)

Bacteriophage	5-Methyl-deoxycytidine, % of C[a]	N[6]-Methyl-deoxyadenosine, % of A[a]	References
T$_2$r + bacteriophage		0.5	Dunn and Smith (1958)
T$_2$r bacteriophage		0.5	Dunn and Smith (1958)
Salmonella c phage		0.4	Dunn and Smith (1958)
Bacteriophage λ	0.25		Gough and Lederberg (1966)
DD$_7$ (host E. coli)	0.5	1.4	Nikolskaya et al. (1968)
T$_1$ THU (host E. coli)	0.16		Klein and Sauerbier (1965)
T$_1$ Hfa UU (host E. coli)	0.28		Klein and Sauerbier (1965)
T$_1$ Hfr UU (pl) (host E. coli)	0.26		Klein and Sauerbier (1965)
T$_1$·B94 (host E. coli)		2.1	Klein and Sauerbier (1965)
T$_1$·B94 (pl) (host E. coli)		2.1	Klein and Sauerbier (1965)
λ rm$_{III}$	pres.	pres.	Lederberg (1966)
λ AB 259	pres.	pres.	Lederberg (1966)
λ cb +	0.3		Ledinko (1964)

TABLE 1 continued

Part B: DNA Samples in Which One of the Parent Nucleosides Is Completely Replaced by a Modified Nucleoside

Organism	Modified Nucleoside	Nucleoside Replaced	References
Bacteriophage XP12 in *Xanthomonas oryzae*	5-Methyldeoxy-cytidine	Deoxycytidine	Kuo et al. (1968)
Bacteriophages T2, T4, T6	5-Hydroxymethyl-cytidine	Deoxycytidine	Wyatt and Cohen (1953); Lehman and Pratt (1960)
Transducing bacteriophage of *B. subtilis*	Deoxyuridine	Thymidine	Takahashi and Marmur (1963a)
Bacteriophages ϕC and SP8 in *B. subtilis*	5-Hydroxymethyl-deoxyuridine	Thymidine	Roscoe and Tucker (1964; 1966); Kallen et al. (1962)

The distribution of methylated nucleosides in DNA of several organisms is presented in Table 1. On the basis of these extensive analytical data it appears that the DNA of most organisms contains a small percentage of methylated nucleosides.

Historically, the first modified component of a nucleic acid reported, 5-methylcytosine, was detected by Hotchkiss (1948) in an acid hydrolyzate of calf thymus DNA. This discovery was later confirmed by Wyatt (1951a). Somewhat ironically, Johnson and Coghill (1925) had earlier reported the isolation of 5-methylcytosine as its picrate from the DNA of avian tubercle bacilli. This finding was later disproved by Vischer et al. (1949), who were unable to detect 5-methylcytosine in a hydrolyzate of DNA from the same source, using sensitive paper chromatographic techniques not available to Johnson and Coghill.

2. DISTRIBUTION OF MODIFIED NUCLEOSIDES IN BACTERIOPHAGE DNA

The distribution of modified nucleosides in the DNA of bacteriophages is somewhat unique: these nucleosides can occur in minor amounts, as they do in the RNA and DNA of bacteria and other organisms, or they can completely replace one of the four major nucleoside components. The latter situation is not known to exist in the nucleic acids of any other organism, including animal and insect viruses. Since the DNA of only a relatively small number of bacterio-phages has been analyzed, it is impossible to know whether this type of replace-ment is a rare or general phenomenon. The substituted nucleosides in these situations are closely related to the replaced nucleoside. For example, 5-hydroxymethyldeoxycytidine replaces deoxycytidine in the DNA of the T-even bacteriophages, 5-methyldeoxycytidine replaces deoxycytidine in the DNA of the XP12 phage and a bacteriophage that infects *Xanthomonas oryzae*, and 5-hydroxymethyldeoxyuridine or deoxyuridine replaces thymidine in the DNA of certain bacteriophages that infect *Bacillus subtilis* (see Table 1 for references).

In the biosynthesis of the bacteriophage DNA that contains a replacement nucleoside, the modified nucleoside is incorporated as the triphosphate into the newly synthesized DNA as if it were the corresponding major nucleoside. This mode of biosynthesis is in distinct contrast to that of other modified nucleosides that are synthesized at the macromolecular level (see Chapter 6).

Some strains of T2 bacteriophage contain small amounts of N^6-methyl-deoxyadenosine (Dunn and Smith, 1958). Trace amounts of methylated components have also been found in other bacteriophages for which *Escherichia coli* serves as host. Klein and Sauerbier (1965) analyzed five strains of T1 phage and found that three contained 5-methyldeoxycytidine and two contained N^6-methyldeoxyadenosine. Several strains of lambda bacteriophage have been analyzed: some strains contain both 5-methyldeoxycytidine and N^6-methyl-deoxyadenosine, whereas others contain only one of these constituents (Gough and Lederberg, 1966; Lederberg, 1966; Ledinko, 1964). The bacteriophage DD_7 also contains both 5-methyldeoxycytidine and N^6-methyldeoxyadenosine (Nikolskaya et al., 1968). The 5-methyldeoxycytidine and N^6-methyldeoxy-adenosine contents of the DNA of different strains of *E. coli* hosts also vary, but there appears to be no direct correlation between the modified nucleoside content of the bacteriophage DNA and that of the host DNA.

There is an experimental pitfall in carrying out and interpreting analyses of DNA viruses with respect to the content of modified nucleosides. Polyoma virus DNA, for example, was at one time thought to contain small amounts of 5-methyldeoxycytidine (Winocour et al., 1965). In an elegant series of experi-ments Kaye and Winocour demonstrated that polyoma virus DNA contains no 5-methyldeoxycytidine, but rather that the virus particle has the capacity to encapsulate some of the host DNA (mouse kidney cells), which contains 5-methyldeoxycytidine (Kaye and Winocour, 1967; Winocour, 1967).

3. PHYSICAL PROPERTIES OF BACTERIOPHAGE DNA IN WHICH A MODIFIED NUCLEOSIDE COMPLETELY REPLACES A MAJOR NUCLEOSIDE

The complete replacement of one of the major nucleosides by a modified nucleoside in the DNA changes the physical properties of a bacteriophage considerably. For example, the melting-out curve of the DNA of a transducing bacteriophage of *B. subtilis*, which contains deoxyuridine instead of thymidine (Takahashi and Marmur, 1963a), has a T_m of 76.5°. This value would correspond to a $G+C$ content of 17.5% if the DNA contained thymidine (Marmur and Doty, 1962). The buoyant density of this DNA is 1.722, which corresponds to a $G+C$ content of 62% on the linear scale of density versus base ratios (Schild-kraut et al., 1962). In actual fact, the chemical analysis for this phage DNA gives values (mole %) of A, 35.9; U, 35.9; G, 13.4; and C, 14.7. These data indicate convincingly that for double-stranded DNA any discrepancy between the melting-out data and the buoyant density data is an indication of the presence of a modified component.

4. SATELLITE DNA LACKS MODIFIED NUCLEOSIDES

The modified nucleosides in tRNA occur at specific locations in the primary sequences, and it is likely that the methylated nucleosides in DNA also occur at specific locations. Moreover, there may be considerable heterogeneity in the distribution of the modified constituents between individual molecules of DNA within a given cell. An improvement in DNA fractionation techniques must be made before such structural features can be investigated in detail; however, an indication of the existence of such heterogeneity stems from studies on the minor or satellite fractions of DNA. Density-gradient ultracentrifugation of DNA samples isolated from plant and animal sources makes possible the separation of a satellite DNA fraction, distinguished from the parent DNA by an altered base ratio. Satellite DNA, detected in DNA preparations isolated from the chloroplasts of the algal flagellate *Euglena gracillis*, possesses an $A+T$ content of 76%, in contrast to 47% for the principal DNA fraction (Ray and Hanawalt, 1964). The principal DNA fraction contains 2.3 mole % of 5-methylcytidine, whereas the satellite DNA fraction has none.

5. METHODOLOGY IN DNA ANALYSES

The modified nucleosides detected in DNA are listed, together with the source of DNA and the amount isolated, in Table 1. It may be helpful in assessing these data to consider briefly the methodology used in the experiments. The simplest and most widely used method is to hydrolyze the DNA sample with a strong acid and chromatograph the released base on paper. The separated bases are eluted, and the amount of each is estimated spectrophotometrically. The errors inherent in paper chromatography are present and are particularly accentuated in the analysis for 5-methylcytosine. The separation between 5-methylcytosine and cytosine is generally poor, and rechromatography is often necessary (see Chapter 3).

The values listed in Table 1 are, in effect, the actual amounts of sample isolated. The limits of detection of the modified base or nucleoside depend on the size of the sample of DNA, but are probably on the order of 0.1–0.2% of the corresponding major base. It is instructive to note that earlier investigators using the standard paper chromatographic exchanges failed to detect 5-methylcytosine in the DNA of bacteria. Doskočil and Šormová (1965), taking advantage of the recently acquired knowledge about the mode of biosynthesis of methylated components of DNA, grew *Bacillus subtilis* in the presence of [^{14}C-methyl]-methionine. They found that more than 90% of the radioactivity incorporated into the DNA was distributed between 5-methyldeoxycytidine and N^6-methyldeoxyadenosine. They subsequently worked up a larger sample of DNA and were able to isolate and characterize 5-methylcytosine.

6. ADDITIONAL MODIFIED COMPONENTS IN DNA?

The question arises as to whether additional modified nucleosides occur in DNA but have simply not been detected by the present methods. Paper chromatography represents a satisfactory method for detecting 5-methylcytosine and N^6-methyladenine, but no one technique can be expected to detect all the possible modified components of nucleic acids. Other techniques have been applied to the analysis of DNA. For example, ion-exchange chromatography permits the use of a relatively large sample, and several groups (Andersen et al., 1952; Hurst et al., 1953; Sinsheimer and Koerner, 1951; Volkin et al.,

1951) have examined DNA by this technique and have detected no additional modified components.

In my own laboratory, I undertook an intensive examination of DNA from an animal source, mouse Ehrlich ascites cells, with the objective of looking for additional undetected components. This particular DNA sample was chosen because of the known hypermethylating enzymic activity in tumor cells (see Chapter 6). Twenty-five grams of DNA was hydrolyzed enzymically, and the constituent nucleosides were subjected to partition chromatography on columns (Hall, 1962). This technique would have permitted the isolation of as little as 1 mg of an unusual component, but only 5-methyldeoxycytidine, in addition to the four major deoxyribonucleosides, was found. These results probably rule out the presence in this DNA of other relatively stable constituents such as methylated nucleosides, but they do not preclude the presence of more labile components.

A small but persistant amount (0.1–0.6 mole %) of amino acids has been found in samples of DNA from bacteria, viruses, birds, and animals by several groups of workers (the literature is summarized by Salser and Balis, 1967). No one has isolated an amino acid-containing nucleoside or nucleotide from DNA as yet, and until this has been done the possibility that the observations are due to nonspecific binding cannot be ruled out.

7. PHYLOGENETIC DIFFERENCES IN PATTERNS OF METHYLATION BETWEEN ANIMAL, BACTERIAL, AND PLANT DNA

The data presented in Table 1 show that, with respect to modified nucleoside content, the DNA of bacteria, plants, and animals falls into three distinct classes. Thus the DNA of animals contains a small percentage of 5-methyldeoxycytidine, while that of plants has a much higher percentage of this nucleoside; the DNA of bacteria, on the other hand, contains N^6-methyldeoxyadenosine in small amounts as well as 5-methyldeoxycytidine. If these distinctions hold true for all members of each phylum, it becomes possible to use this type of analysis to establish phylogenetic relationships. For example, use of this pattern was made in studies of the blue-green alga *Plectonema boryanum* by Kaye et al. (1967). The DNA of this organism contains both N^6-methyldeoxyadenosine and 5-methyldeoxycytidine. This fact suggests that the alga strain resembles bacteria, an observation that supports the classification made by Echelin and Morris (1965) and Schneider et al. (1964) on morphological grounds. In contrast, the DNA of *Euglena gracilis* contains only 5-methyl-deoxycytidine (10% of the cytidine content) (Brawerman et al., 1962), and, therefore, in this regard the algae flagellate more closely resembles plants.

8. HYPERMETHYLATION OF DNA OF BACTERIA IN RESPONSE TO STRESS

The biosynthesis of the methylated components of DNA occurs by direct methylation of the DNA polymer with S-adenosylmethionine as the methyl donor. This biosynthesis mechanism is discussed in more detail in Chapter 6, but one phenomenon deserves comment at this point. In certain bacteria, the number of methyl groups in the DNA molecule increases under conditions of stress. Dunn and Smith (1958) first noted that, when a thymine-requiring strain of *E. coli* (15T⁻) is deprived of thymine or is exposed to thymine antagonists, the N^6-methyldeoxyadenosine level rises from the normal 2.4% (of deoxy-adenosine) to 15%. In some preparations, the analytical data suggested a concomitant decrease in thymidine residues. Although these data were not consistent over several experiments, some authors have assumed incorrectly that the "extra" N^6-methyldeoxyadenosine residues replaced thymidine residues.

This phenomenon was investigated later in more detail by Theil and Zamenhof (1963), and their studies, carried out at a time when the biosynthesis of the methyl constituents of DNA was better understood, have essentially clarified the observed data. The phenomenon is not general; in fact, of three thymine-requiring strains of *E. coli* and one of *B. subtilis*, only one of the *E. coli* strains responded in a manner similar to that of the strain of Dunn and Smith. Using [¹⁴C-methyl]-methionine, Theil and Zamenhof established that when the growth of *E. coli* is inhibited by 5-aminouracil the amount of [¹⁴C-methyl] groups incorporated increases, even after net synthesis of DNA ceases. When the cells are transferred to a normal growth medium, the percentage of methyl groups in the DNA decreases to the previous level. It appears, therefore, that these particular organisms respond to stress by hypermethylating their DNA.

9. BIOSYNTHESIS OF 5-HYDROXYMETHYLDEOXYCYTIDINE

The biosynthesis of 5-hydroxymethyldeoxycytidine occurs at the deoxyribo-nucleotide level (Flaks and Cohen, 1957; Pizer and Cohen, 1962) and involves a new enzyme, deoxycytidylate hydroxymethylase, which does not exist in the uninfected host. The enzyme catalyzes the reaction deoxycytidylate + formaldehyde + tetrahydrofolate $\longrightarrow$ 5-hydroxymethyldeoxycytidylate: Pizer and Cohen (1962) showed that the uninfected organism contains not even one molecule of active enzyme per cell; their studies suggest that the enzyme is synthesized *de novo* on infection by the bacteriophage.

10. GLUCOSYLATION OF 5-HYDROXYMETHYLDEOXYCYTIDINE

Many of the 5-hydroxymethyldeoxycytidine residues in bacteriophage DNA contain a glucosyl or diglucosyl residue. The distribution of the nonglucosyl, monoglucosyl-, and diglucosyl-hydroxymethyldeoxycytidine residues occurs in a pattern distinct to the different T-even phages (Lehman and Pratt, 1960; Lichtenstein and Cohen, 1960). Lehman and Pratt (1960), using α and β glucosidases, established the existence of both anomers in the monoglucosylated samples. Kuno and Lehman (1962) found that the diglucosyl derivative has the gentiobiose configuration [β-glucosyl-(1→6)-α-glucosyl]. These data are summarized in the table (Kornberg, 1962).

	Bacteriophage		
Hydroxymethyl Derivative	T2	T4	T6
	mole %		
Nonglucosylated	25	0	25
α-Glucosyl	70	70	3
β-Glucosyl	0	30	0
β-Glucosyl-α-glucosyl	5	0	72

Kornberg et al. (1959) have shown that the 5-hydroxymethyldeoxycytidine residues of DNA are glucosylated at the polymeric level by direct transfer of glucose from UDPG. The enzymes responsible can be detected in *E. coli* only after infection with the T-even phages. In the case of T4 infection, the synthesis of two enzymes is induced; one enzyme attaches a glucosyl residue in the alpha linkage, and the other in the beta linkage (Josse and Kornberg, 1962). The isolation of these enzymes and their mode of action have been reviewed by Kornberg (1962). The 5-hydroxymethyldeoxycytidine content of T4 DNA, whether glucosylated or not, appears to be critical to the infective process, since this component confers a resistance to nuclease activity that destroys the cytosine-containing DNA of the host (Wiberg, 1967). This result suggests one function for certain of the modified nucleosides, but it is not a general feature of the viral-infective process, since the base composition of the DNA of many bacteriophages qualitatively is identical with that of the host.

The structure of the isolated glucosylated 5-hydroxymethyldeoxycytidine nucleosides has been investigated (Lichtenstein and Cohen, 1960; Sinsheimer, 1954; and Volkin, 1954); the conclusion reached is that the sugar residues are attached at the 5-hydroxyl position. The evidence, however, based mainly on spectral properties, is circumstantial. The observed data would also fit compounds in which the glucosyl residues are attached at the N-3 or N-4 position. In connection with this question, the DNA of phage PBS-2, which does not contain 5-hydroxymethyldeoxycytidine or 5-hydroxymethyldeoxyuridine, contains a reducing sugar thought to be glucose (Takahashi and Marmur, 1963b). In these DNA molecules, the sugar residue appears to be attached to the guanine and cytosine bases.

11. HEXOSE IN THE DNA OF BACTERIOPHAGE SP8

Rosenberg (1965) discovered an interesting fact about the DNA of strains of the *B. subtilis* phage, SP8. The DNA of these strains contains 5-hydroxymethyl-deoxyuridine in lieu of thymidine and also contains a hexose. The DNA of SP8 contains 0.98 mole of D-glucose/mole of phosphorus, and a mutant, SP8T$_s$, contains 0.81 mole of D-mannose/mole of phosphorus. When the DNA of either strain is denatured, it can be separated in the ultracentrifuge into heavy and light strands. Rosenberg concludes that the light strand contains more hexose residues and that therefore in the native DNA the hexose is attached predominantly to one strand.

12. BIOSYNTHESIS OF 5-HYDROXYMETHYLDEOXYURIDINE

The biosynthetic pathway of 5-hydroxymethyldeoxyuridine in *B. subtilis* infected with ϕC bacteriophage appears to follow the same course as that of 5-hydroxymethyldeoxycytidine (Roscoe and Tucker, 1964, 1966). After infection an enzyme appears in the host cell that catalyzes this conversion of deoxyuridylic acid to 5-hydroxymethyldeoxyuridylic acid. In addition, thymidylate synthetase is induced. The SP8 phage (host *B. subtilis*), which also contains 5-hydroxy-methyldeoxyuridine, induces a new enzyme in its host that catalyzes the deamination of 5-methyldeoxycytidylic acid (Nishihara et al., 1967) but not of 5-hydroxymethyldeoxycytidylic acid. The closely related SP3 phage, which contains thymidine but not 5-hydroxymethyldeoxyuridine, does not induce this enzyme.

13. SUMMARY

In summary, a number of general observations can be made. Only 5-methyl-deoxycytidine and N^6-methyladenosine occur in DNA as components that are minor in quantity. There appears to be a distinction among phyla with respect to the distribution of these two DNA components. 5-Methyldeoxycytidine occurs solely in plant and animal DNA, and this nucleoside, as well as N^6-methyldeoxyadenosine, occurs in bacterial DNA. These phylogenetic differences suggest an evolutionary delineation. In some bacteriophages complete replacement of one of the major nucleoside components occurs: 5-hydroxymethyl-deoxycytidine or 5-methyldeoxycytidine replaces deoxycytidine, and 5-hydroxymethyldeoxyuridine or deoxyuridine replaces thymidine. The normal base-pairing ratios are maintained in these substituted DNAs. The DNAs of many different species have been analyzed, and no nucleoside other than a deoxyribonucleoside has ever been detected. The DNA of some bacteriophage, however, contains a hexose attached to the deoxyribonucleotides. Finally, the significance of the modified nucleosides to the DNA has not been established; it has been suggested that the methyl groups may help to provide a species-related individuality to the DNA molecules.

Chapter 6

BIOSYNTHESIS AND METABOLISM OF THE

MODIFIED NUCLEOSIDES

1. INTRODUCTION

A$_{\text{LL THE MODIFIED NUCLEOSIDES}}$ that have been identified are structural modifications of one of the major nucleosides. The first question concerning their biosynthesis is whether they are synthesized by a unique *de novo* synthesis or by the addition of a substituent to an existing major nucleoside. The next question asks whether this process occurs at the nucleoside (-tide) level or at the polymeric level. In view of the fact that DNA is synthesized by means of complementary replication and RNA is synthesized by direct transcription, no known mechanism exists that would allow the modified nucleosides to be incorporated in their proper locations as the nucleic acid polymer is being formed. It follows, therefore, that structural modification occurs after the polymeric chain is formed. This concept is borne out by the known mechanism of biosynthesis of the methylated nucleosides and N^6-(Δ^2-isopentenyl)adenosine, in which the alkyl groups are added to preformed nucleic acid molecules.

Little is known about the metabolic pathways of the modified nucleosides, and in fact the evidence suggests that many of them are not metabolized (except for the removal of the sugar moiety) but, at least in mammalian organisms, are excreted.

2. BIOSYNTHESIS OF THE METHYLATED COMPONENTS OF NUCLEIC ACIDS*

Basic Mechanism for the Methylation of Nucleic Acids. The introduction of methyl groups into DNA and RNA occurs after the primary oligonucleotide sequence has been transcribed. The fundamental mechanism of this process was established on the basis of the pioneering study of Borek and his collaborators (i.e., Fleissner and Borek, 1963; Mandel and Borek, 1963a,b) on a methionine auxotroph of *Escherichia coli*. This organism has an absolute requirement for methionine, but when deprived of this compound it continues to synthesize tRNA, rRNA, and, to a lesser extent, DNA (see Figure 1). The total RNA fraction extracted from cells grown in the absence of methionine lacks the normal number of methyl groups, although the isolated RNA undoubtedly consists of a molecular population of normally methylated and partially methylated molecules. The DNA synthesized after withdrawal of methionine from the medium also lacks the normal complement of methyl groups (Gold and Hurwitz, 1964a,b). The methyl-deficient RNA or DNA serves as a substrate for methylation *in vitro* by a cell-free extract from the same organism; *S*-adenosylmethionine or methionine is the methyl donor.† The addition of the methyl groups *in vitro* seems to be highly specific, since the enzyme system appears to catalyze the addition of the correct number of methyl groups; tRNA or DNA extracted from the organism grown in the presence of methionine does not accept additional methyl groups when exposed *in vitro* to the homologous enzyme system.

The *E. coli* that made these studies possible was isolated by Tatum (1945) from survivors of ultraviolet irradiation and was designated *E. coli* K12 58-166. It required methionine and biotin. Lederberg (1947) derived several variants of the archetype, and in one variant, designated K12 W6, Borek et al. (1955) discovered that the organism loses control over the synthesis of nucleic acid in the absence of methionine. The genetic determinants for this phenomenon were investigated by Stent and Brenner (1961) and Alfoldi et al. (1962), who found

---

*The biosynthesis of the methylated components of nucleic acids has been reviewed in recent articles (Borek and Srinivasan, 1966; Gold et al., 1966; Srinivasan and Borek, 1966). This chapter summarizes the fundamental concepts that have been developed and includes recently published information.

†Fenrych et al. (1968) and Walerych et al. (1966) have reported that methylcobalamin also serves as a donor of methyl groups in the crude *in vitro* methylating system. In all the experiments reviewed in this chapter, methionine or *S*-adenosylmethionine was used as the methyl donor.

that the genetic locus (RC) for control of RNA synthesis can be either under stringent control, designated RC^{str}, or under relaxed control, RC^{rel}. Many strains of *E. coli*, auxotrophic for other amino acids as well as for methionine,

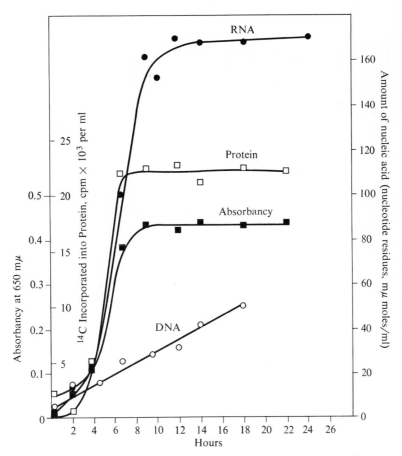

FIGURE 1. Growth of *E. coli* strain 58-161 in the presence of limiting amounts of methionine. The *right* ordinate represents the nucleic acid content of the culture on a basis of nucleotide equivalents; O————O, DNA; ●————●, RNA. Protein (□————□) represents the radioactivity of the hot acid-insoluble material; the growth of the bacteria was followed by measuring the absorbance of the culture (■————■) at 650 mμ. The time of cessation of growth was estimated by the point of intersection of the logarithmic and stationary phases. Data taken from Gold and Hurwitz (1964a).

contain this genetic locus. The phenomenon also occurs in yeast, and Kjellin-Stråby and Boman (1965) describe a methione-requiring mutant of *Saccharomyces cerevisiae* that continues to synthesize tRNA in the absence of methionine.

Specific Receptor Sites in the Nucleic Acid. The inability of the crude enzyme system to add extra methyl groups to homologous normally methylated tRNA *in vitro* suggests a high degree of specificity. The specificity probably takes into account two factors: (1) a specific location on a given nucleoside (i.e., 1-, N^2-, 7-, or $O^{2'}$-methylguanosine); and (2) a specific location in the primary oligonucleotide sequence. Such specificity can be achieved only if there are a large number of methylase systems for each organism. A small beginning has been made toward clarifying this situation. The methylase activity in *E. coli* has been fractionated, and six tRNA methylases and two DNA methylases have been identified (Fujimoto et al., 1965; Gold and Hurwitz, 1964a,b; and Hurwitz et al., 1964, 1965). With respect to tRNA, the isolated enzyme systems from *E. coli* catalyze the formation of 1-methylguanine (two separate enzyme activities), 7-methylguanine, 2-methyladenine, N^6-methyladenine, and N^6,N^6-dimethyladenine.

Sheid et al. (1968) have partially purified an enzyme system in rat liver that catalyzes the formation of 5-methylcytosine in DNA. Kalousek and Morris (1968) detected a DNA methylase in rat spleen nuclei specific for the synthesis of 5-methylcytosine; but, unlike other methylases described, their preparation methylates homologous DNA.

A start toward learning more about the specific receptor sites in tRNA has been made by Baguley and Staehelin (1968), who studied the methylation of *E. coli* tRNA, using purified methylases obtained from rat liver. They observed that an N^2-methylguanine methylase appears to recognize a specific oligonucleotide sequence in the *E. coli* tRNA and that a 1-methyladenine methylase recognizes the specific sequences $Gpm^1ApApUp$ and $Ap(m^1ApAp)Up$. Baguley and Staehelin suggest that the individual methylase enzyme systems recognize specific oligonucleotide sequences and, furthermore, that such accepting sequences are commonly occurring structural elements of many different tRNA molecules. Further evidence for the presence of unique receptor sites in a nucleic acid molecule for individual methylase systems comes from the work of Phillips and Kjellin-Stråby (1967), who obtained evidence for a specific N^2,N^2-dimethylguanine methylase in *S. cerevisiae*. Certain mutants of this organism lack N^2,N^2-dimethylguanine in their tRNA; extracts from mutants containing N^2,N^2-dimethylguanine in the tRNA catalyze *in vitro* the synthesis of this component in the tRNA extracted from the mutants which lack N^2,N^2-dimethylguanine. This suggests that the enzyme system recognizes a strict oligonucleotide sequence in the tRNA for formation of N^2,N^2-dimethylguanine.

A specificity for distinct oligonucleotide sequences is also indicated for the DNA methylases. The methylase induced by T2 phage in *E. coli* will not recognize synthetic polymers such as dAT, dA:dT, dAC:dTC, or dAC:dTG as substrates (Falaschi and Kornberg, 1965). The enzyme exhibits a specificity for adenine residues (i.e., N^6-methyladenine), and these polymers provide all seven possible dinucleotide sequences involving adenine. This suggests that the recognition sequence is longer than two nucleotides, and therefore the particular recognition sequence involved could not be present in these polymers. Synthetic DNA can serve as a substrate, and Gold and Hurwitz (1964b) have shown that the DNA synthesized *in vitro* by DNA polymerase, using T2 phage DNA or *E. coli* DNA as template, accepts methyl groups.

Heterologous Methylation. Crude enzyme preparations from one organism have the ability to methylate *in vitro* the tRNA isolated from another organism. In effect the tRNA becomes hypermethylated (Gold et al., 1963; Srinivasan and Borek, 1963, 1964a). The ability to accomplish hypermethylation varies from organism to organism. The fact that these studies were carried out using crude enzyme preparations, together with the fact that the sites at which the specific tRNA molecules were methylated are not known, makes it difficult to interpret the results. Possibly the fully methylated tRNA fortuitously offers to the heterologous enzyme system correct oligonucleotide acceptor sequences lacking methyl groups. There is no evidence, however, that this interpretation is correct.

Heterologous methylation of normally methylated DNA has also been reported by Gold and Hurwitz (1963) and Borek and Srinivasan (1966). The DNA of several mammalian tissues, for example, serves as a substrate for the DNA methylases of *E. coli* (Kaye et al., 1967). "Hypermethylation" by *E. coli* KB methylase of infectious polyoma DNA does not lower its infectivity (eight methyl groups per double-stranded molecule, m.w. 3×10^6, are added), according to data of Kaye et al. (1967). Similarly, *in vitro* methylation by a heterologous enzyme preparation does not affect the transforming activity of either *Diplococcus pneumoniae* or *Bacillus subtilis* (Gold and Hurwitz, 1964b). The extent of methylation of the heterologous DNA is limited if homologous

BIOSYNTHESIS AND METABOLISM OF THE MODIFIED NUCLEOSIDES

DNA is added to the system (i.e., DNA from the same *E. coli* strain from which the enzyme extract is made). Compounds which bind to DNA also prevent methylation by the *in vitro* methylation system: actinomycin D and proflavin (Gold and Hurwitz, 1964b), lipopolysaccharides isolated from the cell walls of *Salmonella tryphimurium* and *E. coli* (Falaschi and Kornberg, 1965), and heparin (Kaye et al., 1967).

2′-O-Methylribonucleosides. In the earlier studies in this field, the products of enzymic methylation were identified as nucleosides methylated on the heterocycle moiety. Transfer RNA and rRNA, however, contain a substantial portion of 2′-*O*-methylribose. (2′-*O*-Methyl derivatives of the four major nucleosides, 5-ribosyluracil, and N^4-methylcytidine have been identified; see Chapter 2.) The question arises, therefore, whether the mechanism of biosynthesis of 2′-*O*-methylribose follows a course similar to that of the methylated heterocycles. Svensson et al. (1968) obtained a cell-free preparation from *S. cerevisae* that catalyzes the transfer of methyl groups from [^{14}C-methyl]-methionine to the 2′-hydroxyl group of *E. coli* tRNA (this tRNA contains its normal complement of methyl groups); one of the products was identified as 2′-*O*-methyladenosine. Using a similar experimental protocol, Nichols and Lane (1968a) incubated the rRNA isolated from a methionine auxotroph of *E. coli* grown in the absence of methionine with the homologous enzyme system and [^{14}C-methyl]-*S*-adenosylmethionine. Under these conditions methyl groups are incorporated into the RNA; these workers tentatively identified two of the products as 2′-*O*-methylguanosine and 2′-*O*-methylcytidine. Nichols and Lane (1967) and Tamaoki and Lane (1968) also conducted *in vivo* experiments with *E. coli* and L cells, respectively. They grew the organisms in the presence of [^{14}C-methyl]-methionine and obtained radioactive 2′-*O*-methyl derivatives of the four major nucleosides from the isolated rRNA.

Bonney et al. (1971) conducted a similar experiment, using HeLa cells cultured in the presence of [^{14}C-methyl]-methionine. They identified labeled 2′-*O*-methyl derivatives of the four major nucleosides in the tRNA fraction.

These studies suggest that the basic biosynthetic mechanisms for methylation of the nucleoside heterocycle and ribose are identical. These findings have special relevance to the biosynthesis and function of rRNA, since mammalian rRNA contains a relatively large amount (about 80% of the total methyl groups) of 2′-*O*-methyl ribonucleosides, in comparison to nucleosides methylated on the heterocyclic group (Brown and Attardi, 1965; Burdon, 1966).

Norton and Roth (1967) partially purified a phosphodiesterase from *Ana-*

cystis nidulans that specifically cleaves the internucleotide linkages in RNA adjacent to a 2'-*O*-methylribonucleoside, giving 5'-phosphate termini. The enzyme does not attack DNA.

Relationship of Methylation to the Biosynthesis of rRNA. The rRNA of HeLa cells contains 41–44 methyl groups per 28S rRNA molecule and 36–37 methyl groups per 16S rRNA molecule (equivalent to about 13 and 19 methyl groups per 1000 nucleotides for the 28S and 16S molecules, respectively) (Vaughan et al., 1967; Wagner et al., 1967). Although *E. coli* rRNA contains about the same proportions of methyl groups, only about 10% of them are attached to the $O^{2'}$ position of the nucleoside residues, in contrast to the large percentage of $O^{2'}$ methylation in the HeLa cell rRNA (Dubin and Günalp, 1967; Hall, 1964b; Nichols and Lane, 1966b). The question that can be asked is, How essential are the methyl groups to the structure and function of ribosomal RNA? The methylation of ribosomal RNA occurs early in its biosynthesis. According to the results of Greenberg and Penman (1966), methylation of high-molecular weight RNA in HeLa cells grown in culture takes place exclusively on the 45S ribosomal RNA precursor. Methylation apparently occurs close to the growing point of the newly synthesized RNA. The precursor 45S is subsequently converted to the 28S and 16S RNA (Greenberg and Penman, 1966; Zimmerman and Holler, 1967) and then to the ribosomes. When HeLa cells are deprived of methionine, 45S RNA is synthesized, although it becomes methyl-deficient. The 28S RNA is formed, but the 16S RNA resulting from the cleavage of the 45S molecule is lost. The 28S molecules deficient in methyl groups apparently cannot form cytoplasmic 50S ribosomal subunits (Vaughan et al., 1967).

Using a similar experimental design, Sypherd (1968) examined the relationship of the methyl-deficient rRNA to the formation of mature ribosomes in *E. coli* K12 (RC^rel). According to Sypherd, methylation of rRNA occurs at two stages. The first stage occurs early in the synthesis of rRNA and leads to the formation of half-methylated RNA. This methylation step is not obligatory, since it can be bypassed during methionine deprivation. The second stage of methylation takes place just before the final maturation process, which occurs with the addition of protein to the immature particles. Sypherd speculates that the addition of the last few proteins to complete the ribosomal particle renders the rRNA capable of becoming fully methylated and, at the same time, allows the RNA to achieve its final structural configuration. The process of ribosome formation is obviously complex, and the data obtained are insufficient to decide whether the failure of the cells deprived of methionine to form functional ribosomes is due to the lack of methylation or to inability to synthesize some other essential component of the ribosome.

Relationship of Methylation to the Biosynthesis of DNA. Billen (1968) and Lark (1968a,b) studied the mechanism of *in vivo* methylation of the DNA of *E. coli* 15T. The DNA of this strain contains both N^6-methyladenine and 5-methylcytosine (see p. 284, Chapter 3). Their data suggest that the nascent polynucleotide strand is methylated as the DNA is replicated and that, even when the organism is deprived of methionine, DNA synthesis continues without methylation occurring. On the other hand, it appears that replication of the methyl-deficient DNA stops at the point where the normal methylation pattern ceased. On readdition of methionine to the medium, the methyl-deficient DNA is immediately methylated before DNA synthesis resumes. These data support the hypothesis advanced by Srinivasan and Borek (1966) that methylation of DNA may provide a method for the cell to distinguish its own DNA from other similar genetic material. Thus, DNA with a different pattern of methylation might not be replicated. These interpretations also suggest that normal *in vivo* synthesis of DNA may require methylated DNA as the template.

Relationship of Methyl Groups to the Function of tRNA. In order to obtain evidence that bears on the function of methyl groups in tRNA, a number of workers have made use of the methyl-deficient tRNA that can be produced in RC^{rel} *E. coli* mutants. The general approach is to isolate methyl-deficient tRNA and to separate out unique molecular species not present in the tRNA of cells grown in the presence of methionine. A comparison of the properties of the methyl-deficient tRNA with those of the tRNA of the normally grown cells is then made. Capra and Peterkofsky (1966), for example, fractionated methyl-deficient tRNA and detected a new fraction of leucine-tRNA, in addition to the two fractions present in the normal tRNA. The leucine-tRNA from the normally grown cells in the ribosome-binding assay responds to either poly (U,C) or poly (U,G). The new fraction of leucine-tRNA from the methionine-deprived cells responds fully to poly (U,G) and slightly better to poly (U,C). Revel and Littauer (1966) carried out a similar experiment and detected a new fraction of phenylalanine-tRNA from the methionine-deprived cells. This new fraction in the ribosome-binding assay is equally as responsive to poly U, but with poly (U,C) 1 : 2 or poly (U,A) 1 : 1 it is two or three times as responsive as the normal fraction. These results have been questioned by Fleissner (1967), who detected no significant difference in the codon responses of a "methyl-poor" phenyl-alanine-tRNA and phenylalanine-tRNA from normal cells. Since all these experiments have been carried out using specific amino acid-accepting fractions obtained from columns rather than defined molecular species of tRNA, the data are difficult to assess.

Using highly purified tRNA obtained from methionine-deprived *E. coli* (RC[rel]), Shugart et al. (1968a,b) found that four amino acid-accepting activities tested (those of histidine, leucine, phenylalanine, tyrosine) were significantly lower than the level for normal tRNA from the same organism. Methylation of these tRNA fractions by the crude homologous enzyme system *in vitro* completely restored the accepting activity of histidine and phenylalanine species, partially restored that of leucine, and did not restore that of tyrosine at all. Shugart et al. showed that the undermethylated phenylalanine fraction accepts 1.2 moles of methyl groups per mole of tRNA (*E. coli* B tRNA[Phe] contains two methyl groups, m^7G and T; Uziel and Gassen, 1968).

In order to obtain a complete answer to the question of the biological significance of the methyl groups, it will be necessary to know whether the observed differences between the methyl-deficient tRNA and the normal tRNA reported above lie only in the absence or presence of the methyl group(s), or whether differences in the primary sequence arise because of the altered growth conditions of the organism. Although the addition of methyl groups may confer subtle changes on tRNA structure at or near active sites and such structural subtleties may be necessary for efficient binding to other molecules, it should be pointed out that the primitive organism mycoplasma carries out protein synthesis even though its tRNA and rRNA appear to contain very few methyl groups (Hall et al., 1967b; Hayashi et al., 1969).

DNA Methylases Specific to Phage DNA? The data cited in the preceding paragraphs demonstrate that the methylase systems exhibit a strict species specificity. This specificity is reflected in a quantitative change in the methylase activity of host cells on viral infection. Gold et al. (1964) found that after *E. coli* is infected with T2 phage the DNA methylase activity increases one hundred-fold. The methylating pattern actually shifts, according to evidence obtained by Fujimoto et al. (1965); the DNA of *E. coli* K12 contains both 5-methyl-cytosine and N^6-methyladenine. On infection with T2 phage, the methylase activity with respect to 5-methylcytosine remains unchanged. The DNA of T2 phage contains N^6-methyladenine; this enzymic response, therefore, may be in preparation for the synthesis of phage DNA. It is unknown whether this result is due to an increased synthesis of an existing methylase or whether it represents *de novo* synthesis of the enzyme. The methylation of phage DNA directed by the host may be the reason for the host-induced modification of lambda bacteriophage. Arber (1965) noted that lambda grown in a methionine auxotroph of *E. coli* shows a reduced plating efficiency when the host cells are deprived of methionine. He hypothesizes that host specificity in this phage-host relationship may be conferred by the methylation of specific sites in the phage DNA.

One method by which viral infection may cause a change in methylation patterns is illustrated by the study of Gefter et al. (1966), who observed that in *E. coli* infected with T3 phage both the RNA and DNA methylating capacities disappear. This effect is due to the appearance in the infected cells of an enzyme that hydrolyzes S-adenosylmethionine to thiomethyladenosine and homoserine.

Elevated Methylase Activity in Tumor Tissues? The specificity of individual methylating enzymes may be related directly to individual nucleic acid molecules within a given cell. If one considers this possibility in light of the fact that individual tRNA molecules may be protein-specific, a change in distribution of tRNA molecular species due to differential protein synthesis would be reflected in an alteration of methylase activity. In one of the earliest studies directed at this question, Bergquist and Matthews (1962) observed that nucleic acids of certain transplanted tumors in mice have a higher content of methylated components than the nucleic acids of corresponding normal tissues.

Additional data have since been obtained by Tsutsui et al. (1966), who reported that certain rat and mouse tumor tissues appear to have abnormally high levels of methylase activity in comparison to those of rat and mouse liver and brain. In these experiments, the level of methylase activity in each tissue is compared *in vitro* by employing a common substrate, the methyl-deficient tRNA obtained from *E. coli* K12 W6. This substrate is incubated with an excess of [^{14}C-methyl]-S-adenosylmethionine and the crude enzyme extracts under investigation until uptake of the radioactive label ceases, indicating that the substrate has become saturated with respect to the methylating system. The results listed in Table 1 show a marked increase in the capacity of the crude enzyme system from tumor tissues to hypermethylate the substrate tRNA, compared to the enzyme extracts of the liver and the brain.

Using a similar experimental design, Baliga et al. (1965) observed fluctuations in the activity of methylase enzymes as the mealworm *Tenebrio molitor* goes through its stages of metamorphosis. Hancock (1967) reported that the tRNA methylase activity of embryonic rabbit liver is much higher than that of adult rabbit liver, and Simon et al. (1967) found the level of RNA methylase in rat

TABLE 1

tRNA Methylase Activity in Normal and Tumor Tissues. (Data taken from Tsutsui et al., 1966)

The values represent incorporation of radioactive label from [^{14}C-methyl]-methionine into a methyl-deficient tRNA derived from a strain of *E. coli* (RC^{rel}).

Normal Control, cpm/mg tRNA	*Tumor, cpm/mg tRNA*	*Normal Control, cpm/mg tRNA*	*Tumor, cpm/mg tRNA*
Rat liver	Novikoff hepatoma	C-57 mouse liver	Mammary carcinoma
I. 3400	28,000	I. 2200	15,600
II. 2500	25,500	II. 1200	18,000
III. 5200	25,800	DBA × Swiss mouse liver	Melanoma
Regenerating rat liver		I. 8800	13,000
I. 4900		II. 9800	15,300
II. 5700			
C-57 mouse brain	Glioma		
4400	18,300		
	31,800		

brain to be eight times higher in the fetus than in the adult. Hancock et al. (1967) observed that the level of an enzyme that catalyzes the synthesis of *S*-adenosyl-methionine from ATP and L-methionine is highest in fetal mouse and rabbit liver and declines steadily as the animals age.

All of these studies have relied on the quantitative incorporation of [^{14}C-methyl] groups into a common tRNA as an assay and have used the total methylase activity present in a crude extract. For more precise information, however, it would be desirable to identify the nucleoside products as well as the oligonucleotide sequences of the receptor sites in specific tRNA molecules. These data do suggest that the quantitative activity and/or the specificity of the methylase systems can alter during growth and differentiation.

In an extension of these studies, Mittelman et al. (1967b) compared the methylase activity of an SV-40 induced tumor in hamsters with that of the hamster liver. Using normal yeast tRNA as a common substrate, they found that an extract from the tumor tissue hypermethylates the yeast tRNA to a greater extent than extracts from hamster liver, connective tissue, muscle, and thymus (see Table 2). The principal product obtained with the enzyme preparations from all the tissues is 1-methylguanosine. Chemical analysis of hamster liver tRNA shows that this nucleoside comprises 11 mole % of the total nucleo-

BIOSYNTHESIS AND METABOLISM OF THE MODIFIED NUCLEOSIDES

TABLE 2

*Isolation of Specific Methylated Bases, Using Yeast tRNA as a Methyl Acceptor with
Enzyme Preparations Derived from SV-40 Tumor and Normal Tissues. (Data from
Mittelman et al., 1967b)*

	SV-40 tumor, cpm	Muscle, cpm	Connective Tissue, cpm	Liver, cpm	Thymus, cpm
1-Methyladenosine	0	50	0	100	125
N^6-Methyladenosine	1000	0	0	0	0
5-Methyluridine	2500	0	0	0	0
1-Methylguanosine	9000	2000	220	4500	5500
N^2,N^2-Dimethylguanine	880	0	0	0	0

sides of the liver tRNA (Chapter 2). These data raise a related question as to
whether 1-methylguanosine has some special significance to the tRNA of
hamster tissues.

In the protocols of these *in vitro* experiments the crude enzyme extracts are
incubated with the tRNA substrate until maximum incorporation of labeled
methyl groups occurs. This suggests that the methylase enzymes have methylated
all the available receptor sites peculiar to their specificity. In the presence of
0.25–0.35 M ammonium acetate, the intrinsic specificity of the methylases is
apparently suppressed, and 2–4 times as many methyl groups are introduced
into the substrate tRNA (Kaye and Leboy, 1968; Rodeh et al., 1967). Kaye and
Leboy, for example, report that dialyzed extracts of $C_{57}Bl$ mouse leukemic
spleen methylate *E. coli* B tRNA 2–3 times as much as dialyzed extracts of
SWR mouse liver, lung, or spleen. In the presence of 0.35 M ammonium acetate,
the activity of all the enzyme preparations is elevated and the observed differences
in the extent of methylation disappear. These experiments only assess the total
incorporation of methyl groups into the substrate tRNA, and qualitative
differences, if any, are not determined.

In *in vivo* experiments (Bonney et al., 1971; Mittelman et al., 1967a), the
methylation patterns of the tRNA of human normal lymphoblasts and those of
human leukemic lymphoblasts (Burkitt lymphoma) grown in culture were
compared. Normal cells grown in the presence of [^{14}C-methyl]-methionine
incorporate the radioactive label into the 1-methyladenosine residues of the

tRNA. In contrast, the tRNA of the leukemic cells grown under similar conditions incorporates no label into 1-methyladenosine, but rather incorporates a substantial amount of label into the 1-methylguanosine residues, as well as other labeled methylated nucleosides not detected in the tRNA of the normal lymphoblasts.

Insertion of Methyl Groups into Nucleic Acids by Alkylating Agents in vivo. The correlation between methylase activity and the proper functioning of cells poses an intriguing question when the mechanism of action of certain alkylating agents *in vivo* is considered. Alkylating agents can induce tumors in experimental animals, and they can also destroy cancer tissue preferentially (see review by Lawley, 1966). Preliminary experiments to assess whether there is any relationship between the observed biological effects and nucleic acid methylation have centered on the carcinogen dimethylnitrosoamine.

$$\begin{array}{c} CH_3 \\ \diagdown \\ \diagup \\ CH_3 \end{array} N - N \rightarrow O$$

When [^{14}C-methyl]-dimethylnitrosoamine is injected into rodents, a significant amount of the radioactive label is incorporated into both the tRNA and the DNA; the greatest percentage of the radioactivity is located in the former. In both the DNA and the tRNA, practically all the radioactivity is present in the 7-methylguanine residues (Craddock and Magee, 1963; Lawley et al., 1968; Magee and Farber, 1962; Magee and Lee, 1964; Villa-Trevino and Magee, 1966). The N-7 position of guanine is the preferential site of attack for alkylating agents *in vitro* (see p. 371), although Loveless (1969) has suggested that a small amount of alkylation at *O*-6 of guanine residues of DNA may cause the genetic changes. The structure of the actual alkylating species is not known (assuming that dimethylnitrosoamine is metabolized). In view of the known pathway of the biogenesis of the methylated components of nucleic acid, it is doubtful whether a monomer such as guanylic acid is alkylated and then incorporated. In order to exclude this possibility Craddock et al. (1968) injected [^{14}C-methyl]-7-methylguanine into a rat and were unable to detect any incorporation of the radioactive label into the RNA or DNA. It would have been preferable to use labeled 7-methylguanosine, but their data support the concept that the alkylating agent in some ways acts directly on the nucleic acid. The mechanism of the carcinogenic effect, however, is completely obscure.

Summary. The available information is rather meager for an attempt to assess the general biological significance of the methylase systems. Moreover, a great deal of the information has been gained through *in vitro* experiments, and the subtle characteristics of the normal *in vivo* situation are probably obscured. In addition, artifacts in the experimental protocols using crude enzyme preparation may give misleading information. To obtain more precise data, it will be necessary, for example, to work with purified methylase enzymes and single molecular species of nucleic acid of known primary sequence (certainly in the case of tRNA). In these circumstances the products of the reaction should be identified and their location in the primary sequence determined.

In spite of the present experimental shortcomings, the available data indicate that the methylase systems are species-specific and that, with respect to higher organisms, they exhibit different methylating patterns between tissues in different growth states. The rapid advances in nucleic acid technology that are now taking place, particularly in regard to the ability to separate specific tRNA molecules and determine their sequences, will enable the experimentalist to design protocols that will provide more complete answers to the questions that have been raised.

3. BIOSYNTHESIS OF 5-β-D-RIBOFURANOSYLURACIL (PSEUDOURIDINE)*

The nucleoside 5-ribosyluracil represents the only known constituent of RNA in which the glycosylic bond is displaced to another location on the heterocycle ring. If we assume that 5-ribosyluracil, like other modified nucleosides, is formed at the macromolecular level, a rather elaborate mechanism of biosynthesis must be invoked. The available evidence does not as yet give a very clear picture of the mechanism. In the first attempts to establish a mechanism, Hall and Allen (1960) postulated that the transformation of uridine to 5-ribosyluracil occurs via the intermediate 1,5-diribosyluracil. This postulate became more attractive when Lis and Lis (1962) obtained evidence for the presence of 1,5-diribosyluracil in yeast RNA and in extracts of *Penicilliun brevi compastum*. In addition, Pollak and Arnstein (1962) reported that an extract of *E. coli* catalyzes the conversion of uridine to a compound tentatively identified as 1,5-diribosyl-uracil. More recently, doubt has been cast on these conclusions, since the properties of chemically synthesized 1,5-diribosyluracil (Brown et al., 1965; Dlugajczyk and Eiler, 1966a) do not correspond to those reported for the isolated samples.

---

*For an earlier review of the biochemistry of 5-ribosyluracil see Goldwasser and Heinrikson (1966).

The studies of Weiss and Legault-Demare (1965) suggest that RNA is involved in the biosynthesis of 5-ribosyluracil. For example, under conditions sufficient to impair the DNA-directed synthesis of RNA in intact or lysed spheroplasts of *E. coli*, formation of 5-ribosyluracil is also prevented. More directly, when RNA, labeled in its uridine residues, is incubated with lysed spheroplasts in the presence of DNAse or actinomycin, the specific activity of the 5-ribosyluracil residues increases. These authors conclude that the conversion of uridine to 5-ribosyluracil occurs at the polymeric level, although their data would also be consistent with a mechanism involving incorporation of pre-formed 5-ribosyluracil.

In another approach to the problem, Robbins and Hammond (1962) determined that both the ribose and uracil moieties of uridine are incorporated equally well into the 5-ribosyluracil residues of the RNA of growing *Saccharomyces carlsbergensis*. In an extension of these experiments, Robbins and Kinsey (1963) incubated the yeast cells in the presence of 5-fluorouridine labeled in both the ribose and 5-fluorouracil moieties. The fact that labeled 5-ribosyluracil was not detected in the RNA was interpreted as meaning that no transfer of the ribose residue took place, and that the 5-fluoro group blocks an intramolecular rearrangement. Therefore the normal pathway of biosynthesis is probably intramolecular. A similar conclusion was reached by Kusama et al. (1966), who fed uniformly labeled uridine to pyrimidine-requiring *Tetrahymena pyriformis*. 5-Ribosyluracil was formed in the tRNA with conservation of a large part of the uridine ribose moiety. Although these data do not exclude alternative mechanisms, they are certainly consistent with the concept of intramolecular rearrangement. One presumes that this process occurs at the macromolecular level.

Although this point is not firmly established, two additional pieces of indirect evidence suggest that this is in fact the case. Dubin and Günalp (1967) found that both the 16S and 23S RNAs of *E. coli* grown in the presence of chloramphenicol contain a diminished amount of 5-ribosyluracil. Since chloramphenicol interferes with the activity of the enzyme systems responsible for modifying RNA [i.e., the methylating systems (Gordon et al., 1964)], one could argue that the necessary enzyme systems for the formation of 5-ribosyluracil in the RNA at the macromolecular level function poorly under these conditions.

In the second experiment, Ginsberg and Davis (1968) found that although [2-^{14}C]-5-ribosyluracil is incorporated into the RNA of a uracil-requiring mutant of *E. coli*, this incorporation is non-specific. They isolated, for example, the common tetranucleotide, T-ψ-C-G, of the tRNA of the treated cells and found that it contains practically no label. Therefore, under these conditions any incorporation of radioactive label into the RNA can be attributed to the fact that the labeled 5-ribosyluracil is phosphorylated by a kinase (Ko et al., 1964) and incorporated merely as an analog of uridine. These results tend to exclude a mechanism of biosynthesis involving the incorporation of 5-ribosyluracil per se.

A uridine residue of tRNA may not be the direct precursor of 5-ribosyluracil. It is rather curious, for example, that with respect to tRNA sequence homologies, positions 34 and 44 consist of either 5-ribosyluracil, cytidine, or guanosine (Table 3, Chapter 4). Only one sequence, tRNA^{Ala}, contains a uridine in position 37. Does this suggest that cytidine is the precursor of 5-ribosyluracil?

Alternative mechanisms for the metabolism of 5-ribosyluracil may exist in some species. Free 5-ribosyluracil has been detected in the acid-soluble fractions of *T. pyriformis* (Kusama et al., 1966) and *S. cerevisae* (Kuriki, 1964) and in the culture medium of *Agrobacterium tumefaciens* (Suzuki and Hochster, 1964). One might assume that the presence of free 5-ribosyluracil can be attributed to normal breakdown of tRNA. Certainly in mammals free 5-ribosyluracil derived from tRNA is excreted (see p. 312), a fact which means that a definite concentration of free nucleoside (-tide) has to exist in the tissue. However, Heinrikson and Goldwasser (1963, 1964) showed that *T. pyriformis* contains an enzyme that catalyzes the condensation of uracil with ribose 5-phosphate to form 5-ribosyluracil 5'-phosphate. A similar enzyme has been found in *A. tumefaciens* (Suzuki and Hochster, 1966). This enzyme is the only one characterized that catalyzes the formation of the C-C glycosylic bond of 5-ribosyluracil. The relationship, if any, of such enzyme activity to the biosynthesis of the 5-ribosyluracil residues in tRNA is obscure.

4. BIOSYNTHESIS OF 4-THIOURIDINE

4-Thiouridine was first found in the tRNA of *E. coli* (Lipsett, 1965). This component may be peculiar to bacteria; in our own studies, we have not detected it in the tRNA of yeast or mammalian tissue, although other sulfur-containing nucleosides do occur. We identified 2-thio-5-carboxymethyluridine methyl ester in the tRNA of yeast (Baczynskyj et al., 1968), and Carbon et al. (1965) reported the presence of an uncharacterized 2-thiouridine in the tRNA of rabbit liver. Therefore it seems reasonable to assume that the tRNA of many organisms contains sulfur nucleosides.

The biosynthesis of 4-thiouridine has been investigated, using a cell-free enzyme preparation isolated from *E. coli*. Synthesis occurs at the macro-molecular level of tRNA, according to the scheme proposed by Hayward and Weiss (1966), Lipsett and Peterkofsky (1966), and Lipsett et al. (1967).

The essence of the reaction, $[^{35}S]$-cysteine + β-mercaptopyruvate + Mg^{++} + ATP + crude enzyme system extracted from *E. coli* + uracil residue in *E. coli* tRNA——→$[^{35}S]$-4-thiouracil residue in *E. coli* tRNA, involves transfer of a sulfur atom from cysteine to a uracil receptor in the tRNA molecule. Although cysteine is the actual donor of the sulfur atom, β-mercaptopyruvate is required in a catalytic amount to effect the transfer. When Lipsett et al. (1967) incubated $[^{35}S]$-labeled β-mercaptopyruvate and cold cysteine in an *in vitro* assay system, $[^{35}S]$ was not incorporated into the tRNA. The requirement for β-mercaptopyruvate can be met by using pyridoxal phosphate and ATP in the assay; these agents apparently convert a portion of the cysteine to β-mercapto-pyruvate.

The formation of 4-thiouridine at the macromolecular level is analogous to the formation of the methylated components, but some dissimilarities exist in these crude *in vitro* assay systems. The most striking difference lies in the fact that the homologous tRNA extracted from *E. coli* grown under normal condi-tions serves as a substrate for the incorporation of $[^{35}S]$ in the *in vitro* system. In the analogous methylase system, once the tRNA has received its normal complement of methyl groups, the homologous enzyme system does not catalyze further methylation. It would seem that in the thionucleoside system the crude enzyme lacks this specificity. Heterologous tRNA, such as that obtained from yeast and rat liver, also serves as a substrate, according to the data of Hayward and Weiss (1966). In contrast, Lipsett and Peterkofsky (1966) observed no significant incorporation of $[^{35}S]$-sulfur from labeled cysteine when they incubated the tRNA of yeast and rabbit liver with the *E. coli* enzyme preparation. In view of these anomolies and the fact that crude enzyme prepara-tions were used, the possibility that the results are due to an exchange reaction not necessarily related to the *in vivo* synthetic pathway cannot be excluded.

The oxidation state of the thionucleoside component undoubtedly is critical to the structure-function relationship of the tRNA molecule. Lipsett and Doctor (1967) and Doctor et al. (1969) have purified a molecular species of tRNA^{Tyr} from *E. coli* and have shown that it contains two 4-thiouridine residues. In the presence of iodine, this RNA molecule undergoes a slight loss in secondary structure, as shown by a decrease in T_m and an increase in $E(p)$, which suggests that an S-S bond is formed. Lipsett (1967) was able to isolate the bis-*S-S*-(4-

thiouridine) product from the "oxidized" RNA. The tyrosine acceptance capacity of the tRNA is not diminished after iodine treatment. The chemistry of the iodine oxidation reaction is discussed in more detail on p. 379, Chapter 8. Reversible oxidation of the 4-thiouracil residues may represent a cellular mechanism for modulating the activity of species of tRNA containing these residues by changing the conformation.

Carbon [unpublished data, cited by Carbon and David (1968)] could not find evidence for the formation of intermolecular disulfide bonds in iodine-treated tRNA. These data have led to the suggestion that formation of intra-molecular disulfide bonds in tRNA plays a role in conformational changes in tRNA (see Figure 8, Chapter 8). This concept requires more direct evidence for its support; the fact that the four known primary sequences of *E. coli* tRNA contain a single 4-thiouridine (see p. 273) rules out a simple interpretation of the observed data. Finally, it should be pointed out that Lipsett and Doctor (1967) did not find any decrease in tyrosine acceptance activity of their iodine-treated $tRNA^{Tyr}$. One might expect some change in acceptance activity if there were such a drastic change in structure.

5. BIOSYNTHESIS OF N^6-(Δ^2-ISOPENTENYL)ADENOSINE

The mechanism of the biosynthesis of N^6-(Δ^2-isopentenyl)adenosine is discussed in Chapter 7, p. 329.

6. CATABOLISM AND EXCRETION OF THE MODIFIED NUCLEOSIDES

In the normal catabolism of nucleic acid pyrimidines, cytosine and uracil are degraded to β-alanine, and thymine is degraded to β-aminoisobutyric acid. Adenosine and guanosine are degraded to products reusable by the cell. The majority of the modified components of nucleic acids, on the other hand, do not appear to be catabolized to any extent, and in animals are excreted in significant amounts in the urine. The only known exception is N^6-(Δ^2-iso-pentenyl)adenosine, which is degraded to inosine or hypoxanthine in a number of tissues (see Chapter 7, p. 333, for a discussion of the degradation pathway).

Table 3 lists the purine and pyrimidine derivatives detected in human urine. Most samples were obtained from urine collected from individuals fed a caffeine-free, low-purine diet. No direct proof exists that these compounds result from the breakdown of nucleic acids, although the evidence points to such a conclusion. For example, 5-ribosyluracil has been found only in RNA and not in an unbound state in the mammalian cell; therefore, it presumably arises from RNA breakdown. Mandel et al. (1966) demonstrated that the methyl groups of the urinary methylated purines stem from the methyl group of methionine; and since methylation of purines occurs at the RNA level, it appears likely that methylated purines are nucleic acid breakdown products. Treatment of patients with certain drugs known to lead to the breakdown of nucleic acids leads to an increase in the level of the modified nucleosides in the urine (Dlugajczyk and Eiler, 1966b; Park et al., 1962).

The rate and pattern of excretion of nucleic acid components in the urine reflect the metabolic state of the individual, and in certain diseases these parameters change markedly. The rate of excretion of 5-ribosyluracil, for example, rises in patients with gout and in those with leukemia (Adams et al., 1960; Adler and Gutman, 1959; and Dlugajczyk and Eiler, 1966b). One patient with chronic myelocytic leukemia in blastic relapse excreted 416 mg/24 hr. In a study of thirteen patients with different types of acute and chronic leukemia, Park et al. (1962) observed an increase in the urinary level of 1-methylhypoxan-thine and 8-hydroxy-7-methylguanine in all patients, an increase in the level of 7-methylguanine in one-third and an increase in the level of N^6-succinoadenine in one-half.

In assessing this type of data, the possibility that the excretion pattern may

TABLE 3

Excretion of Purines, Pyrimidines, and Nucleosides in Urine

Bases	Amount Excreted per 24 hr by Normal Human Adult, mg	References
Adenine	0.6–1.4	Heirwegh et al. (1967); Mandel et al. (1966); Park et al. (1962); Weissmann et al. (1957a,b)
1-Methyladenine[a]	pres.	Fink and Adams (1968); Mandel et al. (1966)
N^6-Methyladenine	pres.	Fink and Adams (1968)
N^6-Succinoadenine	0.6–0.9	Park et al. (1962); Weissmann and Gutman (1957)

BIOSYNTHESIS AND METABOLISM OF THE MODIFIED NUCLEOSIDES

TABLE 3 continued

Bases	Amount Excreted per 24 hr by Normal Human Adult, mg	References
N^6-(3-Methyl-3-hydroxybutyl)-adenine	trace	Robins et al. (1967)
N-(Purin-6-ylcarbamoyl)-L-threonine	0.2–0.4	Chheda and Mittelman (1967a)
3-Methylcytosine	pres.	Fink and Adams (1968)
Guanine	0.3–1.3	Mandel et al. (1966); Park et al. (1962); Weissmann et al. (1957a)
1-Methylguanine[a]	0.3–0.5	Mandel et al. (1966); Weissmann et al. (1957a,b)
N^2-Methylguanine	0.3–0.6	Park et al. (1962); Weissmann et al. (1957a)
N^2,N^2-Dimethylguanine[a]	trace	Fink et al. (1963); Mandel et al. (1966)
7-Methylguanine[a]	4.4–7.8	Mandel et al. (1966); Park et al. (1962); Weissmann et al. (1957a,b)
7-Methyl-8-hydroxyguanine[a]	0.6–1.8	Mandel et al. (1966); Park et al. (1962); Weissmann et al. (1957b); Weissmann and Gutman (1957)
Hypoxanthine	3.1–10.6	Heirwegh et al. (1967); Mandel et al. (1966); Weissmann et al. (1957a,b)
1-Methylhypoxanthine[a]	0.4–0.9	Heirwegh et al. (1967); Mandel et al. (1966); Park et al. (1962); Weissmann et al. (1957a,b)
Xanthine	5.3–8.7	Mandel et al. (1966); Park et al. (1962); Weissmann et al. (1957a,b)
1-Methylxanthine[b,d]		Weissmann et al. (1957a,b)
7-Methylxanthine[b]		Weissmann et al. (1957a,b)
1,7-Dimethylxanthine[b]		Heirwegh et al. (1967); Weissmann et al. (1957a,b)
Uric acid	390–588	Park et al. (1962)
1-Methyluric acid[b]		Cornish and Christman (1957)
Cytosine	trace	Adams et al. (1960); Heirwegh et al. (1967)
Uracil	4–6	Adams et al. (1960); Adler and Gutman (1959); Heirwegh et al. (1967)
5-Hydroxymethyluracil	pres.	Heirwegh et al. (1967)
5-Acetylamino-6-amino-3-methyluracil[c]		Fink et al. (1964)
Thymine[e]		Adams et al. (1960)

TABLE 3 continued

Nucleosides	Amount Excreted per 24 hr by Normal Human Adult, mg	References
Adenosine	pres.	Fink and Adams (1968); Heirwegh et al. (1967)
1-Methyladenosine	pres.	Fink and Adams (1968)
N^6-Methyladenosine	1.2	Chheda et al. (1968); Fink and Adams (1968)
N-(Nebularin-6-ylcarbamoyl)-L-threonine	0.3	Chheda (1969)
Cytidine	pres.	Fink and Adams (1968)
2'-O-Methylcytidine	1.2	G. B. Chheda, personal communication; Fink and Adams (1968)
Deoxycytidine[f]	pres.	Rotherham and Schneider (1960)
5-Methyldeoxycytidine[f]	pres.	Rotherham and Schneider (1960)
1-Methylguanosine	1.5	G. B. Chheda, personal communication; Fink and Adams (1968)
N^2-Methylguanosine	0.2	Chheda et al. (1969c)
N^2,N^2-Dimethylguanosine	1.8	Chheda and Mittelman (1967b); Fink et al. (1963)
Inosine	pres.	Adams et al. (1960); Heirwegh et al. (1967)
1-Methylinosine	2.1	Chheda and Mittelman (1967b); Chheda et al. (1969c); Fink and Adams (1968)
Uridine	pres.	Heirwegh et al. (1967)
5-Ribosyluracil	45–70	Adams et al. (1960); Adler and Gutman (1959); Eisen et al. (1962); Weissman et al. (1962)
Deoxyuridine[f]	pres.	Rotherham and Schneider (1960)
5-Aminoimidazole-4-carboxamide riboside	pres.	Fink and Adams (1968)

[a] Derived from RNA (Mandel et al., 1966).
[b] Derived from the diet (Weissmann et al., 1957a,b).
[c] Purine breakdown product (Fink et al., 1964).
[d] Guanase catalyzes oxidation of 1-methylguanine to 1-methylxanthine (Hitchings and Falco, 1944).
[e] Thymine has been detected only in the urine of leukemic patients.
[f] Detected only in rat urine.

be influenced by drugs administered to the patient must be considered. For example, Simmonds (1969) found that administration of *allopurinol* resulted in cessation of 5-ribosyluracil excretion.

The possible relationship of the excretion pattern of the modified components of nucleic acids to various diseases has not been explored in any depth. This area appears to be fruitful for research, and as more data become available, significant correlations may be revealed. The presence, absence, or change in the level of certain key nucleic acid components in the urine may become useful as a diagnostic tool, and perhaps such data will signal an important metabolic change before it becomes clinically apparent.

Chapter 7

HYPERMODIFIED NUCLEOSIDES:

N^6-(Δ^2-ISOPENTENYL)ADENOSINE

(AND ANALOGS) AND

N-(NEBULARIN-6-YLCARBAMOYL)-L-

THREONINE*

1. INTRODUCTION

$\mathbf{M}$ OST OF THE MODIFIED nucleosides detected in nucleic acids are relatively simple structural modifications of the various major nucleosides. Some of the modifications, such as attachment of a methyl group, may not appreciably change the chemical properties from those of the parent nucleoside. Other modifications, such as replacement of a hydroxyl group with a sulfur group or the rearrangement represented by 5-ribosyluracil, may make a substantial difference in the chemical and biological properties of the molecule. The basic chemical properties of the modified nucleosides and methods of synthesis are described in Chapter 2; additional information concerning specific chemical reactivities of some of the modified nucleosides is covered in Chapter 8, which is concerned with chemical reactions that can be applied to nucleic acids at the macromolecular level. For the majority of the known modified nucleosides of nucleic acids, this information represents the total information available, although the known chemistry and biology of the parent nucleosides and their analogs are certainly relevant.

---

*The *systematic* names for this compound are: N-[(9-β-D-ribofuranosyl-9H-purin-6-yl)carbamoyl]-L-threonine and N^6-[(1-carboxy-2-hydroxypropyl)carbamoyl]adenosine.

HYPERMODIFIED NUCLEOSIDES

Some of the modified nucleosides recently detected in tRNA possess a relatively elaborate structure, resulting from the attachment of a complex side chain. Not only does the presence of the larger side chain create bulk, but also such side chains contain functional groups (organic chemical definition). For these reasons, such nucleic acid components are called hypermodified nucleosides. Six such nucleosides have now been detected in tRNA: N^6-(Δ^2-isopentenyl)adenosine, *1*; N^6-(*cis*-4-hydroxy-3-methylbut-2-enyl)-adenosine, *2*; N^6-(Δ^2-isopentenyl)-2-methylthioadenosine, *3*; and N-(nebularin-6-ylcarbamoyl)-L-threonine, *24*; methyl, 2-thio-5-carboxymethyluridine, *4*; and 5-carboxy-methyluridine, *5*. The physiological significance of this group of compounds is enhanced by the fact that some, if not all, are located adjacent to the 3' end of the anticodon of tRNA molecules (see Table 2, Chapter 4).

The properties of two of these hypermodified nucleosides, N^6-(Δ^2-isopentenyl)adenosine, *1* (and its ω-hydroxylated and 2-methylthio-analogs, *2* and *3*) and N-(nebularin-6-ylcarbamoyl)-L-threonine, *24*, have been investigated extensively. Since a discussion of their properties is not covered in detail in other chapters, and since they have a significance unique to tRNA structure and function, they warrant a special review.

2. N^6-(Δ^2-ISOPENTENYL) ADENOSINE AND RELATED COMPOUNDS

Molecular species of tRNA containing the Δ^2-isopentenyl group of tRNA have been detected in all organisms investigated. Three Δ^2-isopentenyl-containing nucleosides have been identified: N^6-(Δ^2-isopentenyl)adenosine, *1*; N-(*cis*-4-hydroxy-3-methylbut-2-enyl)-adenosine, *2*; and N^6-(Δ^2-isopentenyl)-2-methylthioadenosine, *3*. The distribution in nature of these compounds is given in Table 1.

The parent nucleoside, *1*, has been found in the tRNA of all classes of organisms: bacteria, yeast, plant, and animal. Compound *2* has been found only in the tRNA of plants, and compound *3* only in that of *E. coli*. In our own studies we have not detected *2* or *3* in the tRNA of mammalian tissues. As more analytical data become available, it remains to be seen whether a phylogenetic segregation of *2* and *3* will hold.

Natural Occurrence of N^6-(Δ^2-Isopentenyl)adenosine and Related Compounds Other than in RNA. A number of compounds with structures related to N^6-(Δ^2-isopentenyl)adenosine have been isolated from natural sources. The free base,

1

2

3

4

5

TABLE 1

Occurrence of N^6-(Δ^2-Isopentenyl)adenosine and Derivatives in Unfractionated tRNA.
(Values in moles/100 moles of total nucleosides)

Source of tRNA	N^6-(Δ^2-Isopentenyl)-adenosine	N^6-(cis-4-Hydroxy-3-methylbut-2-enyl)-adenosine	N^6-(Δ^2-Isopentenyl)-2-methylthio-adenosine	Ref.
Baker's yeast	0.06			(1)
Calf liver	0.05			(1)
Chick embryo	0.03			(1)
Human liver	0.05			(1)
Immature peas	0.003	0.005		(2)
Spinach leaves	0.02	0.01		(2)
Corn kernels				
Immature		0.01		(2)
Mature (seed)		pres.		(3)
Escherichia coli B			pres.	(4,5)
Lactobacillus acidophilus	pres.			(6)
Lactobacillus plantarum	pres.			(6)
Tobacco pith cells grown in culture	pres.			(7)

References
1. Robins et al. (1967)
2. Hall et al. (1967a)
3. McLennan and Hall, unpublished data
4. Burrows et al. (1968)
5. Harada et al. (1968)
6. Fittler et al. (1968a)
7. Chen and Hall (1969)

N^6-(Δ^2-isopentenyl)adenine, *6*, has been isolated from liquid cultures of *Corynebacterium fascians*, a pathogen known to cause abnormal development in plants (Helgeson and Leonard, 1966; Klämbt et al., 1966). The *trans* analog of compound *2*, N^6-(*trans*-4-hydroxy-3-methylbut-2-enyl)adenine, *7*, its ribonucleoside, and its ribonucleotide have been isolated from extracts of *Zea mays* kernels (milk stage) (Letham, 1966a; Miller, 1965). The name zeatin has been assigned to the free base of this compound. Zeatin was identified by Letham et al. (1967) and synthesized by Shaw et al. (1966). Letham (1968) isolated 9-(β-D-ribofuranosyl)zeatin from coconut milk and has shown that it is probably the principal plant growth factor occurring therein. Zeatin and its ribosyl derivative have been isolated from cultures of the puff ball fungus, *Rhizopogon roseolus* (Miller, 1967). Koshimizu et al. (1967) have identified dihydrozeatin, N^6-(4-hydroxy-3-methylbutyl)adenine, *8*, in extracts of immature seeds of *Lupinus luteus*. Isolated compound is optically active with negative rotation.

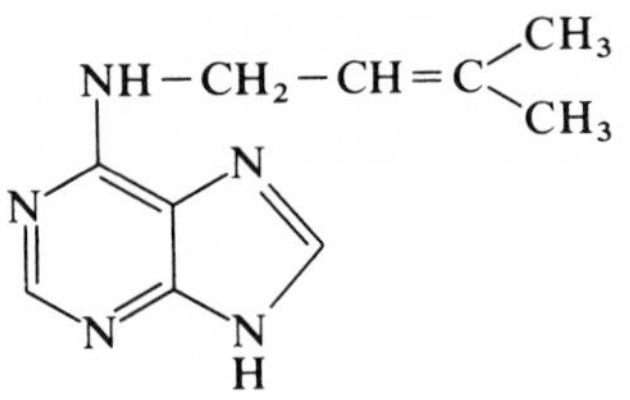

6

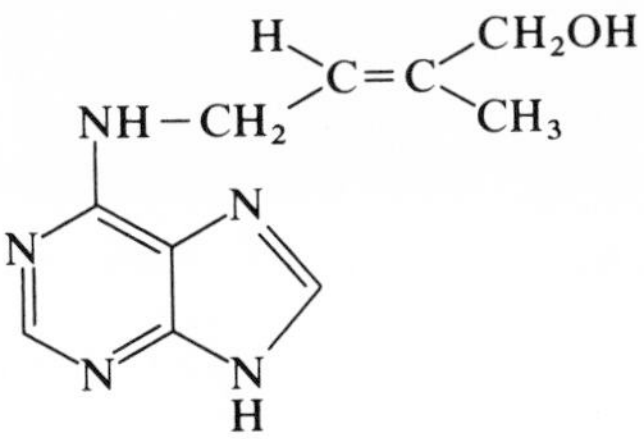

7

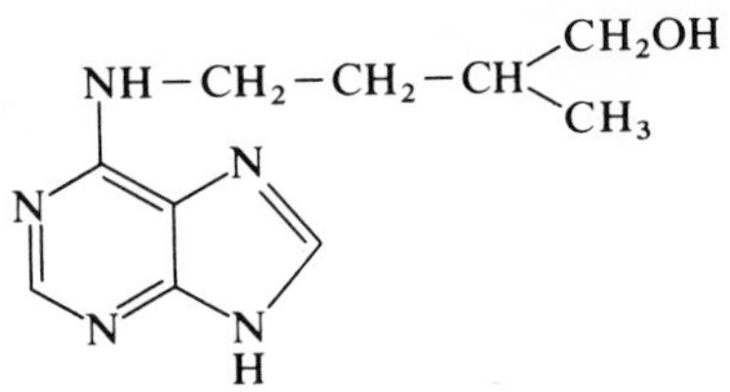

8

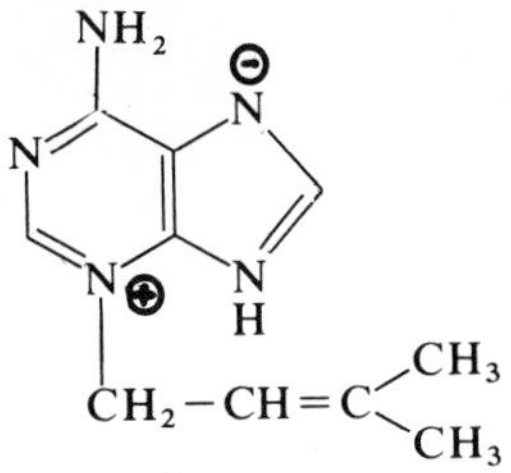

9

The leaves of the honey locust tree, *Gleditsia triacanthos* L., contain an alkaloid material, isolated and identified as 3-(Δ^2-isopentenyl)adenine, *9* (Leonard and Deyrup, 1962). This compound was assigned the name triacanthine. The amount of *9* found varies with the age of the leaf, reaching a maximum in young leaves and decreasing as the leaves age (Rogozinska, 1967). Triacanthine does not act directly as a plant growth factor (cytokinin activity) (Rogozinska et.al., 1964); however, when autoclaved, it is converted into an active substance. These authors speculate that rearrangement of the side chain from the N-3 to the N-6 position may occur, and on this basis Rogozinska (1967) suggested that triacanthine represents an inert storage form for the active cell division factor. This concept is attractive in view of the fact that enzymic degradation of N^6-(Δ^2-isopentenyl)adenosine occurs in plant tissues (Chen et al., 1968). In order to maintain the required nucleoside level in the tissue, a continuous renewal of the supply would be necessary [see p. 335 for a discussion of the biological activity of N^6-(Δ^2-isopentenyl)adenosine]. Triacanthine was also isolated from the leaves of *Holarrhena floribunda* (Janot et al., 1959) and of *Chidlowia sanguinea* (Monseur and Adriaens, 1960), although these two groups of workers did not identify their isolated compounds. Leonard and Deyrup (1962) later showed that these natural products are identical to triacanthine.

Precedence exists for the occurrence in nature of the side chain of compound

2. The plant, *Galega officinalis* L., contains two guanidine derivatives, compounds *10* and *11*. Olomucki et al. (1965) have demonstrated that the hydroxylated form, *11*, occurs in the *cis* configuration.

10   *11*

Synthesis of N^6-(Δ^2-Isopentenyl)adenosine. N^6-(Δ^2-Isopentenyl)adenosine, *1*, can be readily synthesized by two different routes. In one method, the corresponding alkyl side chain in the form of the amine is condensed with 6-chloro-9-(β-D-ribofuranosyl)purine, *14* (Hall et al., 1966). In the second method, the corresponding alkyl side chain in the form of the bromide is used to alkylate adenosine, *12*, directly (Grimm and Leonard, 1967; Grimm et al., 1968; Leonard et al., 1966).

The initial compound formed is 1-(Δ^2-isopentenyl)adenosine, *13*. This compound rearranges to *1* when the N-1-substituted intermediate is heated in aqueous solution at pH 7.5.

The facile rearrangement of the N-1 isomer to the N-6 isomer raises the question as to which one exists in the tRNA, even though the isolated nucleoside has been rigorously identified as the N-6 isomer. This question has a parallel in tRNA structure, since both N-1- and N-6-methyladenosine occur in the nucleic acids of various organisms (Dunn, 1961a). The influence of the Δ^2-isopentenyl side chain on the oligonucleotide structure and the chemical reactivities of the surrounding oligonucleotide segment of the tRNA molecule would vary greatly, depending on whether the side chain is attached at the N-1 or the N-6 position of the adenylic acid component. In order to answer this question, we hydrolyzed tRNA with acid under conditions that would release the purine base and would not permit the N-1→N-6 rearrangement. Under these conditions, only the N-6 isomer is found (Robins et al., 1967). Grimm and Leonard (1967) also obtained data demonstrating that under the conditions used for the isolation of compound *1* from tRNA (Hall et al., 1966; Zachau et al., 1966a,b) the N-1→N-6 rearrangement could not occur.

Synthesis of N^6-(Δ^2-Isopentenyl)-2-methylthioadenosine (Burrows et al., 1968). 2,6-Bis(methylthio)purine, *15*, is refluxed with Δ^2-isopentenylamine to yield N^6-(Δ^2-isopentenyl)-2-methylthiopurine, *16*. A mercuri salt of *16* is prepared and condensed with 1-bromo-2,3,5-O-tribenzoylribofuranose. After deblocking, the desired nucleoside, *3*, is obtained. Compound *16* can also be prepared by condensing 2-methylthio-6-chloropurine with the corresponding alkylamine.

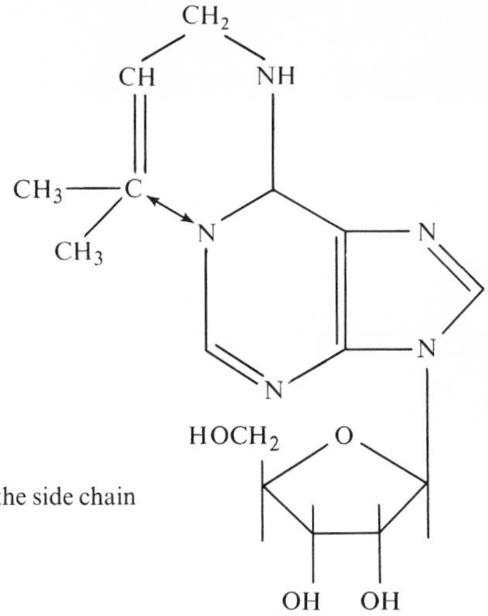

FIGURE 1. Spatial arrangement of the side chain of N^6-(Δ^2-isopentenyl)adenosine.

Chemical Reactions of N^6-(Δ^2-Isopentenyl)adenosine. When N^6-(Δ^2-isopentenyl)adenosine, *1*, is hydrolyzed in dilute acid under conditions that sever the glycosylic bond, the free base is not obtained; rather, the hydrated product, N^6-(3-hydroxy-3-methylbutyl)adenine, *17*, is formed. On continued acid treatment, the hydroxyl group is expelled to form what is presumably a carbonium intermediate that undergoes ring closure to form 3H-7,7-dimethyl-7,8,9-trihydropyrimido-[2,1-i]purine,* *18* (Hall et al., 1966). Compound *18*, although stable to acid, degrades in alkaline solution to form a product that appears to be an imidazole derivative. The ready hydration of the allylic double bond under relatively mild chemical conditions suggests that the N-1 of the purine residue actively promotes reaction of the side chain. The spatial arrangement of the allylic double bond of the side chain and the N-1 position, therefore, may well play an important role in the biological function of N^6-(Δ^2-isopentenyl)-adenosine (Figure 1). The hydration of the double bond of *1* by an enzyme

* The numbering of this ring system is as follows:

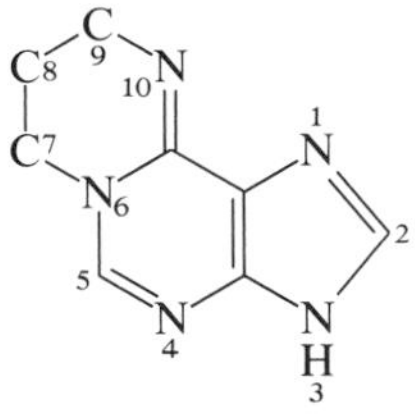

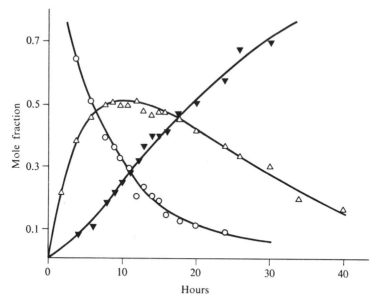

FIGURE 2. Hydrolysis of N^6-(Δ^2-isopentenyl)adenosine in N-hydrochloric acid at 41°. O = mole fraction of nucleoside, *1*; Δ = mole fraction of aglycone, *6*; ▲ = mole fraction of N^6-(3-hydroxy-3-methylbutyl)adenine, *17*. Data taken from Martin and Reese (1968).

possibly present in some tissues to form N^6-(3-hydroxy-3-methylbutyl)adenosine, *21*, underscores this point (see p. 333).

The mechanism of formation of *17* and *18* in acid solution has been studied by Martin and Reese (1968). Their data suggest that in the first step the glycosylic bond is broken and then the free base, *6*, forms a carbonium ion that can accept either a hydroxyl group from the aqueous solution to form *17* or the N-1 proton to form *18*. The relative rates of formation of the products are shown in Figure 2.

The free base of compound *1*, N^6-(Δ^2-isopentenyl)adenine, *6*, can be produced by treating *1* with periodate and then with sodium hydroxide (Hall et al., 1966).

Treatment of compound *1* with dilute aqueous permanganate (pH 7.0) for 10 min at 25° gives the dihydroxylated product, *19*, in 50% yield. In addition to this compound, adenosine (30% yield), is produced. The mechanism of cleavage of the side chain to form adenosine has not been investigated. The mildness of this cleavage reaction has been exploited, however, to remove the

Δ^2-isopentenyl side chain from tRNA without disruption of the rest of the tRNA molecule (see below).

Iodine reacts rapidly with compound *1* in aqueous solution under mild conditions (25°, pH 7.0) to form 7,7-dimethyl-8-iodo-7,8,9-trihydropyrimido-3-(β-D-ribofuranosyl) [2,1-*i*]purine, *20* (Robins et al., 1967). The NMR spectrum for *20* shows two single peaks at δ 1.68 and 1.84 (ratio 60:40); the two peaks together integrate for six protons. These peaks, therefore, represent the methyl groups, and their existence suggests the presence of two isomers. The unsplit nature of each peak indicates the absence of a proton at C-7, which supports the assigned structure, *20*. The fact that the ultraviolet absorption spectra of *20* are similar to those of 1,N^6-dimethyladenosine (Broom et al., 1964) lends additional support to the assignment of the basic structure.

Under these conditions iodine reacts quantitatively with the N^6-(Δ^2-isopentenyl)adenosine residues of tRNA. Furthermore, the reaction is selective; iodine does not react with other known components of tRNA except with the thiolated nucleosides (see p. 379, Chapter 8). The results of two experiments permit the conclusion that the reaction in the tRNA is the same as with free N^6-(Δ^2-isopentenyl)adenosine. In one of them, carried out in our laboratory, Dr. F. Fittler showed that compound *1* cannot be detected in yeast tRNA after the iodine treatment. In the other experiment Kline et al. (1969) showed that the iodinated nucleoside product in the tRNA molecule is identical to that obtained from compound *1* alone.

The ability to selectively immobilize the N^6-(Δ^2-isopentenyl)adenosine residues makes possible an investigation of one aspect of the structure-function relationship of tRNA. Since this nucleoside occurs in yeast tRNA[Ser] adjacent to the anticodon (p. 262), the following question can be asked: What does the iodine modification in the anticodon loop do to the biochemical function of tRNA[Ser]? The data presented in Table 2 show that iodine-treated tRNA does not lose its ability to accept serine residues; on the other hand, the ability of the tRNA[Ser] to bind with the appropriate messenger-ribosome complex is reduced to approximately one-third of the original capacity. If a close fit of the Δ^2-isopentenyl side chain into the tertiary structure of the anticodon loop is assumed, it seems reasonable that a modification of the side chain, particularly one that fixes it into a rigid configuration and also adds a bulky iodine atom, must affect the conformation of the anticodon loop. From the point of view of structure-function relationships, these results suggest that the structural integrity of N^6-(Δ^2-isopentenyl)adenosine is essential for the proper functioning

TABLE 2

Biological Activity of Iodine-Treated tRNA. (Data taken from Fittler and Hall, 1966)

Amino Acid-Accepting Activity[a]	Radioactivity Incorporated	
	Serine, cpm	Phenylalanine, cpm
Iodine-treated tRNA	918	2280
Untreated tRNA	822	2325
No tRNA	43	81

	Aminoacyl tRNA Bound to Ribosomes			
	Untreated		I_2-Treated	
Binding Activity	cpm	μμm	cpm	μμm
Phe with poly U	1730	3.6	1780	3.7
Phe without poly U	142	0.3	156	0.3
Ser with poly (U,C)	676	3.4	238	1.15
Ser without poly (U,C)	132	0.8	71	0.4

[a]The assay was carried out according to the procedure of Hoskinson and Khorana (1965). In each assay 1.7 A_{260} units of unfractionated yeast tRNA was incubated with 6.8×10^{-4} μm of [^{14}C]-amino acid; serine, ^{120}μc/μm, phenylalanine ^{360}μc/μm.
[b]The assay was carried out according to Nirenberg and Leder (1964).

of tRNA molecules containing this nucleoside. This conclusion is convincingly supported by the following experiment.

Effect of N^6-(Δ^2-Isopentenyl)adenosine on Ribosomal Binding. The infection of *E. coli* with the defective transducing bacteriophage $\phi 80$ dsu$_{III}^+$ causes the selective synthesis of suppressor tRNA^{Tyr}. Gefter and Russell (1969) fractionated this tRNA on a reverse-phase column and separated three forms, all of which have the same primary sequence shown on p. 267. The forms differ; although in the extent of modification of the adenosine residue adjacent to the 3′ end of the anticodon, one form contains adenosine in this position, the second contains 2-methylthioadenosine, and the third contains N^6-(Δ^2-isopentenyl)-2-methylthioadenosine. All three forms accept tyrosine at the same rate but differ markedly in their ability to support protein synthesis (see Figure 3). The three forms show the same relative efficiencies in binding to ribosomes in the presence of UAG and UAU, which may account for the observed differences in the rate of protein synthesis. These data clearly indicate that the modification of the adenosine residue adjacent to the 3′ end of the codon is essential for participation of the tRNA molecule in protein synthesis.

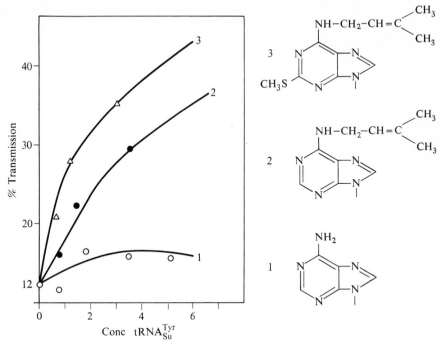

FIGURE 3. *In vitro* suppressor activities of tRNAs containing different modified nucleosides adjacent to the 3' end of the anticodon. The % transmission, as a measure of polypeptide formation coded by f2 sus4A RNA, is plotted against increasing concentration of tRNA. Data taken from Gefter and Russell (1969).

Biosynthesis of N^6-(Δ^2-Isopentenyl)adenosine. Three major questions concern (1) the source of the isopentenyl group; (2) the mode of attachment of the Δ^2-isopentenyl group, particularly whether it is attached to a monomeric precursor or to the already formed tRNA molecule; and (3) the nature of the mechanism that controls the biosynthesis.

The isopentenyl group is derived from mevalonic acid, the source of the five-carbon building blocks of isoprenoid compound. Utilization of mevalonic acid as the precursor of the isopentenyl group in tRNA can be readily demonstrated with the acetate-requiring bacteria, *Lactobacillus acidophilus.* Part of the acetate requirement for this microorganism can be replaced by mevalonic acid, indicating that the acid probably enters the cell intact. When the bacteria is grown in the presence of radioactively labeled mevalonic acid, the tRNA becomes radioactive (Fittler et al., 1968a; Peterkofsky, 1968). All this radio-activity can be accounted for in the N^6-(Δ^2-isopentenyl)adenosine residues of tRNA. It is difficult to experimentally demonstrate this incorporation *in vivo* in other organisms. Bloch and Goodwin (1959) have noted that mevalonic acid penetrates whole cells with difficulty. Fittler et al. (1968a) found that, when Baker's yeast is grown in the presence of radioactively labeled mevalonic acid, the tRNA does not become radioactive. On the other hand, when yeast is grown in the presence of labeled acetate, the precursor of mevalonate, a significant amount of radioactivity is incorporated into the N^6-(Δ^2-isopentenyl)adenosine residues of the tRNA. The acetate label also enters other components of the tRNA, but the specific activity of the N^6-(Δ^2-isopentenyl)adenosine isolated is 50 times that calculated on the basis of random incorporation. Chen and Hall (1969) grew tobacco pith tissue in culture for 4 weeks in the presence of $[2\text{-}^{14}C]$-mevalonic acid and found that a significant amount of radioactivity is incor-porated into a N^6-(Δ^2-isopentenyl)adenosine-like compound in the tRNA. These data strongly indicate that mevalonic acid serves as the precursor of the Δ^2-isopentenyl group of tRNA *in vivo*.

The basic biosynthetic mechanism of N^6-(Δ^2-isopentenyl)adenosine parallels that of the methyl compounds of tRNA; the isopentenyl side chain is attached to an adenylic acid residue in the preformed tRNA molecule. In order to prove this point experimentally by studying the mechanism *in vitro* it was first neces-sary to obtain a tRNA substrate that lacks Δ^2-isopentenyl groups. Since tRNA as isolated would be expected to contain its full complement of such groups, the key technical factor is the ability to remove the Δ^2-isopentenyl group. In our approach to this problem we made use of the reaction of N^6-(Δ^2-isopentenyl)-adenosine with permanganate described on p. 325. Kline et al. (1969) applied this reaction to the tRNA of *L. acidophilus*, in which the Δ^2-isopentenyl groups were labeled with $[^3H]$ or $[^{14}C]$, and showed that about 25% of the Δ^2-isopentenyl side chain could be removed under these conditions. This reaction, therefore, is a useful technique for producing the desired tRNA substrate. It does not appear to cause a significant amount of nonspecific damage to the tRNA molecule, since yeast tRNA treated under these conditions retains about 90% of its amino acid acceptor capability. Fittler and Hall (1966) have shown that alteration of the structure of N^6-(Δ^2-isopentenyl)adenosine residues in tRNA does not interfere with the amino acid-accepting activity (see Table 2).

Studies of the mode of action of the formation of N^6-(Δ^2-isopentenyl)-adenosine *in vitro* were carried out, using a partially purified enzyme system

TABLE 3

Incorporation of $[4\text{-}^{14}C]\Delta^2$-Isopentenyl Groups into the N^6-(Δ^2-Isopentenyl)adenosine Residues of tRNA. (Data from Kline et al., 1969)

The tRNA is incubated with the enzyme system, ATP, and $[4\text{-}^{14}C]\Delta^2$-isopentenyl pyrophosphate for 30 min at $37°$. The tRNA is hydrolyzed, and the radioactivity of the N^6-(Δ^2-isopentenyl)-adenosine is measured. (All the radioactivity is associated with this nucleoside.) The values are the means of at least three separate experiments.

In addition, similar results have been obtained using crude enzyme systems and homologous tRNA from cultured tobacco pith tissue (Chen and Hall, 1969) and rat liver (Fittler et al., 1968b).

tRNA	Enzyme System	Δ^2-Isopentenyl Groups Incorporated into 25 A_{260} Units tRNA (p moles)
Yeast, untreated	Yeast	17
Yeast + $KMnO_4$	Yeast	129
Yeast + I_2, then $KMnO_4$	Yeast	20
Yeast + $KMnO_4$, then I_2	Yeast	121
Rat liver, untreated	Yeast	15
Rat liver + $KMnO_4$	Yeast	9
Escherichia coli B, untreated	Yeast	54
Escherichia coli B + $KMnO_4$	Yeast	45

obtained from yeast and a crude enzyme system from rat liver. These enzyme systems catalyze the transfer of the Δ^2-isopentenyl group from Δ^2-isopentenyl pyrophosphate to the permanganate-treated homologous tRNA as shown in Table 3. The yeast enzyme system does not utilize the permanganate-treated rat liver tRNA as a substrate. The enzyme requires Δ^2-isopentenyl pyrophosphate; the Δ^3-isomer does not work.

An important question concerning the biosynthetic pathway involves the specificity of the reaction. The fact that the enzyme system catalyzes incorporation of Δ^2-isopentenyl groups into the permanganate-treated tRNA and not into untreated tRNA is presumptive evidence that it catalyzes attachment of the isopentenyl group to the adenylic acid residues from which the Δ^2-isopentenyl side chain has been deleted. Furthermore, the nucleoside product of the reaction has been identified as N^6-(Δ^2-isopentenyl)adenosine (Fittler et al., 1968b). Confirmatory evidence for the specific nature of the reaction stems from an experiment in which the N^6-(Δ^2-isopentenyl)adenosine residues of the tRNA are rendered insensitive to permanganate oxidation.

Treatment of tRNA with aqueous iodine solution converts the N^6-(Δ^2-isopentenyl)adenosine residues to compound *20 in situ*, as described above. Since compound *20* no longer contains the allylic double bond, it is insensitive to oxidation under the mild conditions used in these experiments. The rationale behind the experiments, therefore, is to immobilize the Δ^2-isopentenyl group of the N^6-(Δ^2-isopentenyl)adenosine residues by first treating it with iodine and then with permanganate. Yeast tRNA treated first with iodine and then with permanganate does not accept the isopentenyl side chain (Table 3). The two treatments in themselves are not harmful to the tRNA, since reversal of the procedures yields a tRNA sample capable of accepting the isopentenyl group. Because of the selectivity of both the iodine and the permanganate reactions for the N^6-(Δ^2-isopentenyl)adenosine residues, this experiment shows that the enzyme system is specific for the adenosine residues in tRNA which normally would contain a Δ^2-isopentenyl group.

With respect to the nature of the recognition sites in the tRNA molecule, it is instructive that tRNA^{Tyr} and tRNA^{Ser}, the two molecular species of yeast tRNA containing N^6-(Δ^2-isopentenyl)adenosine, possess the common sequence A-i^{6}A-A-ψ-C-U-U in the anticodon loop region. This means that three bases of the anticodon loop and the first four base pairs of the supporting arm are identical in both primary sequences. The corresponding seven-base sequence for rat liver tRNA^{Ser} and *E. coli* Su$_{III}^+$ tRNA^{Tyr} are A-i^{6}A-A-ψm-C-C-A and A-ms^{2}iA-A-ψ-C-U-G, respectively. The striking feature of all these sequences is the occurrence of three consecutive adenosine residues; the middle adenosine contains the Δ^2-isopentenyl side chain. These data all point toward a specific receptor site unique to a small number of tRNA molecular species.

The ability of untreated *E. coli* B tRNA to accept the Δ^2-isopentenyl group catalyzed by both the yeast and the rat liver enzyme systems (Table 3) may be due to the fact that under its growth conditions, a significant proportion of the tRNA molecules normally carrying the Δ^2-isopentenyl group lacks these groups. Goodman et al. (1968) and Gefter and Russell (1969) report the presence of such molecular species of tRNA^{Tyr} in *E. coli*.

Radioactively labeled N^6-(Δ^2-isopentenyl)adenosine was identified in the *E. coli* B tRNA after exposure to the enzyme systems and labeled mevalonic acid, and this product accounted for all the radioactivity in the tRNA (Kline et al., 1969). Although *E. coli* B does not contain N^6-(Δ^2-isopentenyl)adenosine (Fittler et al., 1968a), it does contain N^6-(Δ^2-isopentenyl)-2-methylthioadenosine (Burrows et al., 1968). The fact that N^6-(Δ^2-isopentenyl)adenosine and not

N^6-(Δ^2-isopentenyl)-2-methylthioadenosine is obtained from the tRNA suggests that, if a proportion of the normally Δ^2-isopentenyl-containing molecules in the isolated tRNA lack their Δ^2-isopentenyl groups, these molecules contain adenosine and not 2-methylthioadenosine at the acceptor site.

The central role of mevalonic acid as the precursor of both the isoprenoid compounds and an essential component of tRNA raises the question of the existence of any interlocking metabolic controls*. There is evidence that cholesterol fed to rats suppresses the conversion of β-hydroxy-β-methylglutarate to mevalonate in the liver (Siperstein and Fagan, 1966). Dorsey and Porter (1968) demonstrated that geranyl- and farnesyl-pyrophosphate inhibit the conversion of mevalonic acid to 5-phosphomevalonic acid in an *in vitro* liver system. The cell thus appears to have feedback mechanisms for controlling the availability of mevalonic acid, either by limiting its production or by blocking its conversion to the next intermediate on the pathway. What is the effect of such control mechanisms on the synthesis of the isopentenyl-containing components of tRNA? Does the cell have a regulatory mechanism that ensures a supply of isopentenyl units to tRNA independent of the isoprenoid biosynthesis, and are there interlocking control mechanisms? These questions have particular relevance to the biological function of molecular species of tRNA containing N^6-(Δ^2-isopentenyl)adenosine, since it appears that these molecules will not function without the isopentenyl group.

Enzymic Degradation of N^6-(Δ^2-Isopentenyl)adenosine. During the normal course of catabolism of tRNA, the components are released as mononucleotides, nucleosides, or possibly free bases. The major nucleotide components of tRNA are then metabolized to form reusable products. On the other hand, many of the modified constituents of tRNA (at least in mammals) are excreted, for they can be isolated from urine in the form of the nucleoside or free base (see Table 3, Chapter 6, for a list of the modified nucleosides found in mammalian urine). This observation suggests that many of the modified nucleoside components such as 5-ribosyluracil or the methylated nucleosides are not metabolized by cells. It is quite surprising, therefore, to find that some tissues appear to contain a series of enzymes for the metabolism of N^6-(Δ^2-isopentenyl)-adenosine, *1* (Chen et al., 1968; McLennan et al., 1968), particularly in view of the fact that this is a relatively rare (quantitatively) component of tRNA. These workers have investigated the metabolism of compound *1* in chicken bone marrow and tobacco pith tissue grown in culture. Their data indicate that this

*See Bloch (1965) for a discussion of the pathway of isoprenoid biosynthesis.

nucleoside is metabolized to give a number of products which are further metabolized to inosine, *22*, or hypoxanthine, *23*. Several products are formed in the course of the metabolism, and three of them have been identified. On the basis of these preliminary data, the following degradation scheme has been suggested (Chen et al., 1968; McLennan et al., 1968).

The data indicate that compound *1* is subjected to at least two competing enzymic pathways. In one pathway hydrolysis occurs to form the free base, and in the second the allylic double bond of the side chain is hydrated to form a product tentatively identified as N^6-(3-methyl-3-hydroxybutyl)adenosine, *21*. Compound *21* is subsequently converted to the free base, *17*. Other products have been found in the enzymic reaction mixture but have not yet been identified. It is instructive to note that the free base, *6*, as well as compounds *17* and *21*, exhibits potent cytokinin activity (Hall and Srivastava, 1968). In fact, compound *6* is 20 times as active as the parent nucleoside, *1*, in the tobacco callus tissue assay (Skoog et al., 1967).

Biological Activity of N^6-(Δ^2-Isopentenyl)adenosine. The occurrence of N^6-(Δ^2-isopentenyl)adenosine and/or its derivatives in the tRNA of all species, the fact that it is located adjacent to the 3′ end of the anticodon of some molecular species, and the presence of a series of degradative enzymes in plant and animal tissue that metabolize it, all suggest that this component possesses a degree of significance to the functioning of the cell out of proportion to the relatively small amount present. This significance is even more enhanced by the fact that N^6-(Δ^2-isopentenyl)adenosine exhibits biological activity in several systems. Although at this time no direct evidence exists for a relationship between the occurrence of this nucleoside in tRNA and its biological activity, it will be helpful to consider briefly the nature of the biological activity observed.

This activity is most readily demonstrated in plant systems, since N^6-(Δ^2-isopentenyl)adenosine is a member of a class of compounds, called cytokinins, that exhibit plant hormone activity. Compounds with cytokinin activity have attracted the attention of plant physiologists for several years, and a great deal has been learned about the nature of the hormone activity at the biological level, although little has been uncovered about the mechanism of action.

Cytokinin activity is conveniently assayed using tobacco pith tissue grown in culture, and in this system an active compound promotes cell division and cell differentiation (Linsmaier and Skoog, 1965; Murashige and Skoog, 1962). Cytokinins also evoke biological responses in a number of other systems, for example, stimulation of budding (Benes et al., 1965; Engelbrecht, 1967), the retardation of senescence of excised leaves (Osborne and McCalla, 1961; Srivastava and Ware, 1965), stimulation of the rate of germination of lettuce seeds (Robins et al., 1967; Skinner et al., 1956), and the potentiation of antibiotic activity against microorganisms (Hall and Gale, 1960). Whether the primary locus of action of the cytokinins is the same in all biological systems remains a speculative question. Each of the several known compounds that exhibit cytokinin activity does not have an identical spectrum activity in all these biological test systems. Compounds exhibiting cytokinin activity possess the common structure N^6-(substituted)adenine, and the most active members of this group are the furfuryl, benzyl, and Δ^2-isopentenyl derivatives (see review by Skoog et al., 1967).

Most of the earlier studies in this field were carried out with synthetic compounds such as N^6-furfuryladenine (kinetin) and N^6-benzyladenine. The intense nature of the biological activity of these derivatives led plant physiologists to the conclusion that the activity is an indication of normal hormonal

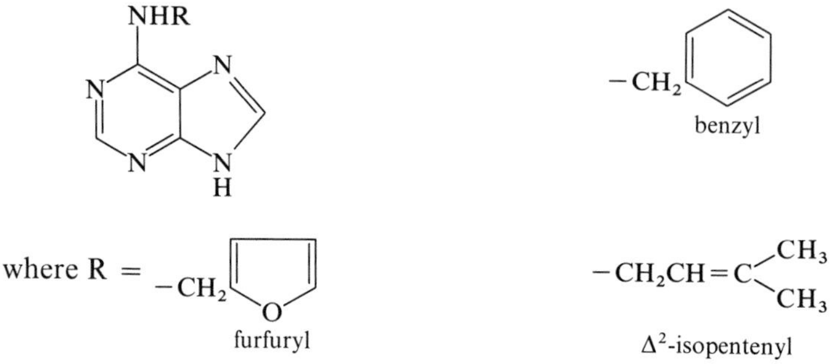

NHR

benzyl

where R =

furfuryl

Δ^2-isopentenyl

N^6-(substituted)adenine derivatives
with cytokinin activity.

activity in plants (see, for example, Mothes, 1967). Therefore they reasoned that there should be naturally occurring cytokinins in plants. Zeatin, 7, found in the immature fruit of several species, and N^6-(Δ^2-isopentenyl)adenosine, 1, and its hydroxylated derivative, 2, detected in tRNA, are the first naturally occurring purine derivatives with cytokinin activity to be identified in plant tissue. These compounds exhibit tenfold or more activity in the standard tobacco callus tissue culture assay system than N^6-furfuryladenine (Leonard et al., 1968; Skoog et al., 1967). The fact that these compounds exhibit a hormone-like activity when added exogenously to experimental systems does not imply, however, that they possess this activity in growing plants.

The mechanism of action of cytokinins in plant tissue is still obscure, and perhaps the most informative data have been obtained in metabolic studies on excised leaves. The antisenescence effect of the cytokinins appears to be related to the maintainence of protein and nucleic acid synthesis. Osborne (1965) reported that the addition of puromycin, which inhibits polypeptide formation, to a culture medium containing detached *Xanthium* leaves reverses the stimulatory effect of N^6-furfuryladenine. On the other hand, addition of actinomycin D, which inhibits DNA-dependent RNA synthesis, does not interfere with the stimulatory action of N^6-furfuryladenine. In another study on excised tobacco leaves, N^6-benzyladenine was found to prevent destruction of chlorophyll a and b; in fact, compared to controls, it maintains photosynthetic activity at a high level (Romanko et al., 1968). These workers also showed that N^6-benzyladenine causes an increase in the rate of nucleic acid and protein synthesis in chloroplasts excised from tobacco leaves.

The effect of N^6-(substituted)adenine derivatives has also been studied in several animal systems. Buckley et al. (1962) reported that N^6-furfuryladenine induces mitotic divisions throughout the digestive tract of the salamander, *Triturus viridescens*. Tetraploid divisions were also reported. This effect follows after the animal has been exposed to the compound for 9 days and is most evident in the duodenum. After the animals are withdrawn from the treatment, mitotic activity gradually decreases, but the number of tetraploid divisions increases (Kevin et al., 1966). These workers also noted that exposure to N^6-furfuryladenine causes absence of the central spindle in the treated tissue and that the spindle reappears in the tissue during the recovery period (in the absence of N^6-furfuryladenine). Grillo and Polsky (1966) also investigated the effect of N^6-furfuryladenine on *Triturus viridescens* and concluded that this derivative does not stimulate mitosis but interferes with karyokinesis, probably by prolonging the actual mitotic time. This effect manifests itself in routine cytological preparation as a higher number of cells than normal in mitosis. In other words, the effect of N^6-furfuryladenine in this system is strikingly similar morphologically to that of colchicine.

These experiments and many others have been conducted with the thought that the cytokinin effect in plants systems might be duplicated in animal systems. Almost all the work has been done with synthetic compounds, such as N^6-furfuryladenine, and the results have generally been negative. These derivatives may not be the appropriate test compounds. Animal systems may require the administration of cytokinin in another form such as the nucleoside; and, perhaps more importantly, animal systems may be more discriminating than plant systems and hence will respond only to the correct chemical structure. If there is a process in animal tissue comparable to the cytokinin process in plant tissue, it will be necessary to characterize the natural components involved in the growth and differentiation of animal cells and to work with these compounds.

The biological activity of N^6-(Δ^2-isopentenyl)adenosine in plant systems and its natural occurrence in both plant and animal cells raise the question of whether this compound has a demonstrable cytokinin activity in animal

TABLE 4

Cells Derived from Human Myelogenous Leukemia (Iwakata and Grace, 1964) Cultured in the Presence of N^6-(Δ^2-Isopentenyl)adenosine for 48 Hours. (Data taken from Grace et al., 1967)

Concentration, μg/ml	Total Cell Count*	Viability, %
0.1	3.9×10^5	70
1	2.3×10^5	40
2	2.9×10^5	16
3	2.5×10^5	10
4	2.3×10^5	10
Control (medium above)	3.2×10^5	70

*Initial input—3.4×10^5, 70% viable.

systems. Some preliminary probing of the biological activity of N^6-(Δ^2-isopentenyl)adenosine in animal cells has been carried out. Grace et al. (1967) reported that this nucleoside is a potent inhibitor of a line of cells derived from human myelogenous leukemia (leukemic myeloblasts) grown in culture (see Table 4). However, there is no significant inhibition in cultured human lymphoblastic leukemic cells (LKID) or Burkitt lymphoma (P-3HR-1) cells at comparable concentrations. This nucleoside is also a potent inhibitor of sarcoma-180 cells. The free base, 6, does not inhibit the human leukemic myeloblast at concentrations up to 50 μg/ml, and this result emphasizes the need to apply the test compound in a suitable form.

The inhibitory activity of N^6-(Δ^2-isopentenyl)adenosine on cultured mammalian cells may be common to other N^6-(substituted)adenosine derivatives. Hampton et al. (1956) reported that N^6-furfuryladenosine inhibits a strain of adult human fibroblasts grown in culture. Under similar conditions this compound has no effect on the strains of cultured cells, HeLa, sarcoma-180, or most fibroblast cells. These data are difficult to interpret. Many adenosine derivatives are toxic to mammalian cells, plant cells, and bacteria (see, for example, a discussion of the nucleoside antibiotics in Fox et al., 1966). It remains to be seen whether the mechanism of action of N^6-(Δ^2-isopentenyl)-adenosine and its analogs in animal cells is related in any way to the mechanism of action in plant cells.

The data obtained in the plant systems prove neither that the cytokinin effect represents a natural process in plants nor that N^6-(Δ^2-isopentenyl)adenosine

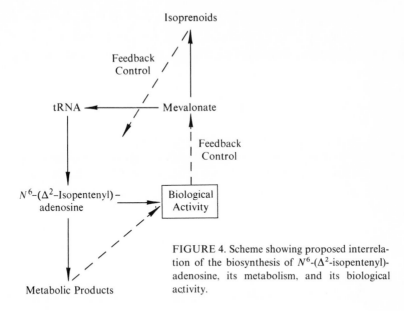

FIGURE 4. Scheme showing proposed interrelation of the biosynthesis of N^6-(Δ^2-isopentenyl)-adenosine, its metabolism, and its biological activity.

or a derivative is a natural regulatory hormone. Nevertheless, the results are suggestive, and if N^6-(Δ^2-isopentenyl)adenosine is a regulatory hormone, its biosynthesis and metabolism in the cell take on additional significance. The relevant observations are correlated in Figure 4. The central concept is that because of the physiological activity of N^6-(Δ^2-isopentenyl)adenosine, its cellular level is critical and can be governed by a balance between the rate of release from tRNA and the rate of degradation by the metabolic enzyme system. A corollary concept to this scheme is that the regulation involves mevalonic acid, which suggests the existence of an overall interlocking control mechanism involving isoprenoids, tRNA, and N^6-(Δ^2-isopentenyl)adenosine.

3. *N*-(NEBULARIN-6-YLCARBAMOYL)-L-THREONINE

Introduction. A number of articles report the presence of amino acids or small polypeptides bound to nucleic acids, apart from the amino acids attached to the acceptor end of tRNA molecules. Ingram and Sullivan (1962) and Akashi et al. (1965), for example, have reported the presence of amino acids bound to RNA which cannot be removed through the use of intensive deproteinizing procedures. Balis et al. (1964) and Olenick and Hahn (1964) have reported the presence of amino acids in highly purified preparations of DNA isolated from a variety of sources. The nature of the amino acid-nucleic acid linkage, however, has not been elucidated. Bogdanov et al. (1962) have found that ribosomal RNA contains amino acids attached to the phosphate residues. Harris and Wiseman (1962) have also reported the presence of small polypeptides attached to the phosphate residue of yeast nucleic acid; the exact nature of such complexes has not been described.

The first amino acid-containing nucleoside, apart from the terminal adenosine of tRNA, that has been definitively characterized is N-(nebularin-6-ylcarbamoyl)-L-threonine,* 24 (Schweizer et al., 1968, 1969). The compound has been detected in the tRNA of E. coli, yeast, and mammalian tissue (p. 109); recently it has been identified in a specific sequence of yeast tRNA^{IIe} adjacent to the anticodon (see p. 260). The level of compound 24 in yeast tRNA is 0.28 mole %, and if it is assumed that only one such residue occurs in a tRNA molecule, about one out of five yeast RNA molecules will contain this compound.

Synthesis. Both the nucleoside, 24, and its free base, N-(purin-6-ylcarbamoyl)-L-threonine, 28, have been synthesized (Chheda, 1969). In one route the isocyanato derivative of threonine, 26 (Mizoguchi et al., 1968), is condensed with 9-(β-D-2',3',5'-tri-O-acetylribofuranosyl)adenosine to yield (25%) the intermediate, 27. Treatment of 27 with hydrogen bromide in trifluoroacetic acid cleaves both the benzyl groups and the sugar moiety to give the free base, 28, in 40% yield. Treatment of 27 with methanolic ammonia removes the acetyl groups to give 29 in good yield. Removal of the benzyl groups of 29 by the use of sodium in liquid ammonia produces a 5% yield of the desired nucleoside, 27. The lability of the ureido side chain to alkali and the lability of the glycosyl bond to acid make the synthesis of nucleoside 27 somewhat difficult.

In an alternative route investigated by Chheda (1969) the azide derivative, 30 (Giner-Sorolla and Bendich, 1958), is converted into the urethane derivative of adenine, 31. This compound condenses smoothly with L-threonine to give a 25% yield of N-(purin-6-ylcarbamoyl)-L-threonine, 28. The chloromercuri derivative of 31 can be prepared, and this intermediate condenses with 2',3',5'-tri-O-acetyl-1-chlororibose to give the blocked intermediate, 32. This intermediate condenses with L-threonine to give a product from which the acetyl groups are removed by treatment with methanolic ammonia to produce the desired nucleoside, 24. The yield of 24 from 32 is 10%.

---

*The analytical data suggest that threonine is in the L form.

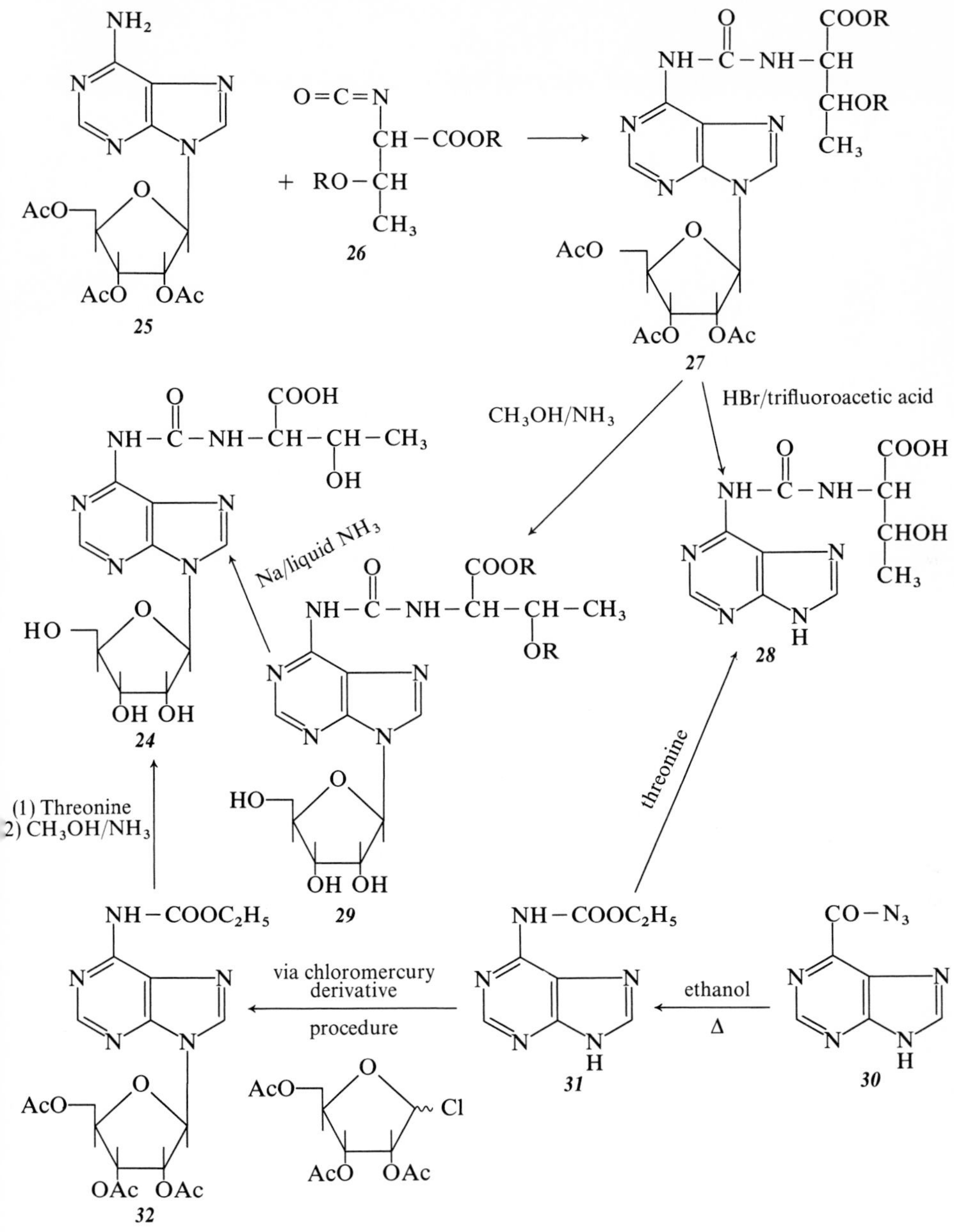

where Ac = acetyl

R = benzyl,

Chemical Reactions. Treatment of *N*-(nebularin-6-ylcarbamoyl)-L-threonine, *24*, in 0.2 *M* ammonium hydroxide for 2 hr at 100° gives adenosine, *12*, and the internal urethane of threonine, *33*. Compound *33*, on treatment with 0.2 *M* sodium hydroxide for 2 hr at 100°, affords free threonine. Compound *24*, on treatment with 1 *M* hydrochloric acid for 15 min at 100°, yields the free base, *N*-(purin-6-ylcarbamoyl)-L-threonine, *28*.

.In light of the results leading to the identification of compound *24* as *N*-(nebularin-6-ylcarbamoyl)-L-threonine, our original results obtained with "aminoacyl nucleosides" isolated from tRNA were reexamined. We had originally reported that a nucleoside obtained from yeast tRNA was an N^6-(α-aminoacyl)adenosine derivative (Hall, 1964a; Hall and Chheda, 1965). The original sample of the nucleoside used in the work described in these two publications is no longer available. We isolated another sample of this nucleoside, following the procedure of Hall (1965). This sample of "N^6-(α-aminoacyl)-adenosine" was hydrolyzed with $1 N$ hydrochloric acid under conditions sufficient to remove the sugar moiety; a product identical to *N*-(purin-6-ylcarbamoyl)-L-threonine was obtained. Analysis of the amino acid content of this nucleoside by quantitative ion exchange chromatography showed that the nucleoside contained only one molecular equivalent of a single amino acid, threonine. This nucleoside gives adenosine on alkaline hydrolysis; therefore, it has the structure *24*.

The original isolated sample (Hall, 1964a), on acid hydrolysis in $6 N$ hydrochloric acid, yielded several amino acids, of which glycine, threonine, and valine predominate. On the basis of the ultraviolet absorption spectral data and chromatographic properties, it is certain that the original sample possessed the general structure *N*-(nebularin-6-ylcarbamoyl)amino acid. It is not certain, however, whether the original isolated sample was a single amino acid analog of compound *24* or a mixture. The other amino acids obtained on acid hydrolysis may have been contaminants or were possibly attached to the isolated sample in some way. Thus, at the present time, we cannot exclude the possibility that other amino acid-adenosine analogs of compound *24* occur in tRNA.

N-(Nebularin-6-ylcarbamoyl)-L-threonine, *24*, actually may exist in tRNA in an esterified form, since Chheda et al. (1969a) report that a small percentage of the isolated sample of *24* lacked an acidic function. Precedence for the occurrence of esterified carboxyl groups in tRNA already exists. Baczynskyj et al. (1968) report the isolation of 2-thio-5-carboxymethyluridine from tRNA as the methyl ester. Gray and Lane (1968) identified 5-carboxymethyluridine in the tRNA of yeast and wheat germ, and from their data it appears that a significant percentage of this compound exists in the tRNA in an esterified form.

344

where R = H or CH$_3$.

4. NOVEL REARRANGEMENT OF N^6-(α-AMINOACYL)ADENOSINE

At the time of the original observation concerning the presence of an aminoacyl-containing nucleoside in tRNA, we started an investigation of N^6-(α-amino-acyl)adenosine derivatives with the objective of learning more about their basic chemical properties. Although such compounds do not represent the exact structure now assigned to compound *24*, any information bearing on the properties of *24* will be helpful in assessing its function in tRNA. At present little is known about the chemical properties of compound *24*; it is conceivable, for example, that in nature it could exist as a hydantoin, *34*, which in turn could undergo ring opening to form an N^6-threonyl derivative of adenosine, *35*.

N^6-Threonyladenine and all other N^6-(α-aminoacyl)adenine derivatives are very unstable in aqueous solution and rapidly undergo a rearrangement to N-(purin-6-yl)amino acids. The nature of this rearrangement was first described by Chheda and Hall (1966), who studied the reaction of a model compound, N^6-glycyladenine.

When a neutral aqueous solution of N-glycyladenine, *36*, is kept at room temperature for several hours or warmed for a few minutes at 100°, it becomes dark purple (the purple color is due to the formation of traces of highly colored impurities), and the compound spontaneously loses the elements of ammonia and undergoes cyclization to form 3-methyl-3-H-imidazo[2,1-i]purine-8-(7H)-one, *37*. Compound *37* in turn is unstable and rearranges to N-(purin-6-yl)-glycine, *38*. The mechanism of this rearrangement was established through a study using [α-^{15}N]-N^6-glycyladenine, *39*, and [N^6-^{15}N]-N^6-sarcosyladenine, *40*. These [^{15}N]-labeled derivatives were allowed to undergo the rearrangement with the following results (Chheda et al., 1969b):

$\overset{*}{N}H-CO-CH_2-NH-CH_3$

40 ·3HBr $\longrightarrow$

$CH_3 \diagdown N \diagup CH_2-COOH$

43

$*N = {}^{15}N$

Compound *39* in aqueous solution affords the cyclic intermediate, *41*, which in turn undergoes ring opening and rearranges to compound *42*. Compound *42* fully retains the atom excess of ${}^{15}N$. These data demonstrate that the α-amino group of the amino acid is retained and hence that the expelled nitrogen must originate from the purine. The sarcosyl derivative, *40*, also spontaneously undergoes conversion to N-methyl-N-(purin-6-yl)glycine, *43*, which fully retains its atom excess of [${}^{15}N$]. Assuming that the rearrangement pathway for the N^6-sarcosyl derivative is the same as that of N^6-glycyladenine, this result shows that the nitrogen atom expelled during the conversion is the N-1 and confirms that the α-amino group of the starting compound becomes the N-6 of the rearrangement product.

The results of these experiments suggest that the mechanism of the rearrangement of N^6-(α-aminoacyl)adenine to N-(purin-6-yl)amino acids is an addition-elimination reaction following the course proposed in the scheme on p. 344.

Chheda and Hall (1969) demonstrated that this rearrangement occurs equally as well when glycine is replaced with other amino acids. Furthermore, it occurs spontaneously when the 9 position of adenine is substituted with a methyl group. These data infer, therefore, that the corresponding N^6-(α-aminoacyl)-adenosine derivatives readily undergo this series of reactions.

Compounds in which the α-amino group of the N^6-(aminoacyl)adenine is protected with an acyl group such as formyl or two alkyl groups [i.e., N^6-(α,α-dimethylglycyl)adenine] are stable. There is no information on the metabolism that N-(nebularin-6-ylcarbamoyl)-L-threonine undergoes in the tRNA molecule, but these data show that if this compound should ever be converted to the N^6-(threonyl) derivative the rearrangement outlined would occur spontaneously in the tRNA molecule.

Chapter 8

CHEMICAL REACTIONS THAT CAN BE
APPLIED SELECTIVELY TO NUCLEIC ACIDS

1. INTRODUCTION

THE ABILITY TO CARRY OUT a specific chemical reaction at a specific location in a nucleic acid molecule becomes highly desirable for many experimental procedures in nucleic acid research. Such reactions can serve as valuable tools, for example, in sequence analysis, investigation of structure-function relationships, and study of physical properties. Chemical reactions can be used to alter the structure of a specific nucleotide in the nucleic acid molecule, and such an alteration may serve as a useful "handle" for subsequent chemical or enzymic studies. Some investigators make effective use of this approach to change the response of nucleic acids to specific nucleases. Other chemical reactions are being developed that could cleave a nucleic acid molecule at a specific point in the primary sequence either directly or in conjunction with an enzyme. This type of nucleic acid technology is still at a rudimentary level, but sufficient development has taken place to indicate its potential value to investigators.

Fundamental to this development is a knowledge of the chemical reactions that individual nucleic components undergo. A vast number of chemical reactions involving the major and modified components of nucleic acids have been described, but few of them were developed with the express intention of applying them to nucleic acids at the macromolecular level. Moreover, many of the reactions are not specific to a single nucleic acid component or are not applicable under mild conditions.

The reaction conditions that do not cause damage to the nucleic acid molecule are fairly restrictive; for example, the reaction should be carried out in aqueous solution for relatively short periods (minutes or, at most, a few hours), at a pH value near neutrality and at temperatures not elevated much above room temperature. These conditions may be stretched, but the risk of side reactions and/or damage to the secondary and tertiary structure of the nucleic acid correspondingly rises. The side reactions may be tolerated, providing they are understood and controlled. Nevertheless, chemists have effectively used a variety of reactions to chemically alter nucleic acid molecules, and in a number of cases the reactions are so specific that a single nucleotide residue in a tRNA molecule can be selectively altered.

The procedures described in this chapter are designed to work in aqueous solution, and the above criteria for mild conditions apply. It is possible that under carefully controlled conditions reactions could be carried out effectively in nonaqueous solutions, and in this event the scope of chemical reactions that could be applied directly to the nucleic acids would be enormously extended.

The secondary and tertiary structures of a nucleic acid molecule can exert a dominant influence on the course of a chemical reaction. This influence sometimes manifests itself in differences in reaction rates between double- and single-stranded DNA or between helical and nonhelical segments of tRNA. A reagent specific for uridine, for example, when reacting with tRNA may attack only uridine residues in the nonhelical regions. In practice, the influence of the secondary structure can be used to enhance the specificity of a reaction. Conversely, some workers have made use of these properties to study the secondary structure of tRNA. There are unknown factors, however, that need to be considered in assessing the influence of secondary structure on the rate and extent of a particular reaction; principally, the attachment of a first molecule of a reagent may disturb the secondary and/or tertiary structure and hence facilitate the entry of the second molecule of reagent.

The format of this chapter presents each chemical reagent under a separate heading. The specificity of the reagents for particular nucleoside components varies from little or none to a specificity for a rare modified component of the nucleic acid. The chemical procedures may be classified in three categories:

1. Reactions specific for one of the major nucleic acid components.

2. Reactions specific for major nucleic acid components that can be combined with an enzymic reaction to sharpen the effective specificity of the enzyme with respect to sites in the primary sequence.

3. Reactions that are exclusive for one of the modified nucleosides.

The selectivity of the reactions discussed in this chapter is summarized in

Table 5, which appears at the end of the chapter. The reactions in category 3 offer perhaps the greatest degree of nucleoside specificity. Most reactions at the present state of the art fall into category 1 and might seem to be of the least value. However, they provide useful information about the behavior of nucleic acid molecules under the particular reaction conditions, and this information may well serve as the basis for the development of more specific reactions.

All the reactions demonstrate clearly that it is possible to manipulate chemically the nucleic acid molecule, and in the case of tRNA, to do this without seriously disturbing the three-dimensional structure. This capability, together with the availability of a variety of specific enzymes, provides the experimenter with an arsenal of techniques for investigation of nucleic acid structure and function.

2. HYDROXYLAMINE

The pyrimidine bases of nucleic acids react with hydroxylamine in aqueous solution (Brown and Schell, 1961; Freese et al., 1961; Kochetkov et al., 1962a,b; Schuster, 1961; and Verwoerd et al., 1961). The rate of reaction depends on the concentration of hydroxylamine and on the pH. The optimum pH is 6.0 for the reaction with cytidine derivatives and 10 for the reaction with uridine derivatives (see Figure 1). Hydroxylamine, although not basic (pK$_a$ 5.96, Bissot et al., 1957),

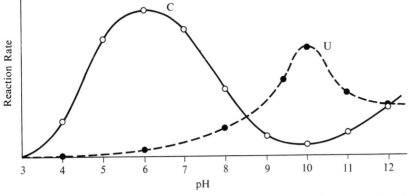

FIGURE 1. pH dependence of the rate of reaction of 6 M aqueous hydroxylamine with cytidine2'(3')-cyclic phosphate (C), and uridine2'(3')-cyclic phosphate (U). The reaction rate is measured as a percentage of the maximum extinction coefficient after a 10 min reaction. Data taken from Verwoerd et al. (1961).

is a powerful nucleophilic agent. The reaction with uridine, *1*, proceeds according to the following scheme (Kochetkov et al., 1963c, 1967c):

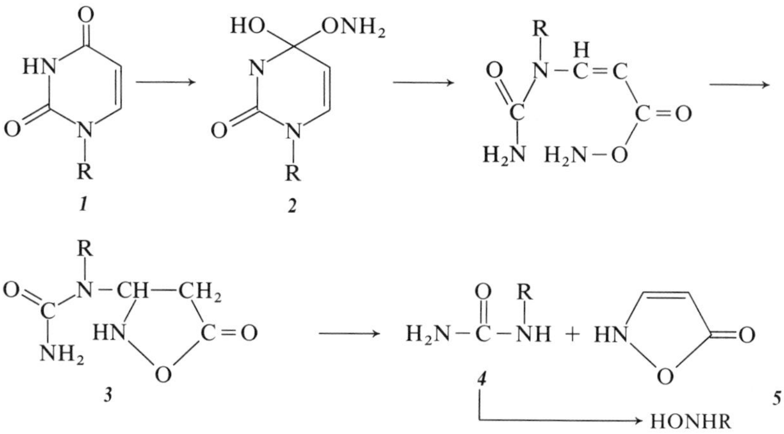

where R = ribose.

Hydroxylamine adds to the α,β-unsaturated carbonyl system, *1→2*, a rearrangement occurs, and a new ring system, 3,4-dihydro-3-ureidoisoxazol-5-one, *3*, forms. Compound *3* eliminates the ribosylureido moiety, *4*. Compound *4*, the primary isolatable product of the reaction, is converted at a slower rate to the ribosyloxime derivative, *5*.

Under optimum conditions (10 M hydroxylamine solution, 10°) it is possible to convert 99% of the uridine residues of RNA to ribosyloxime, *5*, units while only 5% of cytidine residues are changed. The rates of breakdown of uridine residues and the formation of products *4* and *5* on treatment of RNA with hydroxylamine are shown in Figure 2. During the formation of the "deuridylic acid," some adjacent internucleotide bonds split (Budowsky et al., 1968c). The splitting is specific for the reacted uridine residues and offers the possibility of specific cleavage of the polynucleotide chains at these locations.

The nature of the secondary structure of RNA significantly affects the rate of hydroxylaminolysis of the uracil residues. Kochetkov et al. (1967b) showed that the rate of reaction with poly U : poly A is considerably less than that with poly U and therefore suggested that in tRNA the uracil residues in the non-helical regions would react first. The rate of modification of uracil residues in tRNA is affected by the presence of three to five Fe^{+++} ions per 100 nucleotides

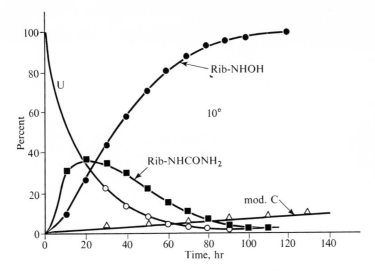

FIGURE 2. Calculation of the amount of uridine, ribosyl urea, ribosyl oxime, and modified cytidine residues in the course of hydroxylamine lysis of RNA. (First three values are in percentages with respect to initial amount of cytidine.) Twenty milligrams/ml RNA, 10 M NH$_2$OH, pH 10. Data taken from Kochetkov et al. (1967c).

(Budowsky et al., 1968b; Turchinsky et al., 1967).

Digestion of the hydroxylamine-treated RNA with pancreatic RNAse results in hydrolysis of the RNA only at the cytidylic acid residues (Kochetkov et.al., 1963c; Verwoerd et al., 1961). In contrast, when the cytosine residues are modified by hydroxylamine, RNAse action on the cytidylic acid internucleotide bonds is not inhibited (Kochetkov et al., 1964).

The reaction of hydroxylamine with cytosine and its derivatives occurs preferentially at pH 6–6.5 (near the pK$_a$ of the reagent). In the initial reaction, two molecules of hydroxylamine react with a cytosine residue, 6, to form a 4,6-dihydroxylamino-2-oxo-5,6-dihydropyrimidine of type 9. Model compounds of type 9 have been prepared in which R = H or deoxyribose (Brown and Schell, 1965; Budowsky et al., 1968b). These compounds are readily oxidized by air and with ferric chloride to give highly colored products; this reaction, incidentally, can be used as a quantitative analytical tool. Compound 9 (R = deoxyribose) undergoes acid-catalyzed elimination of hydroxylamine to give 8 on treatment with 0.1 N hydrochloric acid for 8 min at 100°. The elimination also occurs when the derivative is heated above its melting point.

CHEMICAL REACTIONS APPLIED SELECTIVELY TO NUCLEIC ACIDS

Scheme for the reaction of hydroxylamine with the cytosine moiety of cytosine nucleosides (Brown and Phillips, 1965; Budowsky et al., 1968d; Lawley, 1967):

where R = H, deoxyribose, or ribose.

Brown and Phillips (1965) proposed two routes to compound *9* via intermediate *7* or *8*. They favor intermediate *7* as the principal initial reaction product. In this process, the first molecule of hydroxylamine adds across the 5,6 double bond of cytosine, followed by a rapid exchange at C-4 with the highly nucleophilic reagent. In a detailed examination of the kinetics of the reaction at high concentrations of hydroxylamine at pH 6.5 and 37°, Lawley (1967) found that both intermediates *7* and *8* are produced in a ratio of about 3:1. For 5-(alkyl-substituted)cytosine derivatives, the reaction is much slower and the principal product is the N^4-hydroxy derivative of type *8* (Janion and Shugar, 1965). Treatment of yeast tRNA with hydroxylamine at pH 9.0 or 6.0 under conditions in which 35% of the uracil residues or 23% of the cytosine residues have reacted causes a substantial loss of acceptance capacity for fifteen different amino acids (Cerna et al., 1964). The degree of loss varies considerably among the amino acids.

Hydroxylamine reacts quantitatively with the photohydration product of cytidine, 5,6-dihydro-6-hydroxycytidine, *10*, to form 5,6-dihydro-N^4,6-dihydroxycytidine, *11* (Small and Gordon, 1968). The reaction proceeds in 0.25 M hydroxylamine solution (pH 6.5) at 0° and is essentially complete in 3 hr. Under these conditions cytidine reacts to the extent of 2 % and uridine reacts very little. The photohydration product, *10*, is unstable so that the formation of the stable derivative, *11*, represents a method for quantitating the production of *10* in RNA irradiated with ultraviolet light.

10   *11*

where R = ribose.

Hydroxylamine is highly mutagenic; this effect is believed to be due to modification of the cytosine or 5-(substituted)cytosine residues (Freese et al., 1961; Schuster and Vielmetter, 1961).* Several theories on the mechanism of this process have been advanced. One theory states that hydroxylamine-treated cytosine residues are replicated as if they were uracil residues. Support for this theory stems from studies with poly C. Treatment of poly C with hydroxylamine reduces its capacity to act as a template for poly G synthesis by RNA-dependent RNA polymerase. This capacity is partially restored if ATP (but not CTP or UTP) is added to the reaction mixture. The adenylate residues are incorporated into the synthesized poly G, primarily as single residues flanked on either side by guanylate residues (Phillips et al., 1965; Wilson and Caicuts, 1966).

Many of the studies on mutagenesis have been carried out with T4 bacterio-phage, which contains 5-hydroxymethylcytosine in place of cytosine; therefore the principal reaction product with hydroxylamine would be the N^4-hydroxy derivative. A plausible mechanism of mutagenesis can be deduced, based on the mechanism proposed for the mutagenic effect of alkyl groups attached to the 7 position of guanine residues in DNA. Compounds with structure *8* have an acid dissociation constant approximately 100 times greater than that of an unsubstituted cytosine residue (Fox et al., 1959). Similarly, alkylation of deoxyguanosine residues at N-7 causes a lowering of the pK_a value (Lawley

---

* Phillips and Brown (1967) have reviewed the mutagenic acid of hydroxylamine.

and Brookes, 1962). Nagata et al. (1963) proposed that tautomerization by proton tunneling would be more likely within a 7-alkylguanine-cytosine pair than within the normal guanine-cytosine pair, and this phenomenon could increase the probability of errors in replication. According to this theory, ionization of the acidic hydrogen atom does not occur. Instead, this atom is transferred from its initial site of covalent bonding in the normal base pair to the alternative site of the same hydrogen bond; simultaneous exchange of the proton of the second hydrogen bond to the abnormal position then occurs. Lawley (1967) reasons that a similar situation could arise from an N^4-hydroxycytosine residue, and anomalous pairing such as cytosine-adenosine could result. Most hydroxylamine-induced mutations in T4 bacteriophage are GC to AT transitions (Champe and Benzer, 1962; Freese et al., 1961).

Although hydroxylamine and its derivatives react directly with the cytosine residues under the above conditions, at low concentrations of hydroxylamine ($10^{-1} M$) an indirect reaction predominates (Freese and Bautz-Freese, 1965). Hydroxylamine reacts with oxygen to form hydrogen peroxide and nitroxyl (HNO):

$$NH_2OH + O_2 \longrightarrow HNO + H_2O_2$$

The formed hydrogen peroxide reacts with more hydroxylamine, giving rise to different products such as N_2O and N_2. The chemistry that takes place between this series of products and the nucleic acid molecule remains unknown; the only evidence that a reaction occurs is derived from a biological assay consisting of measurement of the rate of inactivation of transforming DNA.

3. O-METHYLHYDROXYLAMINE

Since the reaction of hydroxylamine lacks absolute specificity for either uracil or cytosine derivatives, Kochetkov et al. (1963a) and Budowsky et al. (1965) examined the action of the O-methyl ether of hydroxylamine on nucleic acids. They reasoned that, if hydroxylamine attacks uridine according to the scheme on p. 350, an alkyl substituent on the hydroxyl group of hydroxylamine should prevent the reaction with uridine. According to their results, uridine does not react at all with O-methylhydroxylamine over the pH range 4.0–10.0. Cytidine, 6, does not react at pH 10.0; but at pH 4.0–6.0 it reacts to form the products 6-methoxyamino-5,6-dihydrocytidine, 12, 6-methoxyamino-5,6-dihydrouridine-O-methyloxime, 13, and uridine-O-methyloxime, 14. Compound 13 is converted readily into 14 at pH values below 4.0 or above 8.0.

where R = ribose.

The effect of this reaction at the polymeric level was studied, using poly-cytidylic acid as a substrate. When this polymer is treated with O-methyl-hydroxylamine for 72 hr at pH 6.0 and 37°, practically all the cytosine residues become substituted (Kochetkov et al., 1963b). The modified polymer is readily digested by the action of pancreatic RNAse. Phillips et al. (1966) used a partially modified poly C containing the 5,6-dihydro-6-methoxyaminocytosine, *12*, residues as a template for the RNA-dependent RNA polymerase system and found that adenylate residues are incorporated into the synthesized poly G.

4. HYDROXAMIC ACIDS

The N-acyl derivatives of hydroxylamine, acethydroxamino acid, and propion-hydroxamino acid, according to studies of Kochetkov et al. (1964), react in aqueous solution at pH 8–10 with nucleosides having an enolizable keto group such as uridine, guanosine, 5-ribosyluracil, inosine, and 4-thiouridine. There is some doubt as to whether this reaction actually proceeds as suggested. In a later publication, Kochetkov et al. (1967d) showed that at the alkaline pH of the reaction hydroxamic acid decomposes to give hydroxylamine and that, at least in the case of uridine, the reaction proceeds according to the scheme shown on p. 350.

5. SEMICARBAZIDE

Incubation of cytidine, 6, in $2 M$ semicarbazide solution (pH 4.2) for 20 hr at 37° affords 4-deamino-4-semicarbazidocytidine, 15 (Hayatsu et al., 1966):

where R = ribosyl.

No reaction with adenosine, guanosine, or uridine occurs under these conditions. Hayatsu and Ukita (1966) extended the reaction to yeast tRNA and found that semicarbazide reacts exclusively with the cytidine residues, although the reaction proceeds more slowly than with cytidine alone. Incubation of yeast tRNA in $3 M$ semicarbazide (pH 7) for 94 hr at 37° results in substitution of all the cytidine residues. Some of the substituted cytidine residues apparently have a tendency to degrade to uridine, since the uridine content of the treated tRNA rises about 25% and the number of substituted cytidine residues correspondingly decreases to about 25% less than the original cytidine content. The authors found that on hydrolysis of the treated tRNA in $0.3 M$ potassium hydroxide for 18 hr at 37° all the substituted cytidylic acid residues are converted to uridylic acid. They noted, however, that the treated tRNA can be hydrolyzed completely to its constituent mononucleotides by means of snake venom diesterase without degradation of any of the substituted cytidylic acid.

Although one might expect some nonspecific cleavage of the internucleotide bonds of tRNA to occur during the lengthy semicarbazide treatment, Hayatsu and Ukita (1966) did not detect any such cleavage. Some loss in secondary structure occurs, however, as indicated by a decrease in hyperchromicity of the treated sample, as well as by the fact that the treated tRNA has an ill-defined melting-out curve (Hayatsu and Ukita, 1966). The treated tRNA also loses amino acid-accepting ability (Muto et al., 1965).

4-Deamino-4-semicarbazidocytidine $2':3'$-cyclic phosphate is hydrolyzed by pancreatic ribonuclease at about one-half the rate of cytidine $2':3'$-cyclic phosphate (Hayatsu et al., 1966).

6. ACYL HYDRAZIDES

Acetyl hydrazide pyridinium chloride (Girard-P reagent), *17*, in aqueous solution (pH 4.2) at a concentration of 2 *M* reacts exclusively with the cytidine residues of tRNA (Kikugawa et al., 1967a,b). The modified tRNA does not appear to suffer any degradation, on the basis of observations that the Svedberg sedimentation coefficient remains at 4S and the product chromatographs homogeneously on a DEAE-cellulose column. The reaction was originally investigated using 1-methylcytosine, *16*, as a model, and the product obtained was identified as 1-methyl-4-deamino-4-acetohydrazidopyridinium cytosine, *18*.

The structure of *18* was confirmed by synthesis from 1-methyl-4-thiouracil and the Girard-P reagent. The reaction between acetyl hydrazide pyridinium chloride and cytidine occurs at the polymeric level in tRNA, and 4-deamino-4-acetohydrazidopyridinium cytidylic acid can be isolated from an enzymic digest of the treated tRNA.

Treatment of tRNA with the Girard-P reagent causes a loss of amino acid-accepting ability, and analysis of the kinetics of the rate of inactivation indicates that inactivation is a one-hit event. Kikugawa et al. also found that the internucleotide bonds adjacent to the modified cytidyl residues are resistant to the action of pancreatic ribonuclease; hence this treatment, coupled with a subsequent specific enzymic degradation, might be useful in structural studies on RNA.

Gal-Or et al. (1967) investigated the reaction of a series of acyl hydrazides with nucleic acids, and their results confirm the conclusion of Kikugawa et al. (1967a,b) that the reaction occurs exclusively with the cytosine residues. They did not fully characterize the reaction products but concluded that for the malonyl-3,4-dicarboxybenzoyl, 3,5-disulfonic benzoyl, and acetyl hydrazides the products result from an addition reaction and are 4-(substituted)-cytosine derivatives. The products resulting from the reaction of these diacidic hydrazides could have a free carboxyl group and therefore would offer the experimentalist an excellent opportunity for devising reactions for additional modification of the nucleic acid structure.

7. CARBODIIMIDE REAGENT

The water-soluble carbodiimide, N-cyclohexyl, N^1-β-(4-methylmorpholinium)-ethyl carbodiimide p-toluene sulfonate, 20, reacts with uridylic acid, 19, 5-ribosyluracil phosphate, guanylic acid, thymidylic acid, and deoxyguanylic acid in water at pH 8 to form derivatives of the type illustrated by the uridylic acid reaction product, 21 (Gilham, 1962; Ho and Gilham, 1967). The reaction with uridylic acid at 30° is complete in 4 hr; the reaction with guanylic acid requires 10 hr for completion. Adenylic acid and cytidylic acid do not react under these conditions.

The adduct formed with 5-ribosyluracil phosphate is much more stable than that formed with guanylic or uridylic acid. Treatment of the adducts with ammonium hydroxide (pH 10–11) at room temperature removes the carbodi-imide residues from guanylic acid and uridylic acid but not from 5-ribosyluracil phosphate (Ho and Gilham, 1967; Naylor et al., 1965). The 5-ribosyluracil phosphate adduct hydrolyzes only in the presence of concentrated ammonium hydroxide. The procedure of treatment with the carbodiimide reagent followed by treatment with dilute ammonium hydroxide, therefore, results in a specific blocking of ribonucleic acid at the 5-ribosyluracil residues.

The internucleotide linkage adjacent to carbodiimide-substituted uridine and 5-ribosyluracil residues is resistant to the action of pancreatic RNAse and of the two exonucleases, spleen and snake venom diesterases (Gilham, 1962; Naylor et al., 1965). The normal hydrolytic progress of the exonuclease halts at the particular internucleotide bond on either side of the modified base. This effect is illustrated by studies on the hydrolysis of a series of dinucleoside phosphates by Naylor et al. (1965), who obtained the data presented in Table 1.

The selectivity of the carbodiimide reaction and the change in susceptibility of the modified nucleic acid residues to enzymic attack present an opportunity for the development of techniques useful in studies on oligonucleotides, as Lee et al. (1965) demonstrated in a study using ribosomal RNA. They treated wheat germ rRNA in 7 M urea solution with the reagent and then with pancreatic RNAse. All of the trinucleotides isolated from the digest were terminated at the 3' end by cytidine. Each of the sixteen possible triplets was obtained. The presence of the 7 M urea presumably diminishes any effect that secondary structure might have on the completeness of this reaction. Ho and Gilham (1967) carried out an identical procedure on tRNA. They did not detect any trinucleotides terminated with the 5-ribosyluracil 3'-phosphate group in the RNAse digest; furthermore, they found a marked reduction in the number of trinucleotides terminated with uridine 3'-phosphate. Comcomitantly, a new nucleotide, UpC, appeared in the digest.

These data indicate that a large percentage (but not all) of the uridine residues in tRNA are blocked under the reaction conditions. Thus the secondary and tertiary structures appear to play some role in the extent of the reaction. It should also be kept in mind that addition of the reagent causes some loss of secondary structure, as demonstrated by the fact that the treated tRNA loses hyperchromicity and the ability to accept amino acids (Girshovich et al., 1966; Knorre et al., 1966).

CHEMICAL REACTIONS APPLIED SELECTIVELY TO NUCLEIC ACIDS

TABLE 1

Percentage Hydrolysis of Dinucleoside Phosphates[a]. (Data from Naylor et al., 1965)

Dinucleoside Phosphate	Snake Venom Diesterase	Spleen Diesterase
UpA[b]	100 (U, pA)	94 (Up, A)
U̅pA	25 (U̅, pA)	47 (U̅p, A)
ψpA	88 (ψ, pA)	69 (ψp, A)
ψ̅pA	45 (ψ̅, pA)	35 (ψ̅p, A)
CpU	93 (C, pU)	61 (Cp, U̅)
CpU̅	0	0
Cpψ	53 (C, pψ)	23 (Cp, ψ)
Cpψ̅	0	0
UpC	94 (U, pC)	100 (U̅p, C)
U̅pC	40 (U̅, pC)	80 (U̅p, C)

[a] Ten absorbancy units (260 mμ) of each dinucleoside phosphate was incubated with snake venom phosphodiesterase (0.025 mg) for 11 hr at 37° in 0.02 M Tris buffer (1.2 ml) at pH 8.0, or with spleen phosphodiesterase (0.15 unit) for 2 hr at 37° in 0.02 M ammonium acetate (1.2 ml) at pH 6.0. The products of the enzyme digests are included in parentheses. In the spleen diesterase digests, adenosine is usually obtained as inosine because of presence of deaminase.

[b] UpA = uridine-(3′→5′)-adenosine; U̅pA = UpA with a blocking group on the uridine moiety.

Ivanova et al. (1967) have also applied the carbodiimide reaction to a study of tRNA. They treated tRNA with the reagent at pH 8.0 for 12 hr at 40°, and in the subsequent workup of the digest could detect no oligonucleotides ending in uridine 3′-phosphate. These results suggest that all the uridine residues in the tRNA had been substituted under their conditions.

Augusti-Tocco and Brown (1965) found that the carbodiimide reagent reacts only with non-hydrogen-bonded bases. This fact may explain why complete substitution of the uridylic acid residues appears to have occurred in the experiment of Ivanova et al. (1967) with ribosomal RNA, whereas in Ho and Gilham's (1967) experiments with tRNA unsubstituted uridylic acid residues remained. These results do not coincide exactly with those of Ivanova et al. (1967), but under the conditions used in the workup of the reaction the latter workers could have missed detecting a small percentage of oligonucleotides ending in uridine 3′-phosphate.

In this experimental protocol, before the RNAse digest can be resolved by means of ion-exchange chromatography, it is necessary to remove the carbodi-imide blocking group. This means that all traces of RNAse must be eliminated from the reaction mixture, or the liberated Up(Np) oligonucleotides will be hydrolyzed.

Brostoff and Ingram (1967) have carried out a more exact analysis of the reaction between the carbodiimide reagent, 20, and tRNA. They prepared the [14C]-methiodide analog. Treatment of unfractionated yeast tRNA with this reagent in 0.2 M magnesium chloride–0.01 M borate solution, pH 8.0, for 4 hr at 38° results in the addition of 6 moles of reagent per mole of tRNA (77 nucleotides). The reaction actually proceeds very rapidly in the initial stages; 2 moles of reagent are attached per molecule of tRNA within the first 60 sec. The reaction reaches a maximum limit of 6–7 moles of reagent.

These workers treated purified yeast tRNA[Ala] with the reagent and analyzed the oligonucleotides produced by digestion with pancreatic RNAse and T_1 RNAse. They concluded that the reagent attacks only residues in the anti-codon loop, as shown in Figure 3. They were able to show definitely that the

FIGURE 3. The sequence of yeast alanine-tRNA arranged in the cloverleaf structure proposed by Holley. The arrows point to the areas of the sequence that react with the reagent. The bases enclosed in boxes do not react. Data taken from Brostoff and Ingram (1967).

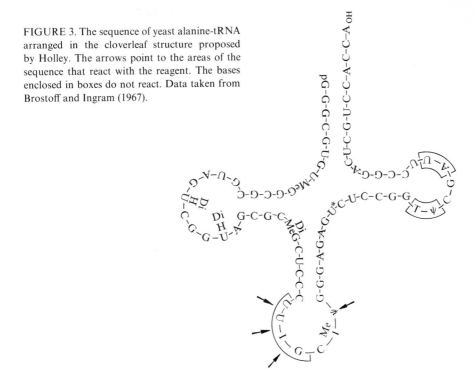

T-ψ-C-G-A-U sequence in the right-hand loop had not reacted. Nelson et al. (1967) (see p. 379) showed that pyrimidines in this particular sequence are less susceptible than those in the anticodon loop to bromination by N-bromo-succiniimide. Yoshida et al. (1968a,b) (see p. 369) found that the T-ψ-C-G tetra-nucleotide of tRNA[Ala] resists the attack of acrylonitrile. These data suggest that the right-hand loop lies buried in the tertiary structure and consequently is less available for chemical reaction.

The use of the carbodiimide reagent as a probe for secondary structure may have some limitation. Girshovich et al. (1968) studied the reaction of the reagent with yeast tRNA[Val] in the presence of $0.01-0.05\,M$ magnesium ion at $15-30°$, conditions which should promote stabilization of the secondary structure. They observed that the extent of the reaction is proportional to the starting concentration of the reagent and also that the rate of the reaction is independent of reagent concentration. It thus appears that the three-dimensional structure undergoes changes during the reaction.

Brownlee et al. (1968) treated 5S rRNA with the carbodiimide reagent. This RNA does not contain any 5-ribosyluracil residues, so that the reaction occurs principally with uridine residues. Ribonuclease A cleaves adjacent to unmodified pyrimidines, and therefore a ribonuclease digest of the treated RNA yielded a new series of oligonucleotide fragments containing carbodiimide substituents. These particular oligonucleotides are readily separated from the unmodified fragments on electrophoresis because of their extra positive charge. The overall technique thus offers an additional weapon in the arsenal of techniques for RNA sequence analysis.

Reaction of tRNA[Ala] with the carbodiimide reagent causes a decrease in alanine-accepting capacity (Brostoff and Ingram, 1967). After a 60 sec reaction the molecule retains 71% of acceptor ability, and after 24 hr, 51% of acceptor capacity still remains.

Salganik et al. (1967) employed the carbodiimide reagent, 20, in an interesting fashion to measure the amount of denaturation in DNA samples. Reaction of the reagent with partially denatured DNA results in modification of only the single-stranded regions, so that on subsequent enzymic hydrolysis of the sample with DNAse and snake venom diesterase the denatured regions remain as large oligonucleotides.

8. GLYOXAL AND OTHER 1,2-DICARBONYL COMPOUNDS

Guanosine and N^2-methylguanosine, *22*, in aqueous solution at pH 4.0 react with a large excess of glyoxal, giving rise to products with structure *23* (Broude et al., 1967; Shapiro and Hachmann, 1966; Staehelin, 1959). The reaction proceeds to completion in 3 hr at 66° or in 6 days at room temperature. Under these conditions the reaction appears to be specific for guanine and N^2-methylguanosine, although, according to Nakaya et al. (1968), this specificity is a matter of reaction rates, since the other three major nucleosides also react to some extent at high concentrations of glyoxal. The glyoxal adduct of guanosine is stable in aqueous solution below pH 6.0, but at pH 7.8 and 25° it decomposes completely within 2 hr to give guanosine (see also Figure 4).

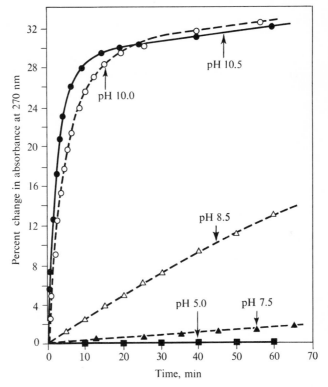

FIGURE 4. Stability of glyoxalguanosine 5'-phosphate at various pH values. A solution of glyoxal (1 O.D.$_{270}$ nm unit in 0.03 ml of water) was incubated at 37°. Data taken from Mitchel (1968).

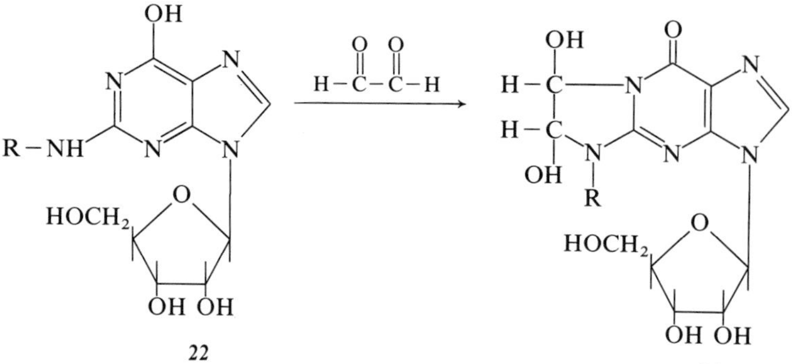

22   23

where R = H or CH$_3$.

Shapiro and Hachmann (1966) investigated a series of dicarbonyl reagents and found that kethoxal* (β-ethoxy-α-ketobutyraldehyde) and ninhydrin (1,2,3-indantrione monohydrate) also react quantitatively with guanosine to give single products with structures 24 and 25, respectively:

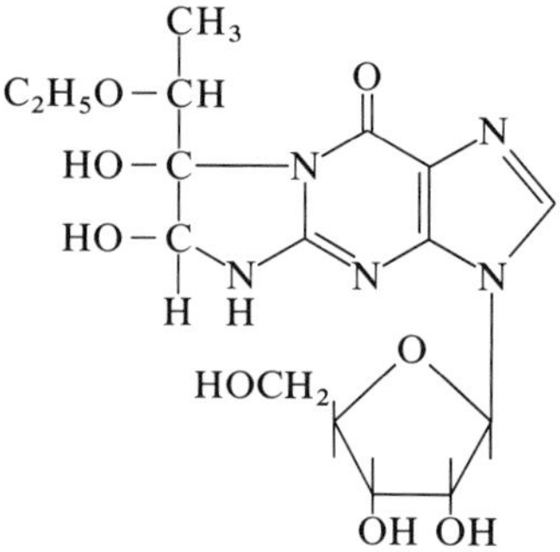

24 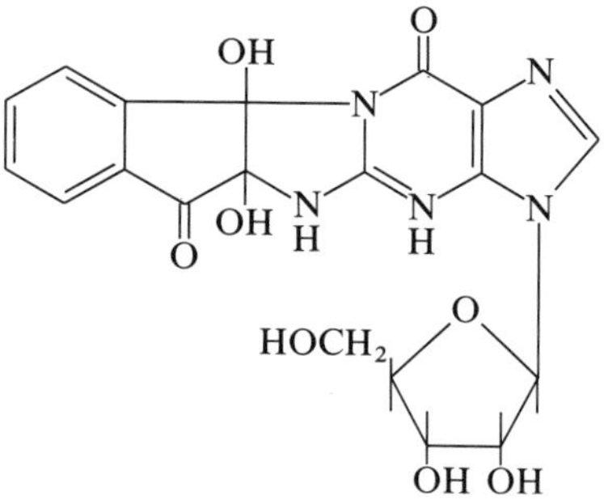   25

The adduct of ninhydrin, as well as that of glyoxal, is converted readily to xanthosine by the action of nitrous acid.

Ninhydrin, unlike glyoxal, also reacts readily with cytosine derivatives to form an adduct with structure 26 (Shapiro and Agarwal, 1968). Adenine and uracil derivatives do not react. The cytosine reaction involves formation of a new C-C bond and is not reversed by alkaline treatment. Cytidine 2′,3′-cyclic phosphate is converted completely to the adduct in 18 hr at room temperature. The adduct is not resistant to the action of pancreatic RNAse.

* Kethoxal is the registered trademark of the Upjohn Co.

26

where R = ribose.

Treatment of glyoxalguanosine 5′-phosphate with periodate produces the diformyl derivative, *27*, which loses one formyl group to afford N^2-formyl-guanosine 5′-phosphate, *28* (Shapiro et al., 1967). The formyl group, which renders guanosine resistant to deamination by nitrous acid, is readily removed by treatment with dilute ammonium hydroxide. The formyl group can also be reduced with LiAlH$_4$ to yield the N^2-methyl derivative (Shapiro et al., 1969). Mitchel (1968) treated glyoxalated tRNA with periodate and noted a decrease in resistance of the treated tRNA to attack by RNAse T$_1$.

27   **28**

Complete reaction of all the guanosine residues of tRNA with kethoxal causes no disruption of primary structure; however, the kethoxalated tRNA loses its capacity to accept amino acids (Litt and Hancock, 1967; Mitchel, 1968). This capacity is restored to 50% of the original after removal of 95% of the kethoxal groups by incubation of the reacted tRNA in 0.1 M Tris buffer (pH 7.6) for 18 hr at 37°. Litt and Hancock (1967) suggest that the reaction of kethoxal with tRNA proceeds faster in unpaired regions of the tRNA molecule. However, the secondary structure of the molecule appears to be seriously disrupted by the addition of kethoxal, so that the reaction conditions would have to be controlled carefully in order to achieve any degree of selectivity in this regard. Nevertheless, the secondary structure of nucleic acids may play an important role in the glyoxal reaction. Nakaya et al. (1968) found that, under conditions in which 72% of the deoxyguanine residues of denatured calf thymus DNA react, none of the deoxyguanosine residues in native calf thymus DNA react.

The action of nucleases on RNA in which all the guanyl residues are substituted with a glyoxal group has been investigated. The activity of the guanyl-ribonuclease from *Actinomices aureoverticillatus* is not influenced (Broude et al., 1967). RNAse T_2 cleaves less than 10% of the residues that it would normally cleave (Mitchel, 1968). Similarly, the activity of RNAse T_1 is strongly but not completely inhibited (Broude et al., 1967; Kochetkov et al., 1967b; Mitchel, 1968; Whitfeld and Witzel, 1963).

Mitchel (1968) introduced an innovation by adding borate ions to the glyoxal-treated tRNA. The borate complex with the glyoxal-guanyl adducts affords virtually 100% protection to the guanylic internucleotide bonds from attack by RNAse T_1 and snake venom phosphodiesterase (exonuclease). The glyoxal-borate group attached to guanosine 5'-phosphate prevents attack by snake venom 5'-monophosphatase.

Since glyoxal does not react with inosine (RNAse T_1 cleaves internucleotide bonds adjacent to this component), the specificity of RNAse T_1 can be significantly enhanced. Kochetkov et al. (1967a,b) treated yeast tRNA^{Val} with glyoxal and subjected the treated tRNA to RNAse T_1 digestion. They obtained two large oligonucleotide fragments. Yeast tRNA^{Val} contains an inosine component in the anticodon loop (see p. 264, Chapter 4), and one might presume that a specific cleavage had taken place at this location. Unfortunately, Kochetkov and his colleagues did not report the results of end group analyses or other characteristics of the produced oligonucleotides. Nevertheless, a technique for producing a highly specific and quantitative cleavage at inosine or 1-methyl-guanosine residues in tRNA now appears to be available.

9. ACRYLONITRILE (CYANOETHYLATION)

Acrylonitrile in aqueous solution reacts with nucleosides possessing a dissociable hydrogen (Chambers, 1965; Chambers et al., 1963; Ofengand, 1965; Yoshida and Ukita, 1965a). The reaction illustrated for 5-ribosyluracil, *29*, involves a Michael-type condensation according to the following scheme:

29   30

31

where R = ribose.

The initial reaction occurs preferentially at the N-1 position to yield *30*. Introduction of a second cyanoethyl group at the N-3 position, giving rise to *31*, occurs more slowly at a rate equivalent to that of the reaction of uridine with acrylonitrile.

At pH 11.5, acrylonitrile reacts with adenosine, cytidine, guanosine, inosine, 5-ribosyluracil, and uridine; but at lower pH values the acidic dissociation of most of these nucleosides is repressed, and at pH 8.8 the reaction becomes relatively selective for inosine (pK_a 8.75) and 5-ribosyluracil (pK_a 8.9). Figure 5 shows the rate of reaction of acrylonitrile with inosine, 5-ribosyluracil, and uridine at pH 8.6.

The susceptibility of cyanoethyl-substituted 5-ribosyluracil derivatives to pancreatic RNAse activity was tested by Chambers (1965), who prepared the 1-cyanoethyl and 1,3-dicyanoethyl derivatives of 5-ribosyluracil-2′:3′-cyclic phosphate. The monosubstituted derivative is rapidly cleaved by RNAse, whereas the disubstituted derivative, even after several days' incubation, remains unchanged.

The mild conditions of the cyanoethylation reaction enable it to be used for the selective blocking of the inosine and 5-ribosyluracil residues in RNA. Yoshida and Ukita (1965a) incubated unfractionated tRNA in an aqueous

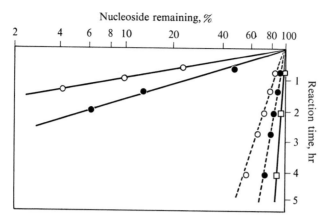

FIGURE 5. Rates of cyanoethylation of inosine, 5-ribosyluracil, and uridine at pH 8.5 and 7.5. Each nucleoside was incubated at 37° in 2.2 M acrylonitrile in 0.1 M phosphate buffer containing 2.7 M dimethylformide. Data taken from Yoshida and Ukita (1968a). ○——○, inosine, pH 8.5; ○---○, inosine, pH 7.5; ●——●, 5-ribosyluracil, pH 8.5; ●---●, 5-ribosyluracil, pH 7.5; □——□, uridine, pH 8.5.

solution containing 2.7 M dimethylformamide and 2.2–2.4 M acrylonitrile (pH 8.6) for 48 hr at 37°. Under these conditions, 90% of the 5-ribosyluracil and inosine residues and 8% of the uridine residues are modified; this means that 12 out of every 200 nucleoside residues in the tRNA are cyanoethylated, of which 8 are 5-ribosyluracil, 1 is inosine, and 3 are uridine. There is no evidence for cleavage of internucleotide bonds during the reaction. These workers found that the presence of 2.7 M dimethylformamide in the reaction medium doubles the rate of cyanoethylation of 5-ribosyluracil and inosine but has little effect on the rate of reaction with adenosine or uridine.

The effect of the dimethylformamide illustrates how the solvent environment can mediate the course of the cyanoethylation of tRNA. Yoshida and Ukita (1965b, 1966, 1968a,b) found that the rate of reaction of acrylonitrile with 5-ribosyluracil residues of tRNA in the aqueous dimethylformamide solution can also be influenced by sodium chloride or magnesium chloride. When the reaction is carried out in 0.01 M phosphate buffer containing dimethylformamide, all the 5-ribosyluracil residues are cyanoethylated in 50 hr. When this reaction medium contains 0.5 M sodium chloride, about one 5-ribosyluracil residue per tRNA molecule reacts within 10 hr and the remaining residues react very slowly. In the presence of 0.002 M magnesium chloride the rate of reaction is

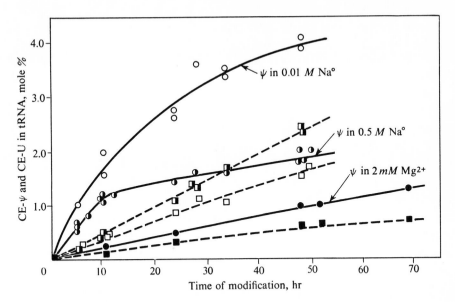

FIGURE 6. Rates of cyanoethylation of 5-ribosyluracil and uridine residues in tRNA under various conditions. Yeast tRNA was treated with 2.2 M acrylonitrile in 0.01 M phosphate buffer, 2.7 M dimethylformamide (pH 8.6) in the absence (at 40°) or in the presence (at 42°) of salt. Data taken from Yoshida and Ukita (1968b). ○——○, CE-ψ, 0.01 M phosphate; □---□, CE-U, 0.01 M phosphate; ◑——◑, CE-ψ, 0.5 M NaCl–0.01 M phosphate; ◧---◧, CE-U, 0.5 M NaCl–0.01 M phosphate; ●——●, CE-, 2 mM MgCl$_2$ in 0.01 M triethanolamine; ■---■, CE-U, 2 mM MgCl$_2$ in 0.01 M triethanolamine.

slow, and even after 70 hr only about 0.5 of a 5-ribosyluracil residue per tRNA molecule reacts. These reaction rates are illustrated in Figure 6. The sodium chloride and magnesium chloride ions stabilize the secondary and tertiary structures of tRNA, causing a more compact molecule, so presumably some of the 5-ribosyluracil residues become inaccessible to the reagent. Yoshida and Ukita (1968b) determined that the one 5-ribosyluracil residue in the tRNA that reacts in the presence of sodium chloride is not the residue in the common G-T-ψ-C-G sequence. From our knowledge of tRNA sequences, it would appear that the other 5-ribosyluracil residues occur only in the anticodon region; this leads to the inference that under the conditions of the reaction (high salt) a 5-ribosyluracil residue in the anticodon loop reacts selectively.

Although most 5-ribosyluracil residues of tRNA become resistant to cyanoethylation in the presence of 0.002 M magnesium ions, the inosine residues are rapidly modified by the reagent (Yoshida et al., 1968a,b). This property enables a selective modification of the anticodons containing an inosine residue. For example, yeast tRNA^{Ala} (see p. 259 for sequence) was treated with acrylonitrile in the presence of 0.002 M magnesium chloride for 10 hr. One mole of inosine residues and 0.4 mole of the 5-ribosyluracil residues in the anticodon loop per mole of tRNA^{Ala} were cyanoethylated. This treated tRNA was fully active in the alanine acceptor assay but did not bind to ribosomes in the presence of GpCpU, GpCpC, and GpCpA. These latter data confirm the work of Fittler and Hall (1966) (see p. 327), who found that the anticodon does not seem to be essential for the recognition or binding with the aminoacyl synthetase.

The effect of the cyanoethylated-inosine residue on the codon-anticodon interaction is also demonstrated in the reverse situation. Sekiya et al. (1967) blocked the inosine residue of the trinucleoside phosphate, GpApI, with acrylonitrile. This modification eliminated the ability of the triplet to stimulate binding of E. coli tRNA^{Glu} to ribosomes.

Rake and Tener (1966) have extended the studies of the solvent effect on the interaction of unfractionated tRNA and acrylonitrile by carrying out the reaction in a mixture of dimethyl sulfoxide, 1 M dimethylaminoethanol, and varying amounts of water. The percentage of water in the solvent controls the degree of substitution of the 5-ribosyluracil residues. In 10% aqueous solution, 0.3 of a residue per tRNA molecule reacts; in 25% water, 1.2 reacts; in 35% water, 2 react; and in 50% water, all 5-ribosyluracil residues react (an average of 3 per tRNA molecule).

The amino acid-accepting capacity of the tRNA is sensitive to cyanoethylation. According to Rake and Tener (1966), cyanoethylation of one 5-ribosyluracil residue per tRNA molecule reduces the acceptor activity about 70%; cyanoethylation of two or three residues, about 85%. These workers did not determine which 5-ribosyluracil in the tRNA reacts first, but in view of the results of Yoshida and Ukita (1968b) this residue probably is located in the anticodon loop region. The reduction in amino acid-accepting capacity may be partially or wholly due to damage to the secondary structure. Rake and Tener (1966) have commented that a possible function of 5-ribosyluracil in tRNA is the maintenance of secondary structure; hence the cyanoethylation of even one 5-ribosyluracil residue would perturb the molecule.

These data clearly show that cyanoethylation can be used as an effective tool to modify selected sites of RNA molecules without damaging the molecules.

10. ACID HYDROLYSIS; SELECTIVE REACTION OF A BASE RESIDUE IN tRNA^{Phe}

The following experiment illustrates that, when the structure of a tRNA molecule and the chemistry of its components are understood, a very general reaction can be used to produce a highly selective result. Yeast tRNA^{Phe} contains an unidentified nucleoside, Y, adjacent to the 3' end of the anticodon (p. 261). This rather hydrophobic nucleoside is more acid labile than, for example, adenosine or guanosine. Thiebe and Zachau (1968) incubated an aqueous solution of yeast tRNA^{Phe} at pH 2.9 for 2 hr. They showed that under these conditions a substantial proportion of the base of Y is expelled. The treated tRNA loses its ability to bind with the messenger-ribosome complex but does not lose its amino acid-accepting ability. Yeast tRNA^{Phe} also contains the acid-labile nucleoside 7-methylguanosine, but under conditions that cause hydrolysis of 70% of Y no hydrolysis of 7-methylguanosine occurs.

Philippsen et al. (1968) carried this reaction one step further. The treated tRNA molecule contains a ribose moiety in the place of Y that presumably possesses a C-1 aldehyde group. This group labilizes the adjacent phosphodiester bond, and when the tRNA is treated with an amine (in aqueous solution, pH 8–10), the molecule is quantitatively split in half. The cleavage probably occurs at the phosphodiester bond *beta* to the aldehyde group in a reaction analogous to the mechanism of cleavage of internucleotide bonds first described by Whitfeld and Markham (1953).

It is interesting that when the two halves of the tRNA^{Phe} molecule are recombined they accept 50–60% phenylalanine compared to unsplit tRNA^{Phe} that had been acid-treated. Bayev et al (1967a) have also reported that tRNA^{Val} split in the anticodon loop by T$_1$-RNAse regains full acceptor activity on recombination of the two fragments (see also Chapter 4, p. 279).

11. ALKYLATION (METHYLATION)

Alkylation of nucleic acids has attracted considerable attention because of the mutagenic, carcinogenic, and antitumor effects produced in biological systems (see, e.g., the review by Lawley, 1966). In studying the effect of alkylating reagents at the biological level, reaction of a single base residue may be sufficient to evoke an observable response. Since we are more concerned in this chapter with the use of alkylation as a chemical tool, the quantitativeness and selectivity of the reaction become important criteria.

All nucleosides have sites susceptible to alkylation on both the heterocycle and the sugar. In any alkylation reaction all sites of a given nucleoside compete for the reagent, but the rates of reaction are generally different enough so that, in practice, a single site can be alkylated. The reaction conditions are critical, and a reagent may preferentially alkylate one site under one set of reaction conditions and another site under a different set. Mixed nucleosides in an oligonucleotide present multiple sites for alkylation; however, as in the single-nucleoside reaction, rates vary considerably and with a judicious manipulation of conditions a fair degree of selectivity may be achieved. Therefore, in view of the multiplicity of the possible alkylation reactions of nucleic acids, this section will present first a discussion of the alkylation of single nucleosides and then a discussion of the reactions as applied to nucleic acids.

All the major nucleosides can be alkylated in good yield using methods suitable for preparative organic chemistry, although such methods are not necessarily suitable for the alkylation of nucleic acids. These reactions, nevertheless, demonstrate the favored site of alkylation in each nucleoside.

Diazomethane in methanolic solution methylates uridine in good yield to give 3-methyluridine (Miles, 1956). Dimethyl sulfate in N,N-dimethylformamide solution also methylates cytidine at the 3 position to give 3-methylcytidine (Brookes and Lawley, 1962). Treatment of adenosine or deoxyadenosine in N,N-dimethylformamide solution with either dimethyl sulfate or methyl p-toluenesulfonate affords 1-methyladenosine or 1-methyldeoxyadenosine (Brookes and Lawley, 1960; Jones and Robins, 1963). Treatment of deoxy-guanosine with diazomethane in methanol yields a mixture of O^6 and N^7 monomethyl derivatives (Friedman et al., 1965). This result contrasts with treatment in aqueous solution (described below) in which the O^6 derivative is not obtained.

Methylation of inosine dissolved in dimethyl sulfoxide or guanosine dissolved in N,N-dimethylacetamide with methyl iodide affords the 7-methyl derivative in each case (Jones and Robins, 1963). When the reaction in dimethyl sulfoxide solution is carried out in the presence of potassium carbonate, the corresponding 1-methyl derivatives are obtained (Broom et al., 1964). Both 1-methylguanosine and 7-methylguanosine can be methylated further under these conditions to yield 1,7-dimethylguanosine. Alkylation at the N-1 position of guanosine and inosine in the presence of carbonate can be attributed to the removal of the N-1 proton, which makes this site the most electrophilic center and hence the pre-ferential site of alkylation. The N-7 position of guanosine is normally the most electrophilic center (Miles et al., 1963; Pfleiderer, 1961; Tsuboi et al., 1962); it follows that a nucleophilic reaction such as alkylation would occur at this position.

Under conditions feasible for the treatment of nucleic acids, methylation by a number of reagents readily occurs. In a series of model experiments, Haines et al. (1964) compared the extent of reaction under identical conditions between diazomethane and each of the major ribonucleosides in aqueous solution. The following products were obtained: 3-methyluridine and 7-methylguanosine (80–90%), 3-methylcytidine (ca. 40%), and 1-methyladenosine (ca. 10%). The order of reactivity differs markedly for reaction with dimethyl sulfate or methyl esters of other strong acids, that is, guanosine > adenosine > cytidine; uridine is virtually inert (Lawley, 1961).

Treatment of tRNA or double-stranded DNA with an excess of methyl methanesulfonate or diazomethane in aqueous solution results in extensive methylation (Kriek and Emmelot, 1964; Lawley and Brookes, 1963; Zakharyan et al., 1967). The methylated products are listed in Table 2. These data clearly show that the N-7 position of guanosine is the favored site of alkylation in both RNA and DNA. Alkylation of the adenine residues of DNA at the N-3 position rather than at the N-1 position, as in RNA, may be due to participation of the N-1 position in hydrogen bonding. Methylation of deoxyadenosine per se affords the N-1 derivative exclusively (Haines et al., 1964; Jones and Robins,

TABLE 2

Methylated Bases of Diazomethane-Treated Nucleic Acids: Mean with Standard Deviation for Three Experiments with DNA. (Data from Kriek and Emmelot, 1964)

Base	Salmon Sperm DNA		Yeast sRNA	
	Mole % of Base Present	*Relative Amount*[a]	*Mole % of Base Present*	*Relative Amount*[a]
7-Methylguanine	20.8 ± 1.0	81 (86)[b]	35	65 (64)[b]
3-Methyladenine	3.6 ± 0.3	19 (9)[b]	0	0 (0)[b]
1-Methyladenine	0	0 (5)[b]	24	27 (21)[b]
1-Methylcytosine	0	0 (0)[b]	0	8 (14)[b]

[a] Sum of methylated bases taken as 100.

[b] Methylation by methyl methanesulfonate according to Lawley and Brookes (1963).

CHEMICAL REACTIONS APPLIED SELECTIVELY TO NUCLEIC ACIDS

1963). The secondary and tertiary structures clearly have an effect on the course of methylation. Zakharyan et al. (1967), for example, found that 0.01 M magnesium chloride in the buffer significantly limits the extent of methylation of the guanine residues.

The glycosylic bond of 3-methyldeoxyadenosine and 7-methyldeoxyguanosine groups of DNA is labile, and in aqueous solution at neutral pH the bases are rapidly released. The 3-methyladenine residues are expelled 4–5 times as rapidly as the 7-methylguanine residues (Kriek and Emmelot, 1964). The glycosylic bonds of the 7-methylguanosine, *32*, residues in RNA, on the other hand, are stable below pH 7; but at slightly higher pH values the imidazole ring opens rapidly to give 2-amino-4-hydroxy-5-*N*-methylcarboxamido-6-(*N*-β-D-ribofuranosyl)aminopyrimidine, *33* (Haines et al., 1962). Compound *33* has a stable glycosylic bond.

32   *33*

where R = D-ribose.

Although the heterocyclic units of nucleic acids are the most susceptible to alkylation, the phosphate group and the 2'-hydroxyl group of ribose could also, in theory, be alkylated. Brimacombe et al. (1965) and Griffin et al. (1967) addressed themselves specifically to this question. They treated poly U with diazomethane or dimethyl sulfate under conditions that resulted in methylation of 70 % of the uracil residues. No cleavage of the internucleotide bonds occurred; cleavage might have been expected if phosphate groups were esterified to give the unstable phosphate triesters. The methylated polymer was hydrolyzed and the hydrolyzate examined for the presence of 2'-*O*-methyl nucleotides, none of which was found. The same workers treated uridylyl-(3'→5')-adenosine and adenylyl-(3'→5')-uridine with a 200-fold excess of diazomethane and found in each case that methylation occurred exclusively at the N-3 position of the uridine residue.

These results are contradicted by those of Holy and Scheit (1967), who treated adenylyl-(3'→5')-uridine with diazomethane; they reported that, in addition to the N-3 position of the uridine residues, the N-1 position of the adenosine residues and 33 % of the 2'-hydroxyl groups of the uridine moiety are methylated. Furthermore, they observed that a significant percentage of the internucleotide bonds was broken. The conditions used in these experiments must have been quite unique, since the evidence obtained by the other workers suggests that the heterocyclic residues of oligonucleotides can be methylated without any methylation of the 2'-hydroxyl or phosphate groups.

In a later paper, Scheit and Holy (1967) reported that treatment of uridylyl-(3'→5')-inosine with dimethyl sulfate at pH 7 results in methylation of the N-7 and/or N-1 positions of inosine. Under the conditions of the reaction, the hypoxanthine residue undergoes ring opening with the formation of 3-methyl-5-(N-formyl-N-methyl)-6-aminopyrimidine-4-one, *34*, and 5-(N-formyl-N-methyl)-6-aminopyrimidine-4-on, *35*, derivatives. There is no evidence of methylation of the phosphate group, since no fission of the internucleotide bonds occurs.

34   35

Methylation of nucleotide homopolymers has been used as a technique to obtain information about the secondary structure and the biological properties of oligonucleotides. Treatment of poly A or poly C in aqueous solution with dimethyl sulfate results in methylation of approximately 50% of the base residues, without any degradation taking place (Brimacombe et al., 1965). The extent of methylation can be increased to 100% of the base residues when tri-n-butylamine is included in the reaction medium (Michelson and Pochon, 1966). Introduction of methyl groups at the N-1 position of adenine residues in poly A abolishes the ability of the polymer to stimulate glycine incorporation in the polypeptide synthesizing system *in vitro* (Michelson and Pochon, 1966). This effect is probably due to the loss of hydrogen-bonding capability of the N-1 position. A similar effect results when the methyl groups are introduced at the N-1 position of guanine residues of poly G. Pochon and Michelson (1967) prepared this polymer by first synthesizing 1-methylguanosine 5′-diphosphate and then polymerizing it by the use of polynucleotide phosphorylase. Poly m^{1}G is incapable of hydrogen bonding with poly C and, in addition, shows a lack of an ordered secondary structure at pH 7 in 0.15 M sodium chloride. [Poly G under these conditions possesses a stable secondary structure (Pochon and Michelson, 1965).] Poly m^{7}G, on the other hand, can still hydrogen-bond with poly C, although the stability of the copolymer is considerably less than that of poly C : poly G.

The effect of a methyl group on the normal complementary base-base pairing is also illustrated by the fact that substitution of 3-methyluridine for uridine in the triplet GpUpU completely eliminates its template activity in the tRNA-ribosome binding assay (Grünberger et al., 1968).

Methylation of nucleic acids can bring about a change in susceptibility to enzymic attack. Methylation of poly G to produce poly m^{7}G, for example, makes the polymer completely resistant to hydrolysis by rattlesnake (*Crotalus atrox*) venom (Michelson and Pochon, 1966). The fact that RNA is preferentially methylated at the N-7 position of the guanosine residues was exploited by Brownlee et al. (1968) in their study of the primary sequence of 5S rRNA. The 5S RNA was partially methylated and then completely digested with ribonuclease T$_1$. Since the methylated guanosine residues are resistant to the action of the enzyme, these workers obtained a different spectrum of oligonucleotide fragments from that of the unmethylated RNA. As a consequence, a number of overlapping sequences useful in the reconstruction of the primary sequence were obtained.

Strauss and Robbins (1968) observed a change in susceptibility of methylated DNA to nuclease action. Coliphage T7 DNA was methylated with methyl methanesulfonate to give an alkali-stable product containing sixteen added methyl groups per molecule. This methylated DNA is attacked by an extract from *Micrococcus lysodeikticus*, which causes single-strand breaks at or near the site of methylation. Extracts from *Bacillus subtilis* cause a similar result. Nonmethylated native DNA is not attacked by either extract. The authors believe that this enzymic activity is different from that of the enzyme that catalyzes deletion of ultraviolet-damaged sections of DNA, since the crude extract of *B. subtilis* has no effect on ultraviolet-treated DNA. The essential point suggested by this study is that methylation alone changes the structure of DNA sufficiently to make it recognizable by a nucleolytic enzyme.

12. BROMINATION

Bromine in aqueous solution (pH 5.7) at room temperature reacts rapidly with uridylic and cytidylic acids, slowly with guanylic acid, and not at all with adenylic acid (Yu and Zamecnik, 1963a,b). The mechanism of the reaction, as illustrated for uridine, *1*, appears to consist of electrophilic attack of Br^+ at the C-5 position to give a transient intermediate, *36*, followed by the addition of Br^- or OH^- at C-6 to give *37* or *38* (Wang, 1959a,b; Yu and Zamecnik, 1963b). Compounds *37* and *38*, when refluxed in water for 25 min, give rise to 5-bromouridine, *39*, and some uridine. In the case of guanosine, bromine attacks initially at the C-8 position, but the 8-bromoguanosine derivative undergoes further oxidation by bromine with destruction of both rings of the heterocycle (Shapiro and Agarwal, 1966).

Weil et al. (1964) and Weil (1965) studied the bromination of tRNA in the form of a quaternary ammonium salt dissolved in *N,N*-dimethylformamide. If the reaction is allowed to proceed briefly at room temperature, the oxidized tRNA contains 8-bromoguanine, 5-bromocytosine, and 5-bromouracil residues. The bromination can be carried out equally as well in aqueous solution (Yu and Zamecnik, 1963a). The rate of bromination of tRNA in either aqueous or organic solvent depends on the amount of bromine and the reaction time; therefore, by judicious choice of conditions the extent of bromination can be carefully controlled. In the presence of a minimal amount of bromine only one cytidine and one uridine residue per eighty base residues of the tRNA react. From the ultracentrifugation patterns and melting-out curves of the treated tRNA, it appears that the low level of bromination does not affect secondary structure except at the point of attack of the bromine atoms (Weil et al., 1964; Yu and Zamecnik, 1963a).

It is particularly important for the development of the chemical technology of nucleic acids that the physical properties of the tRNA brominated in dimethylformamide solution are not significantly changed. This suggests that the nonaqueous solvent does not cause nonspecific damage to the tRNA secondary structure. Moreover, although there is some loss of amino acid-accepting activity, under the reaction conditions the degree of loss is propor-

CHEMICAL REACTIONS APPLIED SELECTIVELY TO NUCLEIC ACIDS

where R = ribose.

tional to the degree of bromination, indicating that the loss of activity is not due to solvent interactions.

The secondary and tertiary structures of tRNA influence which of the nucleoside residues is available for attack by the bromine. Nelson et al. (1967) demonstrated this effect by an elegant study on the nature of the bromination of yeast tRNA^Ala. The tRNA^Ala was brominated with N-bromosuccinimide under conditions that led to the incorporation of four atoms of bromine per molecule of RNA. The general location of the bromine atoms in the primary sequence was determined with the results shown in Figure 7. These results indicate that the T-ψ-C-G loop is protected in some way by the tertiary structure of the molecule. Similarly, this tRNA loop is resistant to cyanoethylation

(p. 369). These results show how a rather general reaction can be employed to modify specific residues of tRNA by taking advantage of the three-dimensional structure.

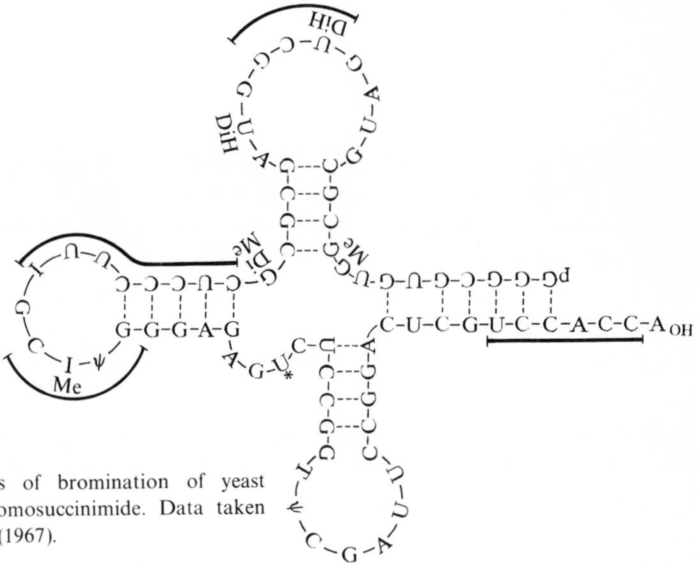

FIGURE 7. Sites of bromination of yeast tRNA^{Ala} by N-bromosuccinimide. Data taken from Nelson et al. (1967).

13. IODINATION

Iodine in dilute aqueous solution at pH 7.0 reacts rapidly with N^6-(Δ^2-isopentenyl)adenosine (Robins et al., 1967) and 4-thiouridine (Lipsett, 1965) (possibly also with other sulfur-containing nucleosides), but not at all with the major nucleosides and other known modified nucleosides. Iodine does react with other components of nucleic acids, but only under relatively severe conditions, such as heating at 100° in aqueous solution at pH 5.3. [Under these conditions, cytosine residues are preferentially oxidized (Jones et al., 1966).] The reaction at room temperature is very mild and has no effect on the major nucleoside components, as indicated by the fact that on exposure to aqueous solution the infectivity of tobacco mosaic virus RNA (Brammer, 1963) or the activity of transforming DNA (Hsu, 1964) is not diminished. The reactions of iodine with N^6-(Δ^2-isopentenyl)adenosine and 4-thiouridine, therefore, are among the most selective chemical reactions available for specific modification of nucleic acid structure.

40   *41*

The reaction of iodine and N^6-(Δ^2-isopentenyl)adenosine is discussed on p. 327, and this section will deal only with the reaction with 4-thiouridine. The tRNA of *E. coli* contains a relatively large proportion of 4-thiouridine (about 6% of the uridine content), as well as small quantities of other thiolated nucleosides (see Chapter 2). 4-Thiouridine, *40*, is readily oxidized to form the disulfide derivative, *41*. The reaction proceeds more rapidly at pH 8–9 than at pH 7 and is readily reversed by the action of thiosulfate. Although the disulfide derivative has not been isolated from iodine-treated tRNA, the oxidation reaction presumably proceeds when 4-thiouridine is part of tRNA. Carbon et al. (1965) observed that unfractionated *E. coli* tRNA, on treatment with iodine, loses its ability to accept lysine and phenylalanine, but not arginine and valine; further more, the inactivation is reversed by treatment with thiosulfate. The authors interpreted these findings as being due to the presence of 4-thiouridine in tRNA^{Lys} and tRNA^{Phe}. They carried these experiments further and fractionated the tRNA^{Lys} into two fractions, only one of which was inactivated by the iodine treatment. This heterogeneity with respect to iodine sensitivity within an amino acid-accepting group of tRNAs was confirmed by Goehler et al. (1966) and Goehler and Doi (1968). They fractionated the amino acid-accepting ability for serine, and for lysine of *B. subtilis* tRNA. Treatment with iodine totally inactivated all tRNA^{Lys} species, but only partially inactivated the tRNA^{Ser} species. The valine-accepting activity of *B. subtilis* tRNA, like that of *E. coli*, is not lost by the iodine treatment.

The reversibility of the iodination reaction suggests that the susceptible tRNA molecules have two thio-nucleosides near each other. The evidence for this possibility is conflicting, however. Lipsett and Doctor (1967) isolated a species of *E. coli* tRNA^{Tyr} that contains two residues of 4-thiouridine, and Lipsett (1967) isolated bis-(4-thiouridylic acid) from iodine-oxidized tRNA. This evidence suggests, therefore, that intramolecular formation of disulfide bonds could occur (see Figure 8); furthermore, Carbon [unpublished data cited

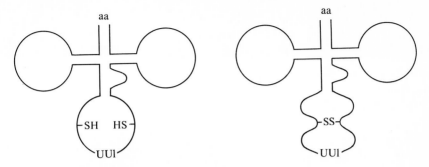

FIGURE 8. A model to illustrate a change in conformation of the anticodon site of oxidized tRNA. As suggested by Goehler and Doi (1966).

by Carbon and David (1968)] could not find any evidence for intermolecular disulfide bonds in iodine-treated tRNA. These observations may have another interpretation and do not necessarily suggest the concept that some *E. coli* tRNA molecules have two 4-thiouridine residues. The four known primary structures (p.273) of *E. coli* tRNA all have one 4-thiouridine residue.

Escherichia coli tRNA contains N^6-(Δ^2-isopentenyl)-2-methylthioadenosine (p. 320), and perhaps some of the results obtained with iodination of this tRNA can be attributed to reaction with this nucleoside. In practice it should be relatively easy in most instances to determine whether a 4-thiouridine or N^6-(Δ^2-isopentenyl)adenosine derivative had reacted. The reaction with 4-thiouridine is characterized by (1) the reversibility of the reaction and (2) the fact that iodine is not incorporated into the product. In contrast, the reaction with N^6-(Δ^2-isopentenyl)adenosine is characterized by (1) nonreversibility of the reaction and (2) the fact that iodine is incorporated into the product.

14. NITROUS ACID

Adenosine, cytidine, and guanosine are deaminated by nitrous acid to yield inosine, uridine, and xanthosine, respectively. Figure 9 shows the comparative rate of reaction under identical conditions. The reaction with guanosine, *22*, also produces 2-nitroinosine, *42*, in addition to xanthosine, *43* (Shapiro and Pohl, 1968). The formation of 2-nitroinosine is favored by increasing the nitrite ion concentration to 8 *N*. Other nucleophiles can substitute in the 2 position; when guanine is treated with nitrous acid in the presence of fluoroboric acid, 2-fluorohypoxanthine, *44*, is produced (Montgomery and Hewson, 1960).

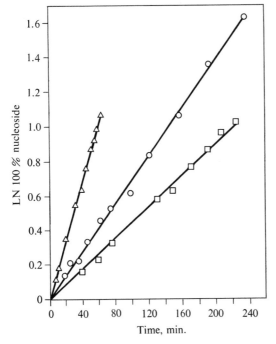

Nitrous acid is mutagenic (Horn and Herriott, 1962; Sinsheimer, 1960), an effect of considerable interest to biologists. Nitrous acid deaminates the three bases of tobacco mosaic virus (TMV) RNA (Schuster, 1960; Schuster and

FIGURE 9. Plot of pseudo-first-order kinetics for the reaction of adenosine (O), cytidine (□), and guanosine (△) with 0.57 M sodium nitrite in 3.1 M acetic acid-sodium acetate buffer, pH 3.75, 37.5°. Data taken from Shapiro and Pohl (1968).

Vielmetter, 1961) at about the same rate and also deaminates the three bases of
E. coli tRNA (Carbon, 1965), at approximately the same rate. The rates of
mutation and inactivation of TMV-RNA follow first-order kinetics, indicating
that one deamination can produce a mutation or cause a lethality. In the
reaction of DNA with nitrous acid, the guanine residues react fastest; it has
been suggested that this is because the amino groups of guanine are not hydrogen
bonded and hence are more available to the reagent (Schuster, 1960; Vielmetter
and Schuster, 1960). The ratio of deamination of adenine in native DNA is
about 30 times as slow as the rate of deamination of free adenine. Treatment of
DNA with nitrous acid also causes cross-linking between the two strands
(Geiduscheck, 1961; Luzzati, 1962).

The wide range of the nitrous acid reaction limits its usefulness with respect
to specific chemical modification of nucleic acids. In combination with other
techniques, however, the reagent can be of value. Carbon and Curry (1968), for
example, treated tRNA^{Gly} (presumed anticodon CCU) with nitrous acid. Even
though the reaction with nitrous acid would proceed randomly, a certain
percentage of the tRNA molecules should undergo a change in their anticodon
to CUU or UCU, in which case they would respond to the codon GAA (glutamic
acid) or AGA (arginine). Although there was no clear-cut chemical evidence
that such a deamination had indeed been produced in the anticodon of the
treated tRNA^{Gly}, the authors detected incorporation of glycine into a polypeptide in response to the code word AGA [poly (A,G)].

15. PERMANGANATE OXIDATION

Treatment of N^6-(Δ^2-isopentenyl)adenosine in dilute aqueous solution (pH 7)
with a slight excess of permanganate for 10 min cleaves the Δ^2-isopentenyl side
chain to give adenosine in 30% yield (Robins et al., 1967). This reaction can be
applied to tRNA with little damage to the structure, judging by the fact that the
amino acid acceptor activity decreases only about 15% (Fittler et al., 1968b;
Kline et al., 1969). At the same time about 20% of the Δ^2-isopentenyl groups
are removed. This reaction is described in more detail in Chapter 7, p. 325.

The mild permanganate treatment may cause some nonspecific oxidation.
Hayatsu and Ukita (1967) concluded that treatment of tRNA with 26.4 *mM*
permanganate solution, pH 6.7, for a few minutes at 0° causes oxidation of a
limited number of pyrimidine residues. They base this conclusion on a loss of

ultraviolet absorption of the treated tRNA and suggest that, of the four major nucleosides, only the pyrimidines react under these conditions.

This oxidation reaction of the pyrimidine nucleosides is a complex reaction, which has been studied in some detail by Howgate et al. (1968). Treatment of thymidine with excess permanganate for 19 hr at 37° (pH 8.5) produces 5,6-dihydroxy-5,6-dihydrothymidine, which is then oxidized and hydrolyzed to give *N*-(2-deoxy-D-ribofuranosyl)-*N'*-pyruvoylurea. Further oxidation and hydrolysis produces 2-deoxy-D-ribofuranoyslurea, which in turn gives rise to urea and 2-deoxy-D-ribose.

16. SODIUM BOROHYDRIDE REDUCTION

Uridine, *1*, and uridylic acid in aqueous solution (pH 10.0) are reduced rapidly to their respective 5,6-dihydro derivatives, *45*, by sodium borohydride in the presence of light from a low-pressure mercury vapor lamp (2537 Å) (Cerutti et al., 1965, 1968b). The reaction is mild enough to apply to nucleic acids, and when tRNA is treated under these conditions, only the uridine residues are reduced (Table 3). The treatment causes some alterations in secondary structure, presumably in the regions where the uridine residues are reduced (Adman and Doty, 1967). The reaction can be used as a technique for the general labeling of RNA with tritium. Kirkegaard and Bock (1968),* in a light-catalyzed reaction, treated RNA with [³H]-sodium borohydride in aqueous solution, pH 8.5, for 3 min at room temperature. Under these conditions, 3–5% of the uridine residues in the RNA is reduced. This technique may be particularly valuable for

TABLE 3

Photoreduction of Uridine in tRNA; Base Composition of Baker's Yeast tRNA after Exposure to Light in the Presence of Sodium Borohydride. (Data from Cerutti et al., 1965)

Irradiation Time, min	Up, %	Cp, %	Ap, %	Cp, %
0	24.9	22.3	25.0	27.7
60	14.3	21.8	25.1	27.5
120	9.0	21.4	25.2	27.5
180	6.5	20.5	25.0	27.4
240	5.6	21.1	25.3	27.3

*See also Leppla et al. (1968).

radioactive labeling of the RNA of organisms that cannot be conveniently grown in a labeled medium.

In a light-independent reaction sodium borohydride in aqueous solution reductively cleaves 5,6-dihydrouridine, *45*, to yield *N*-(β-D-ribofuranosyl)-*N*-(γ-hydroxypropyl)urea, *46* (Cerutti and Miller, 1967; Cerutti et al., 1968b). The rate of hydrogenolysis of dihydrouridine is considerably slower than the rate of the photoreduction of uridine, a fact that makes it possible to isolate the 5,6-dihydrouridine from a reaction starting with uridine. No significant reduction of uridylic acid by sodium borohydride takes place in the dark.

where R = ribosyl or (5'-phosphate)ribosyl.

The mild conditions (aqueous solution, pH 8, room temperature) under which 5,6-dihydrouridine is reductively cleaved make the sodium borohydride reaction ideal for selective reduction of 5,6-dihydrouridine residues in tRNA. Cerutti and Miller (1967) demonstrated that it is possible to use the reaction for this purpose. They treated unfractionated yeast tRNA with [³H]-sodium borohydride and, after an alkaline hydrolysis of the treated tRNA, isolated [1-³H]-*N*-ribosyl-2',3'-phosphate-3-ureidopropanol-1. These experiments do not indicate which 5,6-dihydrouridine residues in the tRNA sequences are being reduced or how extensive the reaction is. In a preliminary study, Cerutti (1968) investigated the effect of sodium borohydride treatment on the serine and valine acceptance of unfractionated yeast tRNA (see Chapter 4 for the primary sequences showing the location of the 5,6-dihydrouridine residues). Serine-accepting ability was not affected, but valine-accepting ability decreased to a level of 50% of the original in 3 hr.

Molinaro et al. (1968) repeated these experiments with unfractionated yeast tRNA and obtained slightly different results. Under conditions that resulted in cleavage of about 90% of the 5,6-dihydrouridine residues, the alanine-, serine-, and valine-accepting activity decreased only about 20%. There was no significant difference in results between the serine- and the valine-accepting abilities. These authors also treated yeast tRNA[Ser] and observed a decrease of about 20% in accepting ability, as well as a very slight reduction in ribosome-binding activity. Taking into consideration the complexity of these experimental protocols and the possibility that some nonspecific damage occurs, one might infer that the integrity of the 5,6-dihydrouridine residues is not essential for the biological parameters measured.

In contrast, the conversion of uridine residues to 5,6-dihydrouridine residues in tRNA can have an effect on the amino acid-accepting activity. Adman and Doty (1967) found that reduction of one-half of the uridine residues of unfractionated yeast tRNA is sufficient to inactivate alanine acceptance. These particular data are too preliminary to draw any conclusions, but on the basis of theoretical considerations the reduction of a uridine residue to a dihydrouridine residue would have a marked effect on the secondary and tertiary structures. The 5,6-dihydrouracil base has a shortened π-electron system and, unlike uracil, is nonplanar. Such a change in physical conformation undoubtedly affects the normal stacking and hydrogen-bonding capabilities of uridine-containing oligonucleotides. This effect on hydrogen bonding is illustrated by studies with oligonucleotide triplets containing 5,6-dihydrouracil in place of a uracil residue. AphUpG will not stimulate binding of tRNA[Met] in the ribosome-binding assay (Lee et al., 1967), and neither GpUphU nor GphUpU stimulates binding of tRNA[Val] (Smrt et al., 1966).

Sodium borohydride also reacts readily with some of the modified nucleosides in the light-independent reaction. It reduces N^4-acetylcytidine to N^4-acetyl-3,4,5,6-tetrahydrocytidine (Miller and Cerutti, 1967), and 4-thiouridine, 40, to N-ribosyl-2-oxohexahydropyrimidine, 47 (Cerutti et al., 1968a). Also, it reduces 1-methyladenosine to a product that can be readily reoxidized to produce 1- or N^6-methyladenosine (Macon and Wolfenden, 1968). These authors report that 3-methylcytidine is rapidly reduced with concomitant loss of ammonia. On the basis of the susceptibility of 5,6-dihydrouridine, 4-thiouridine, and N^4-acetyl-cytidine to sodium borohydride reduction, Cerutti et al. (1968a) developed a convenient assay for the presence of these three modified nucleosides in RNA by treating the sample with [^{3}H]-sodium borohydride.

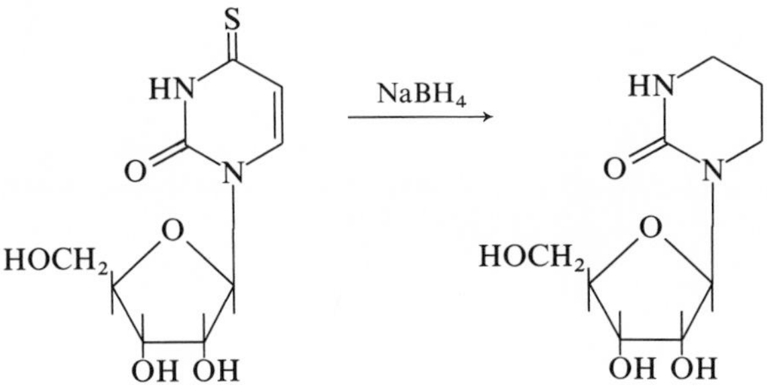

The photoreduction of thymidine, *48*, by sodium borohydride, like the reaction with uridine, proceeds in two stages. First, hydrogenation of the 5,6 double bond occurs, and, second, the dihydro compound is reductively cleaved in a light-independent reaction. In contrast to the reduction of uridine, 5,6-dihydrothymidine cannot be isolated as a reaction product. The major product obtained is N^3-[2-deoxy-β-D-*erythro*-ribofuranosyl]-3-ureido-2-methylpropan-1-ol, *49* (Ballé et al., 1966; Kondo and Witkop, 1968). The reaction proceeds stereospecifically to yield the *erythro* configuration:

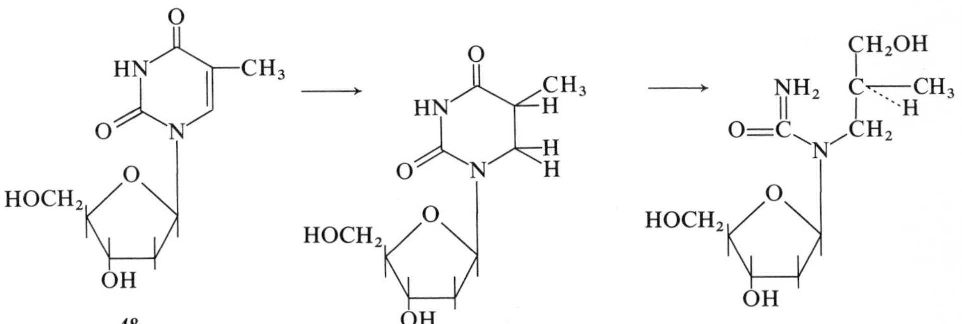

The thymine dimer produced in DNA` by ultraviolet irradiation can be reductively cleaved by sodium borohydride. Treatment of the thymine dimer,

50, in aqueous solution at room temperature for 4–10 hr produces a mixture of compounds *51* and *52* (Kunieda and Witkop, 1967):

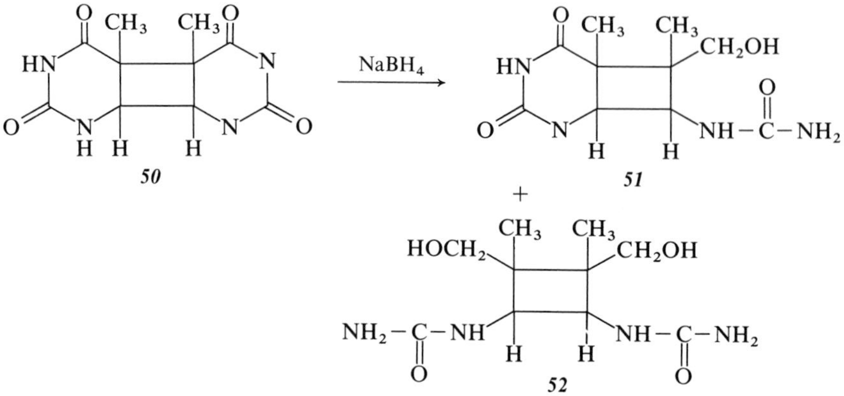

In view of the mildness of the reaction conditions, this technique should be applicable to the reduction of thymine dimers in DNA.

17. CYANOGEN BROMIDE

Treatment of 4-thiouridine 5'-phosphate, *53*, with cyanogen bromide in carbonate buffer solution (pH 8.9) at 20° rapidly converts it to the thiocyanato derivative, *54* (Saneyoshi and Nishimura, 1967):

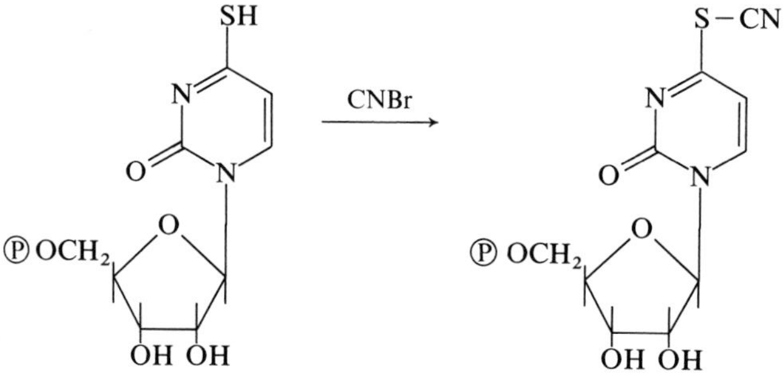

This reaction appears to be highly specific; the common nucleotides 5'-AMP, 5'-GMP, 5'-UMP, and 5'-IMP do not react at all. 4-Thiouridine is a prominent modified component of *E. coli* tRNA (p. 273); accordingly, Saneyoshi and

Nishimura investigated the effect of cyanogen bromide on the biological activity of this tRNA. They found that the amino acid-accepting capacity and ribosome-binding capacity for some, but not all amino acids, are reduced (Table 4).

TABLE 4

Effect of Cyanogen Bromide on the Biological Activity of E. coli tRNA. (Data from Saneyoshi and Nishimura, 1967)

Amino Acid	Acceptor Activity, % of control	Binding to mRNA Ribosome Complex, % of control
Alanine	100	
Arginine	100	100
Glutamic acid	10–20	
Glycine	100	100
Histidine	100	0
Isoleucine	100	100
Lysine	10–20	
Methionine	100	
Phenylalanine	60–70	
Threonine	100	
Tyrosine	60–70	0[a]
Valine	60–70	100

[a] The treated [^{14}C]-tRNA^{Tyr} was not bound to ribosomes in the presence of poly (A$_5$,U), but the binding was stimulated in the presence of poly (U$_4$,G).

18. 2-AMINO-*p*-BENZENEDISULFONIC ACID

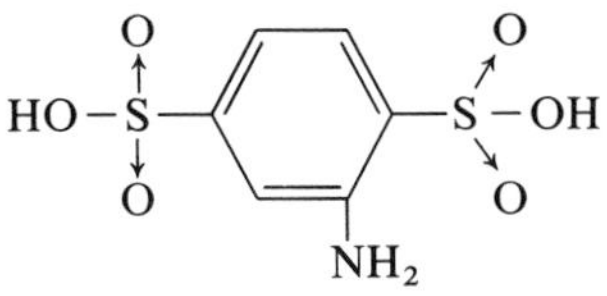

55

2-Amino-*p*-benzenedisulfonic acid, *55*, at pH 9.0 reacts with the guanine, adenine, and cytosine residues of DNA and RNA (Moudrianakis and Beer, 1965a,b). The adenine and cytosine adducts are acid-labile, and the reagent is detached from these residues when the treated DNA is precipitated from an acid solution (Erickson and Beer, 1967). The guanine derivative appears to be more stable because coupling occurs at C-8; consequently, the overall procedure provides a technique for preferentially labeling the guanine residues of DNA or RNA. The selectivity of the overall procedure is slightly better for DNA than RNA; for example, in the case of RNA, when 60% of the guanine residues has reacted, about 10% each of the adenine and cytosine residues has also reacted. The treated DNA can be stained with uranyl acetate, and the fixed heavy uranium atoms can then be visualized in the electron microscope. This technique has been developed for the purpose of studying base sequences in nucleic acids by means of electron microscopy (Moudrianakis and Beer, 1965a,b).

19. MERCURY DISODIUM p-HYDROQUINONE DIACETATE

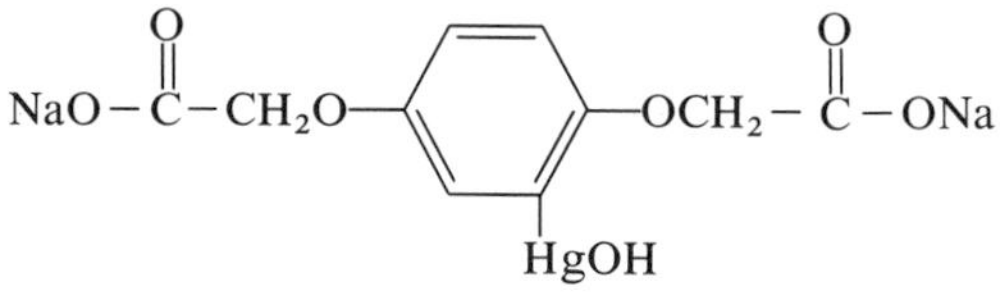

56

Simpson (1964) has shown that methylmercury hydroxide in aqueous solution (pH 8–9) binds much more strongly to guanosine and uridine than to adenosine and cytidine. The methylmercury ion binds to the N-3 and the N-1↔N-7 positions, respectively, of uridine and guanosine. This binding appears to be characteristic of the mercury ion; Fiskin and Beer (1965) found that another derivative, mercury disodium p-hydroquinone diacetate, 56, at pH 8.5, binds preferentially with uridine and guanosine. The reagent also binds with poly U, with more than one molecule of reagent binding per two uridylate residues.

20. 8-AMINO-1,2,6-NAPHTHALENE TRISULFONIC ACID

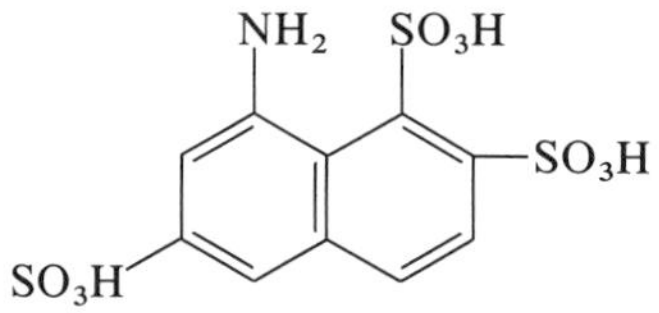

57

This reagent, 57, in the diazotized form reacts preferentially with the adenosine residues of DNA (Beer and Moudrianakis, 1962).

21. PERPHTHALIC ACID

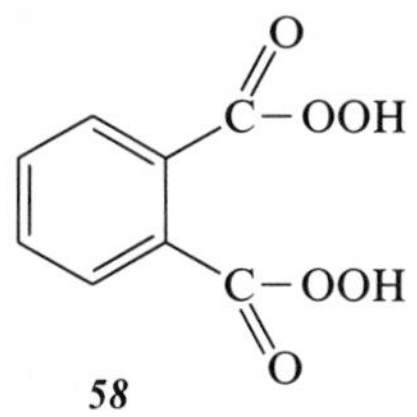

58

Perphthalic acid, *58*, in aqueous solution, pH 7, and at room temperature oxidizes adenine and cytosine derivatives to give the respective *N*-oxide derivatives, *59* and *60* (Cramer et al., 1963; Cramer and Seidel, 1963):

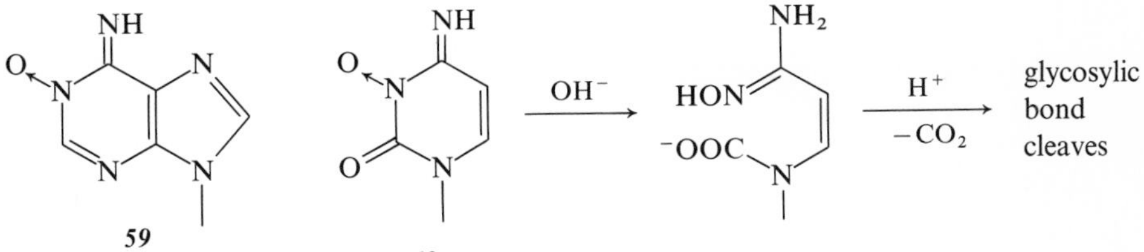

Cramer and Seidel (1964) extended the reaction to the oxidation of synthetic polyribonucleotides. Poly A can be 86% oxidized in a few minutes with a tenfold excess of perphthalic acid. The extent of oxidation of poly (A,U) is much less. Poly C is about 50% oxidized in 2 hr by a fivefold excess of the reagent. Uridine and guanosine residues do not react with perphthalic acid under these conditions.

In the presence of alkali, the N-3 oxide of cytidine, *60*, undergoes degradation with final splitting of the glycosylic bond in the presence of acid (Seidel, 1967). Seidel treated a model compound d(pCpT) with the reagent and then exposed the oxidized derivative to 0.3 *M* potassium hydroxide solution for 30 hr at 40°. This product was dissolved in 50% acetic acid solution and left for 20 hr at 40° in order to cleave the glycosylic bond. A second treatment of this derivative with alkali resulted in a beta elimination of the sugar residue giving rise to thymidine. This series of reactions, however, is somewhat vigorous for nucleic acids, especially for RNA.

22. *N*-2-FLUORENYLHYDROXYLAMINE AND *N*-ACETOXY-2-ACETYLAMINOFLUORENE

61   62

These reagents, *61* and *62*, respectively, appear to react preferentially with the guanine residues of DNA and RNA (Kriek, 1965; Miller et al., 1966). The reaction proceeds in aqueous solution (pH 4–5 for *61*; pH 7 for *62*) for 3–6 hr at 37°. These workers did not identify the reaction products but based their conclusion that the reagents react with guanine on the fact that the guanylate content of the treated nucleic acids decreases. These compounds are carcinogenic.

23. OSMIUM TETROXIDE

Thymidine, uridine, and cytidine are oxidized by osmium tetroxide at 0° in ammonium chloride buffer (pH 8.7) (Burton, 1967; Burton and Riley, 1966). The rate of reaction with thymidine is about 45 times as fast as that with cytosine nucleosides. The rate with uridine is intermediate. The reaction requires ammonia unless it is carried out at high temperatures. Burton suggests that the active oxidant is an uncharged complex formed between one molecule of the tetroxide and two molecules of unprotonated ammonia. The nature of the reaction is illustrated by the oxidation of 1-methyluracil, *63*, to give 5,6-dihydro-1-methyl-4,5,6-trihydroxypyrimidine-2, one, *64*.

63   64

4-Thiouridine is readily oxidized by osmium tetroxide in sodium bicarbonate buffer at pH 9 (Burton, 1967). The reaction product has not been identified. When the reaction is carried out in the ammonium chloride buffer, the major product is cytidine. Oxidation with periodate under similar conditions also yields cytidine (Ziff and Fresco, 1968; see Section 24). Burton believes that in the latter case the sulfur atom is oxidized to $-SO_2H$, which serves as a leaving group. This nucleophilic substitution reaction then proceeds at a much faster rate than hydroxylation of the 5,6 double bond.

In studies on poly U, Burton (1967) found that osmium tetroxide oxidizes about 95% of the uridylate residues in 60 min at 0° in an ammonium chloride buffer, pH 9.6. The reaction is almost completely inhibited when poly A is added to the mixture. Yeast tRNA was treated for 4 hr under the same condition. The cytosine, uracil, and thymine residues were oxidized to the extent of 12%, 20%, and 33%, respectively. From these observations Burton concludes that oxidation is limited by the secondary structure and that the groups that do react are probably in the single-stranded regions.

The effect of secondary structure on the reaction is particularly noticeable for DNA. Native double-stranded DNA does not react at all with the reagent, but the thymine residues of denatured single-stranded DNA are readily oxidized (Beer et al., 1966). The fact that the reaction rate is much faster with thymidine than with deoxycytidine makes the reaction somewhat selective.

24. PERIODATE OXIDATION

65   66

Treatment of 2',3'-O-isopropylidine-4-thiouridine, 65, with 0.01 M sodium periodate oxidizes it to the corresponding 2-oxypyrimidine-4-sulfonate, 66 (Ziff and Fresco, 1968). The reaction proceeds with pseudo-first-order kinetics (t1/2 = 15 sec at 35°, pH 7.0). The reaction product, 66, is susceptible to nucleophilic attack. Treatment of 66 with sodium hydroxide (pH 12.0) under mild conditions yields 2′,3′-O-isopropylidineuridine, and treatment with ammonium chloride (pH 8.5, 35°) results in formation of the cytidine derivative. This latter reaction is analogous to the nucleophilic substitution by NH_2. When 4-thiouridine is oxidized with osmium tetroxide in the presence of ammonium chloride, cytidine is the main product.

The conditions of the periodate oxidation are mild enough to apply to tRNA. A major, but not necessarily troublesome, side reaction would be the oxidation of the terminal nucleoside. Ziff and Fresco, in a proof note in their article, report that they have successfully converted 4-thiouridylate residues in tRNA to N^4-methylcytidylate residues.

25. N-ETHYLMALEIMIDE

Reagent 67 reacts selectively and quantitatively with 4-thiouridylic acid, 53, or with the 4-thiouridine residues in the tRNA of E. coli (Carbon and David, 1968). The reagent does not react with the 2-thiopyrimidine residues or with the other bases in tRNA. The use of $[^{14}C]$-N-ethyl-maleimide permits selective labeling of the 4-thiouridine residues of tRNA.

26. CHEMISTRY OF 5-(β-D-RIBOFURANOSYLURACIL) (PSEUDOURIDINE)

5-Ribosyluracil, 29, represents the only identified constituent of nucleic acids possessing a C-C glycosylic bond; because of this unique structure, it displays properties that warrant special mention in this chapter. The chemistry of 5-ribosyluracil has been reviewed by Chambers (1966), and in this section the chemical reactions that might be exploited for purposes of selective reactions in RNA will be discussed.

The C-5 attachment of the ribose means that the C-1′ is in an alpha position to the C-5/C-6 double bond, and this configuration imparts an allylic character to the C-1′. The consequential carbonium ion character of C-1′ may explain the facile isomerization of 5-ribosyluracil (Cohn, 1960; Shapiro and Chambers, 1961), which Tomasz et al. (1965) attributed to the formation of an open-chain intermediate, 68. They postulated that this would recyclize in four ways to give the anomeric forms 69, 70, 71, and 29:

Assuming that *68* can exist, it is reasonable to hypothesize that water could add across the double bond in a completing reaction to form a hydrated product of type *72*. This derivative distinguishes itself from all other ribonucleosides by the presence of the C-1'-, C-2'-vicinal dihydroxy grouping; therefore, on this basis, the 5-ribosyluracil residues in oligonucleotides should be sensitive to periodate oxidation. Tomasz et al. (1965) investigated this possibility, using 5-ribosyluracil 3'-phosphate as a model compound. The compound is oxidized by periodate, but unfortunately the conditions (pH 8.9, 50°, 23 hr) required to force the reaction are somewhat severe. The major products of the reaction are 5-formyluracil, *73*, 5-carboxyuracil, *74*, and sugar breakdown products. The 5-carboxyuracil does not result from the oxidation of 5-formyluracil but possibly is produced via direct oxidation of the hydroxylated intermediate, *72*, at C'. This oxidation gives compound *75*, which, in turn, could undergo further oxidation and hydrolysis to form 5-carboxyuracil.

Periodate oxidation of 5-ribosyluracil-3′,5′-diphosphate under the same conditions produces 5-formyluracil, 5-carboxyuracil, erythrose-4-phosphate, and inorganic phosphate. Cleavage of one of the phosphate bonds suggests that these conditions might be suitable for cleaving tRNA at the point of location of a 5-ribosyluracil residue; however, in practice, the method suffers severe limitations. The foremost limitation is that periodate destroys a large percentage of all the uracil residues of tRNA. Tomasz et al. (1965) treated *E. coli* tRNA for 50 hr at 50° and pH 5.0 with 84 moles of periodate per mole of tRNA and, after hydrolysis, recovered 70% of the original uridylic acid residues and only 55% of the pseudouridylic acid residues. This reaction obviously is not suitable for selective cleavage of tRNA.

Tomasz and Chambers (1964) observed that on exposure to light (253.7 mμ) pseudouridylic acid undergoes photolytic oxidation. The mild nature of the

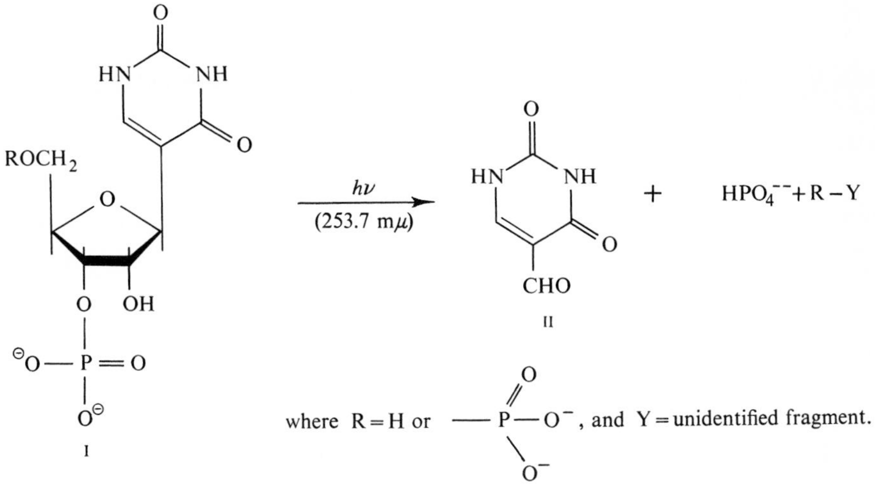

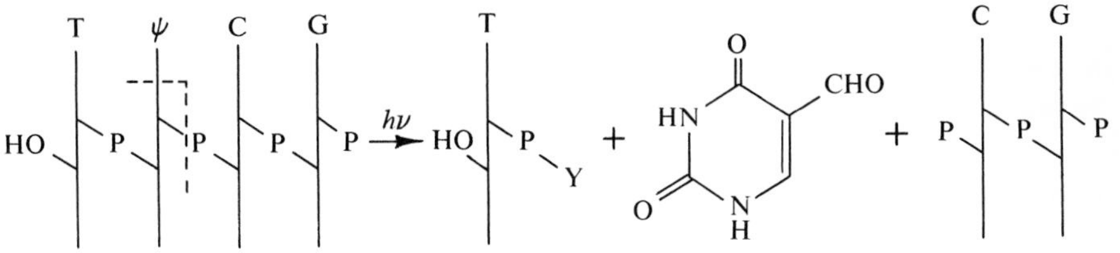

FIGURE 10. Photolysis of the tetranucleotide TpψpCpGp. Data taken from Tomasz and Chambers (1966).

reaction led to its investigation as a means of selectively cleaving oligonucleo-tides which contain a pseudouridylic acid residue. The tetranucleotide TpψpCpGp, common to yeast tRNAs (Zamir et al., 1965), was photolyzed; the results are shown in Figure 10. The photolytic action also causes a loss of approximately 20% in ultraviolet absorption, indicating the destruction of heterocyclic residues. This destruction is attributable to rapid nucleophilic addition to the 5,6 double bond of the pyrimidines (McLaren and Shugar, 1964), which results in hydration (Moore and Thomson, 1955) or dimerization (Beukers and Berends, 1960). Moreover, photolysis of 5,6-dihydrouridine represents another factor to take into account if one wishes to apply this reaction to tRNA. The photolytic reactions occurring under these conditions, therefore, appear to be too varied to employ on a multifunctional molecule such as tRNA.

27. PHOTO-OXIDATION OF 4-THIOURIDINE

Irradiation of 4-thiouridine in air-saturated t-butanol at 330 mμ leads to rapid and near quantitative conversion to uridine (Pleiss et al., 1969). No reaction occurs in the absence of oxygen. The pseudo-first-order rate constant for the formation of uridine is 0.08 min^{-1}. When the reaction is carried out in the presence of a mixture of ammonia and air (1:1), a mixture of cytidine and uridine (2:1) is obtained. The initial product of this reaction appears to be the sulfonate which suggests that the oxidation mechanism is analogous to that obtained with periodate (see Section 24).

Irradiation of 4-thiouridine in water leads to a complex mixture of products resulting from the addition of water to the 5,6 double bond and other reactions (see Section 28 on the photolysis of pyrimidines).

Pleiss et al. (1969) prepared the hexadecyltrimethylammonium salt of E. coli tRNA and irradiated it in t-butanol in the presence of air or ammonia-air. The 4-thiouridine content of the tRNA disappeared at a rate comparable to the rate of reaction of 4-thiouridine per se. They did not detect any breakdown of the tRNA under these conditions.

The major nucleosides do not react under these conditions, a fact which makes the reaction highly selective for 4-thiouridine.

This reaction offers the unique possibility of selectively converting a modified nucleoside to a major nucleoside and thus provides a convenient experimental tool for investigating the structure-function relationships of tRNA.

28. PHOTOCHEMISTRY

The pyrimidine components of nucleic acid undergo photoreactions that result principally in hydration of the 5,6 double bond or dimerization. When a nucleic acid sample is irradiated, the course of the photochemical reaction does not proceed completely randomly but is influenced by the nature of the secondary structure. For example, Schulman and Chambers (1968), working with purified tRNA[Ala], found that certain segments of this molecule are more vulnerable than others to a chemical reaction; furthermore, as might be expected, they observed that the ionic environment that affects secondary structure also influences the course of the reaction.

The photochemistry of nucleic acid components is complex, and an adequate discussion is beyond the scope of this book. Three excellent reviews, however, cover the subject:

A. D. McLaren and D. Shugar, *Photochemistry of Proteins and Nucleic Acids*, Macmillan, New York, 1964.

R. B. Setlow, *The Photochemistry, Photobiology and Repair of Polynucleotides*, Vol. VIII in *Progress in Nucleic Acid Research and Molecular Biology*, J. N. Davidson and W. E. Cohn, Eds., Academic Press, New York, 1968.

A. Wacker, *Molecular Mechanisms of Radiation Effects*, Vol. I in *Progress in Nucleic Acid Research and Molecular Biology*, J. N. Davidson and W. E. Cohn, Eds., Academic Press, New York, 1963.

TABLE 5

Reagents That React Selectively with Nucleic Acids under Mild Conditions: Principal Base Residues Attacked[a]

Reagent	Page No. in Text	pH	Base Residue								
			A	G	C	U	ψ	I	i⁶A	s⁴U	hU
Hydroxylamine	349	6.0			+						
		10.0				+					
O-Methylhydroxylamine	354	5.0			+						
Semicarbazide	356	4.2			+						
Acyl hydrazides	357	4.2			+						
Carbodiimide reagent	358	8.0		+		+	+				
Glyoxal	363	4.0		+							
Acrylonitrile	366	8.8				+	+	+			
Diazomethane	371	6.8	+	+							
Bromine	377	5.7		+	+	+					
Iodine	379	7.0							+		
Nitrous acid	381	3.8	+	+	+						
Permanganate	383	7.0							+	+	
Sodium borohydride	384										
Light		8.0				+					
Dark		8.0								+	+
Cyanogen bromide	388	8.9							+		
2-Amino-p-benzenesulfonic acid	389	9.0	+	+	+						
Mercury disodium-p-hydroquinone diacetate	390	8.5			+		+				
8-Amino-1,2,6-naphthalene trisulfonic acid	390			+							
Perphthalic acid	391	7.0	+		+						
N-2-Fluorenylhydroxylamine	392	5.0		+							
Osmium tetroxide	392	8.7			+	+					
Periodate	393	7.0								+	
N-Ethylmaleimide	394									+	

[a] A, adenosine; G, guanosine; C, cytidine; U, uridine; ψ, 5-ribosyluracil; I, inosine; i⁶A, N^6-(Δ^2-isopentenyl)adenosine; s⁴U, 4-thiouridine; hU, 5,6-dihydrouridine.

Chapter 9

GENERAL SUMMARY

U NTIL RECENTLY each nucleic acid was thought to consist of four
basic nucleoside structures. This view has changed markedly in the
last 15 years, and at present nine deoxyribonucleosides (five are modified
nucleosides) and thirty-nine ribonucleosides (thirty-five are modified nucleosides) are known to occur in DNA and RNA, respectively. In actual fact, DNA
and RNA can be thought of as consisting of four major nucleosides; all the
additional components are structural modifications of one of the eight basic
nucleoside structures. It is becoming apparent that these modified components
confer unique properties critical to the biological functioning of nucleic acid
molecules, and this concept adds a new dimension to the study of nucleic acid
structure and function.

The types of structural modifications found in this class of nucleic acid components are varied. The most common form is the addition of a methyl group
to one of the major nucleosides; more than twenty methylated components of
DNA and RNA are known. Other relatively simple structural alterations consist
of the replacement of an amino group with a sulfur atom (4-thiouridine), the
replacement of an amino group with a hydroxyl group (inosine), the reduction
of the 5,6 double bond of uridine (5,6-dihydrouridine), the attachment of an
acetyl group (N^4-acetylcytidine), and the rotation of a pyrimidine base on its
ribosyl pedestal (5-ribosyluracil). In addition to these types, modified nucleosides exist in which a more complex alteration occurs, for example, the attachment of a carboxymethyl group (5-carboxymethyluridine, 5-carboxymethyl-
2-thiouridine), the attachment of a Δ^2-isopentenyl group [N^6-(Δ^2-isopentenyl)-
adenosine], and the attachment of a carbamoylthreonine group [N-(nebularin-
6-ylcarbamoyl)-L-threonine].

No one molecular species of nucleic acid contains all the known modified nucleosides. In fact, the DNA and RNA of no one organism contain all the known modified nucleosides and, further, a certain amount of species-specific distribution may exist. The nucleic acids of a relatively limited number of organisms have been analyzed with respect to modified nucleoside content, although the organisms studied represent viruses, bacteria, yeast, plant tissues, and animal tissues. On the basis of these results it is possible to draw some general correlations concerning the pattern of distribution of the modified nucleoside. 5-Hydroxymethylcytosine and 5-hydroxymethyluracil, for example, occur only in the DNA of certain bacteriophages. 5-Methylcytosine occurs in the DNA of plant and animals, whereas N^6-methyladenine, as well as 5-methylcytosine, occurs in the DNA of bacteria. 4-Thiouridine has been found in the tRNA of *E. coli* but not in that of mammalian or plant tissue. N^6-(Δ^2-Isopentenyl)adenosine occurs in the tRNA of animals, plants, and bacteria, but its two derivatives, N^6-(Δ^2-isopentenyl)-2-methylthioadenosine and N^6-(*cis*-4-hydroxy-3-methylbut-2-enyl)adenosine, have been found only in the tRNA of *E. coli* and plant tissue, respectively. A noticeable difference occurs in *O* and *N* methylation in the rRNA of *E. coli* and mammalian cells. In *E. coli* rRNA methylation occurs predominantly on the heterocyclic residues of the nucleosides; in mammalian cell rRNA, on the 2'-*O* position of the ribosyl moiety. Undoubtedly, as the nucleic acids of additional organisms are investigated, modified nucleosides and/or distribution patterns of modified nucleosides peculiar to the nucleic acids of different organisms will be discovered.

Not all classes of nucleic acids contain modified nucleosides. The 5S ribosomal RNA and mRNA are not known to contain them. The DNA, in comparison to the rRNA and tRNA, is modified relatively little, and the modifications that occur do not affect the complementary hydrogen-bonding characteristics. Therefore, it seems that the species of nucleic acid responsible for encoding and transmitting the genetic message, such as DNA and mRNA, are not extensively structurally modified. On the other hand, the species involved in the functional activities necessary to synthesize proteins, such as rRNA and tRNA, are extensively modified.

With respect to the function of the modified nucleosides, it is impossible to treat them as a single class of nucleic acid components. Because of their structural variety, they obviously perform many different functions. It is difficult to assess the exact function of each of the modified nucleosides, although theoretical considerations as well as a number of general experimental observations permit some predictions. The methylated nucleosides, for example, represent a subtle change in the shape of the parent nucleosides. Apart from this, the methyl substituent changes the hydrogen-bonding characteristics and definitely alters the base-stacking characteristics. In other words, the attachment of one methyl group at a strategic location in the nucleic acid molecule could cause a significant change in its three-dimensional structure. Destruction of the aromaticity of the uracil residue as represented by 5,6-dihydrouridine similarly would alter drastically the neighbor-neighbor stacking interaction normally associated with uridine and its neighbors; this would undoubtedly have a considerable influence on the conformation of the region of the molecule containing 5,6-dihydrouridine residues. The substitution of an oxygen atom by a sulfur atom not only changes bulk at that particular location but also provides the nucleic acid molecule with an opportunity to form S-S bonds. Such bonds, particularly if intramolecular, would have a profound influence on the configuration of the molecules. Thus, in different ways, a minor alteration in structure of one of the nucleic acid components can have a significant influence on the three-dimensional structure of the whole nucleic acid molecule and may give that molecular species a unique character.

Some of the modified nucleosides that occur in tRNA result from the attachment of a relatively large side chain containing a functional group (organic chemistry definition). These "hypermodifications," apart from adding considerable bulk to the nucleoside, contain a functional group such as carboxyl, hydroxyl, or allylic double bond. The presence of a highly reactive group at a unique location in the tRNA molecule undoubtedly has considerable influence on the reactivity of these nucleic acid molecules.

The hypermodified nucleosides have particular relevance to the ability of the tRNA molecule to respond to the correct codon. In all the tRNA sequences that have been determined a hypermodified nucleoside, if present, occurs at a common position in the anticodon loop, adjacent to the 3′ end of the anticodon. The anticodon region in itself represents a structurally well-defined segment of the tRNA molecule. It consists of a single-stranded loop containing seven nucleotides (as based on the cloverleaf model). The triplet oligonucleotide representing the anticodon, located in the middle of the loop, is bracketed by a uridine residue on the 5′ side and the hypermodified nucleoside on the 3′ side. In terms of the importance of the hypermodified nucleoside to the functioning of the anticodon, it is significant that this nucleoside provides a functional group (i.e., hydroxyl, carboxyl, etc.) in a structurally unique configuration. Evidence supports the view that this modified component is essential to the anticodon-codon interaction of some tRNA molecules. Since the ribosome completes the binding trinity, it is conceivable that the hypermodified group is also concerned with attachment of the tRNA molecule to the ribosome-binding site. Whatever the exact mechanism may be. it is apparent that the hypermodified nucleoside plays a significant role in the translation mechanism.

Some of the structural modifications, particularly in tRNA, appear to be common to all molecular species and may be critical to the general functions of tRNA. For example, methylated guanosine residues, 5-methylcytidine, 5-methyluridine, 5-ribosyluracil, and N^6-methyladenosine occur at specific locations in each of the known tRNA sequences (see Figure 2, Chapter 4). It is interesting that the methylated guanosine residues and the 5-methylcytidine occur at the place in the primary sequence linking two helical regions, and they may enable the primary sequence to adopt its correct confirmation at these points. The dihydrouridine residues are concentrated in one general section of the primary sequence, and 5-methyluridine and 5-ribosyluracil occur together in a specific sequence common to all tRNA molecules. These specific structural features may represent common binding sites of the tRNA molecule with the ribosomes and/or with the aminoacyl synthetase, but they do not necessarily involve the specificity of the binding reactions.

The possible significance of modified nucleosides to nucleic acid structure, particularly that of tRNA, may be assessed better in the light of current concepts of tRNA function. Evidence is now accumulating that tRNA is involved in activities more complex than the passive carrying of an amino acid. It appears that regulation of differential protein synthesis may occur via not only a transcription mechanism but also a translation mechanism; in such a mechanism, tRNA undoubtedly would occupy a central place. In support of this concept, evidence has been obtained that certain tRNA molecules may be associated with specific protein molecules, and in this event cells would contain perhaps hundreds of molecular species of tRNA. Therefore, if there is a changing requirement for the synthesis of various cell components during the life of the cell, only a selection of the potentially available tRNA molecules would be active at any given instant. One method for the cell to create the correct spectrum of tRNA molecules would be to destroy unnecessary molecules and to synthesize necessary ones. Alternatively, a molecular mechanism could be invoked in which individual molecular species of tRNA are quantitatively activated-deactivated or their protein-related specificity is changed qualitatively. In this respect it is also conceivable that the aminoacyl synthetase and/or coding specificity of the tRNA molecule could be subtly altered to suit the changing requirements of specific protein synthesis. These changes, presumably reversible, could be accomplished by reversible modification at key locations in the tRNA molecule. Such a molecular mechanism would enable tRNA to play a critical role in the selection of proteins to be synthesized at a given time in the life cycle of a cell.

In light of these concepts, it is interesting that, although the basic structure of tRNA appears to be common to all organisms, when one goes up the phylogenetic scale, the extent of modification of the tRNA structure increases. This fact may well relate to the increasing complexity of cellular activity in higher organisms.

The mechanism of biogenesis of the modified components of nucleic acids as presently understood supports in a general way the postulates advanced above, for it appears that the alterations in nucleic acid structure occur after replication or transcription and not by incorporation of modified nucleosides. The experimental evidence shows that the alkyl substituents (methyl and Δ^2-isopentenyl) are added to the preformed nucleic acid molecule, and it is to be expected that all the other modified components are synthesized likewise at the macromolecular level. Such a mechanism raises two possibilities with respect to the timing of the modifying reactions. Modification of the primary structure could occur while the nucleic acid molecule is being synthesized or after synthesis is complete. In the case of rRNA and DNA, some evidence indicates that addition of methyl groups occurs at the time of the synthesis of the molecule. It does not necessarily follow that all the modified nucleosides are formed at a similar stage in the biosynthesis of the nucleic acid molecule. In the case of tRNA, this question has some theoretical significance because, if the structural modifications occur during transcription of the molecule, it can be presumed that the molecule still lacks its secondary structure at this stage. On the other hand, if the modification takes place at a later stage, after the tRNA molecule has assumed its fundamental secondary and tertiary conformations, this fact could have considerable bearing on the specificity of the enzymes involved in formation of the modified constituents. In the first instance, the primary sequence alone would dictate the enzyme specificity with respect to the particular modifications. In the second instance, a combination of primary and secondary structures might provide the specificity.

Experimental evidence suggests that the attachment of modifying substituents can take place *in vitro*, using tRNA that retains its secondary and tertiary structures. With respect to the hypermodified nucleosides, it appears that in some cases two, three, or more enzymes would be necessary to complete the structural modification; this would lead to some crowding if it occurred at the time of transcription. In view of the diversity of the structural modifications and their correspondingly diverse functions, the time of synthesis is probably not consistent for all modified nucleosides, and therefore various modifications occur at different times in the synthesis cycle and possibly at different locations in the cell.

The presence of enzymes in the cell which modify tRNA after it has assumed its three-dimensional structure raises the possibility that certain types of structural alterations may occur at any time during the existence of the tRNA molecule and that these alterations are important components of the mechanism by which the tRNA molecule functions. Consideration has already been given to the concept that reversible modification of tRNA structure may play a critical role in the control of the biological activity of this molecule. This idea raises the possibility that foreign agents could change the normal activity related to modification of the tRNA molecule. For example, certain drugs might inhibit such reactions on a very selective basis. Another means of interfering with the normal modification reactions is viral infection, which may change the spectrum of enzymic modifications of tRNA molecules. This could result from either *de novo* synthesis of tRNA or from the alteration of existing tRNA molecules. In any event, the infecting virus introduces a foreign genome, and, significantly, part of the new genome is concerned with the synthesis of viral-specific tRNA molecules.

Little is known about the metabolism of the modified nucleosides after they are released by catabolism of the nucleic acids. It appears that many modified nucleosides, at least in animals, are excreted intact or at most have the sugar moiety removed. Plant and animal tissues, however, contain enzymes that metabolize one of the modified components, N^6-(Δ^2-isopentenyl)adenosine, to give a number of products. This metabolic pathway seems unique with respect to the known modified nucleosides and may relate in some way to a biological function of this component apart from its presence in the tRNA.

N^6-(Δ^2-Isopentenyl)adenosine in plant systems stimulates cell growth and cell differentiation. This phenomenon, known as the cytokinin effect, is a biological property of several N^6-(substituted)adenine derivatives. The fact that N^6-(Δ^2-isopentenyl)adenosine exhibits this activity may be fortuitous or may in some way be related to a natural hormone function. If this is the case, the tRNA could represent an essential component in the biosynthesis of N^6-(Δ^2-isopentenyl)adenosine per se. If N^6-(Δ^2-isopentenyl)adenosine does have a hormone activity, its metabolic enzyme system becomes critical to this activity. Therefore, the level of free N^6-(Δ^2-isopentenyl)adenosine in the cell could be easily regulated by the rate of breakdown of the tRNA molecules containing this component and the activity of the metabolic enzyme system. A novel feature of this model is that mevalonic acid is the precursor of the Δ^2-isopentenyl side chain, and therefore regulation of the availability of this side chain for tRNA biosynthesis may be related to the control of isoprenoid biosynthesis. The point has been made that the Δ^2-isopentenyl group is essential to the function of certain tRNA molecules. Therefore, an important relationship can be envisaged between the control of the availability of mevalonate, the function of certain tRNA molecules, and the growth stimulatory activity of a nucleic acid component.

Apart from their biological significance, the modified nucleosides offer a number of advantages in the investigation of nucleic acid structure and function. For example, in the sequence work on rRNA, the modified nucleoside pair $m_2^6Apm_2^6Ap$, which occurs in the 16S RNA molecule, serves as an excellent anchor point for conducting sequence studies. With respect to tRNA, many of the modified nucleosides, because of unique chemical properties, can serve as loci for application of specific chemical reactions to modify tRNA structure. Such chemical alterations can be exploited for studies on nucleic acid structure. In conjunction with enzymic hydrolysis of tRNA they can be used to enhance significantly the specificity of a number of nucleases.

In summary, the modified nucleosides provide nucleic acids with a large number of diverse structural components. These components appear to perform a variety of essential functions, ranging from the subtle perturbation of the three-dimensional structure of the nucleic acid molecule to qualitative and quantitative modulation of its biological activity.

APPENDIX

Paper Chromatographic Data

The Rf values were obtained on Whatman No. 1 paper run in descending fashion.

		Solvent Systems				
	A	*B*	*C*	*D*	*E*	*V*
	1BuOH 86	2PrOH 2	2PrOH 680	2PrOH 7	EtoAC 4	BuOH 5
	NH₃ 5	1%(NH₄)₂̄	conc. HCl 170	H₂O 2	1PrOH 1	HOAC 3
Compound	H₂O 14	SO₄ 1	H₂O 144	NH₃ 1	H₂O 2	H₂O 2
Bases						
1 Adenine	0.30	0.64	0.24	0.46	0.20	0.66
2 1-Methyladenine	0.23	0.44	0.15	0.63	0.02	0.57
3 2-Methyladenine	0.51	0.67	0.34	0.58	0.23	0.64
4 N^6-Methyladenine	0.53	0.72	0.35	0.65	0.37	0.73
5 N^6,N^6-Dimethyl-adenine	0.65	0.78	0.43	0.72	0.62	0.73
6 N^6-(Δ^2-Isopentenyl)-adenine	0.85	0.91	0.76	0.89	0.84	0.87
7 N^6-(cis-4-hydroxy-3-methylbut-2-enyl)-adenine	0.71	0.74	0.58	0.68	0.54	0.77
10 Cytosine	0.24	0.52	0.39	0.49	0.06	0.54
11 3-Methylcytosine	0.41	0.68	0.38	0.59		0.59
13 5-Methylcytosine	0.31	0.57	0.33	0.51	0.08	0.59
14 5-Hydroxymethyl-cytosine	0.15	0.50	0.39	0.43	0.03	0.47
17 Guanine	0.00	0.00	0.10	0.12	0.00	0.18
18 1-Methylguanine	0.24	0.47	0.11	0.67	0.08	0.53
19 N^2-Methylguanine	0.13	0.47	0.30	0.62	0.11	
20 N^2,N^2-Dimethyl-guanine	0.06	0.50	0.21	0.43	0.12	0.69

APPENDIX

Solvent Systems

	Compound	A 1BuOH 86 NH$_3$ 5 H$_2$O 14	B 2PrOH 2 1%(NH$_4$)$_2^-$ SO$_4$ 1	C 2PrOH 680 conc. HCl 170 H$_2$O 144	D 2PrOH 7 H$_2$O 2 NH$_3$ 1	E EtoAC 4 1PrOH 1 H$_2$O 2	V BuOH 5 HOAC 3 H$_2$O 2
21	7-Methylguanine	0.08	0.47	0.18	0.28	0.08	0.56
22	Hypoxanthine	0.09	0.64	0.19	0.51	0.12	0.50
23	1-Methylhypoxanthine				0.45		
24	Uracil	0.20	0.63	0.67	0.48	0.27	0.58
25	3-Methyluracil	0.50	0.76	0.84	0.71	0.60	0.68
26	5-Methyluracil (thymine)	0.46	0.65	0.72	0.58	0.44	0.66
28	5-Hydroxyuracil	0.03	0.57	0.49	0.19	0.21	0.75
29	5-Hydroxymethyluracil	0.14	0.59	0.59	0.54	0.14	0.50
	Nucleosides						
35	Adenosine	0.30	0.57	0.29	0.53	0.19	0.53
36	1-Methyladenosine	0.17	0.46	0.27	0.46	0.00	0.51
37	2-Methyladenosine	0.37	0.63	0.41	0.60	0.24	0.69
38	N^6-Methyladenosine	0.46	0.91	0.35	0.70	0.30	0.69
39	N^6,N^6-Dimethyl-adenosine	0.68	0.78	0.47	0.70	0.69	
40	2'-O-Methyladenosine	0.47	0.76	0.39	0.81	0.32	0.72
41	N^6-(Δ^2-Isopentenyl)-adenosine	0.83	0.87	0.83	0.87	0.85	0.88
45	N-[9-(β-D-ribofur-anosyl)purin-6-ylcarbamoyl]-L-threonine	0.00	0.38	0.31	0.28	0.00	0.53
46	Cytidine	0.12	0.57	0.35	0.44	0.04	0.39
47	3-Methylcytidine	0.42	0.54	0.43	0.63	0.00	0.48
49	5-Methylcytidine	0.17	0.61	0.33	0.47	0.05	0.54
50	2'-O-Methylcytidine	0.28	0.64	0.53	0.62	0.09	0.57
51	N^4-Acetylcytidine	0.19	0.67	0.42	0.48	0.24	0.63
53	Guanosine	0.03	0.46	0.16	0.21	0.04	0.32
54	1-Methylguanosine	0.16	0.56	0.27	0.51	0.18	0.42
55	N^2-Methylguanosine			0.40	0.42	0.21	
56	N^2,N^2-Dimethyl-guanosine	0.14	0.59	0.37	0.48	0.19	
57	7-Methylguanosine		0.51	0.17	0.29	0.09	
58	2'-O-Methylguanosine	0.21	0.64	0.37	0.46	0.30	
59	Inosine	0.03	0.72	0.17	0.35	0.04	0.46
60	1-Methylinosine				0.51		0.53

Solvent Systems

	Compound	A 1BuOH 86 NH$_3$ 5 H$_2$O 14	B 2PrOH 2 1%(NH$_4$)$_2^-$ SO$_4$ 1	C 2PrOH 680 conc. HCl 170 H$_2$O 144	D 2PrOH 7 H$_2$O 2 NH$_3$ 1	E EtoAC 4 1PrOH 1 H$_2$O 2	V BuOH 5 HOAC 3 H$_2$O 2
61	Uridine	0.08	0.61	0.62	0.41	0.19	0.42
62	3-Methyluridine	0.45	0.77	0.79	0.67	0.47	0.60
63	5-Methyluridine	0.21	0.65	0.62	0.48	0.27	0.55
64	2'-O-Methyluridine	0.20	0.69	0.72	0.54	0.41	0.65
65	5-(β-D-ribofuranosyl) uracil (pseudouridine)	0.00	0.52	0.41	0.26	0.04	0.35
67	2-Thio-5-carboxy-methyluridine, methylester	0.07	0.66		0.40	0.64	0.65
68	4-Thiouridine	0.07	0.70	0.74	0.44	0.61	0.44
69	5-Hydroxyuridine	0.10	0.55	0.52	0.17	0.12	0.47
72	5,6-Dihydrouridine	0.15	0.56	0.55	0.47	0.08	0.40
74	Deoxyadenosine	0.36	0.64	0.22	0.58	0.22	0.62
75	N^6-Methyldeoxyadenosine	0.52	0.72	0.45	0.70	0.38	0.66
76	Deoxycytidine	0.25	0.61	0.50	0.59	0.06	0.58
77	5-Methyldeoxycytidine	0.35	0.63	0.58	0.63	0.10	0.61
79	Deoxyguanosine	0.08	0.55	0.13	0.32	0.05	0.50
80	Deoxyuridine	0.20	0.67	0.70	0.53	0.25	0.61
81	5-Methyldeoxyuridine (thymidine)	0.39	0.75	0.73	0.68	0.36	0.62
82	5-Hydroxymethyl-deoxyuridine	0.11	0.66	0.68	0.48	0.18	0.49

BIBLIOGRAPHY

Adams, A., Lindahl, T., and Fresco, J. R. (1967), *Proc. Natl. Acad. Sci. (U.S.)* **57**, 1684.

Adams, W. S., Davis, F., and Nakatani, M. (1960), *Am. J. Med.* **28**, 726.

Adler, M., and Gutman, A. B. (1959), *Science* **130**, 862.

Adler, M., Weissmann, B., and Gutman, A. B. (1958), *J. Biol. Chem.* **230**, 717.

Adman, R., and Doty, P. (1967), *Biochem. Biophys. Res. Commun.* **27**, 579.

Akashi, S., Murachi, T., Ishihara, H., and Goto, H. (1965), *J. Biochem.* (Tokyo), **58**, 162.

Albert, A., and Brown, D. J. (1954), *J. Chem. Soc.* 2060.

Alegria, A. H. (1967), *Biochim. Biophys. Acta* **149**, 317.

Alfoldi, L., Stent, G. S., and Clowes, R. C. (1962), *J. Mol. Biol.* **5**, 348

Amos, H., and Korn, M. (1958), *Biochim. Biophys. Acta* **29**, 444.

Anand, N., Clark, V. M., Hall, R. H., and Todd, A. R. (1952), *J. Chem. Soc.* 3665.

Andersen, W., Dekker, C. A., and Todd, A. R. (1952), *J. Chem. Soc.* 2721.

Anderson, C. D., Goodman, L., and Baker, B. R. (1959), *J. Am. Chem. Soc.* **81**, 3967.

Anderson, N. G. (1962), *Anal. Biochem.* **4**, 269.

Anderson, N. G., Green, J. G., Barber, M. L., and Ladd, F. C. (1963), *Anal. Biochem.* **6**, 153.

Anderson, N. G., and Ladd, F. C. (1962), *Biochim. Biophys. Acta* **55**, 275.

Andrews, K. J. M., and Barber, W. E. (1958), *J. Chem. Soc.* 2768.

Antonov, A. C., Favorova, O. O., and Belozerskii, A. N. (1962), *Dokl Akad Nauk SSSR* **147**, 1480.

Apgar, J., and Holley, R. W. (1962), *Biochem. Biophys. Res. Commun.* **8**, 391.

Arantz, B. W., and Brown, D. J. (1968), in *Synthetic Procedures in Nucleic Acid Chemistry*, Vol. 1, W. W. Zorbach and R. S. Tipson, Eds., Interscience–John Wiley and Sons, New York, p. 55.

Arber, W. (1965), *J. Mol. Biol.* **11**, 247.

Armstrong, D. J., Burrows, W. J., Skoog, F., Roy, K. L., and Söll, D. (1969), *Proc. Natl. Acad. Sci. (U.S.)* **63**, 834.

Asai, M., Miyaki, M., and Shimizu, B. (1967), *Chem. Pharm. Bull.* (Tokyo) **15**, 1856.

Augusti-Tocco, G., and Brown, G. L. (1965), *Nature* **206**, 683.

Augusti-Tocco, G., Carestia, C., Grippo, P., Parisi, E., and Scarano, E. (1968), *Biochim. Biophys. Acta* **155**, 8.

414

BIBLIOGRAPHY

Baczynskyj, L., Biemann, K., and Hall, R. H. (1968), *Science* **159**, 1481.

Baczynskyj, L., Biemann, K., Fleysher, M. H., and Hall, R. H. (1969), *Can J. Biochem.* **47**, 1202.

Baddiley, J., Buchanan, J. G., Hardy, F. E., and Stewart, J. (1959), *J. Chem. Soc.* 2893.

Baddiley, J., Lythgoe, B., McNeil, D., and Todd, A. R. (1943a), *J. Chem. Soc.* 383

Baddiley, J., Lythgoe, B., and Todd, A. R. (1943b), *J. Chem. Soc.* 386.

Baguley, B. C., and Staehelin, M. (1968), *Biochemistry* **7**, 45.

Baker, B. R., Joseph, J. P., and Schaub, R. E. (1954), *J. Org. Chem.* **19**, 631.

Baliga, B., Srinivasan, P. R., and Borek, E. (1965), *Nature* **208**, 555.

Balis, M. E., Salser, J. S., and Elder, A. (1964), *Nature* **203**, 1170.

Ballé, G., Cerutti, P., and Witkop, B. (1966), *J. Am. Chem. Soc.* **88**, 3946.

Bank, A., Gee, S., Mehler, A., and Peterkofsky, A. (1964), *Biochemistry* **3**, 1406.

Barnett, W. E., and Brown, D. H. (1967), *Proc. Natl. Acad. Sci. (U.S.)* **57**, 452.

Batt, R. D., Martin, J. K., Ploeser, J. M., and Murray, J. (1954), *J. Am. Chem. Soc.* **76**, 3663.

Bayev, A. A., Fodor, I., Mirzabekov, A. D., Axelrod, V. D., and Kazarinova, L. Ya. (1967a), *Mol. Biol. (USSR)* **1**, 859.

Bayev, A. A., Venkstern, T. V., Mirzabekov, A. D., Krutilina, A. I., Li, L., and Axelrod, V. D. (1967b), *Mol. Biol. (USSR)* **1**, 754.

Bayev, A. A., Venkstern, T. V., Mirzabekov, A. D., and Tatarskaya, R. I. (1963), *Biokhimiya* **28**, 931.

Beaven, G. H., Holiday, E. R., and Johnson, E. A. (1955), in *The Nucleic Acids*, Vol. 1, E. Chargaff and J. N. Davidson, Eds., Academic Press, New York, p. 493.

Beer, M., and Moudrianakis, E. N. (1962), *Proc. Natl. Acad. Sci. (U.S.)* **48**, 409.

Beer, M., Stern, S., Carmalt, D., and Mohlhenrich, K. H. (1966), *Biochemistry* **5**, 2283.

Behrend, R., and Roosen, O. (1889), *Ann.* **251**, 235.

Benes, J., Veres, K., Chvojka, L., and Friedrich, A. (1965), *Nature* **206**, 830.

Bergmann, F., and Dikstein, S. (1955), *J. Am. Chem. Soc.* **77**, 691.

Bergquist, P. L., Burns, D. J. W., and Plinston, C. A. (1968), *Biochemistry* **7**, 1751.

Bergquist, P. L., and Matthews, R. E. F. (1962), *Biochem. J.* **85**, 305.

Beukers, R., and Berends, W. (1960), *Biochim. Biophys. Acta* **41**, 550.

Bhat, C. C. (1968), in *Synthetic Procedures in Nucleic Acid Chemistry*, Vol. 1, W. W. Zorbach and R. S. Tipson, Eds., Interscience–John Wiley and Sons, New York, p. 200.

Biemann, K., and McCloskey, J. A. (1962), *J. Am. Chem. Soc.* **84**, 2005.

Biemann, K., Tsunakawa, S., Sonnenbichler, K., Feldmann, H., Dütting, D., and Zachau, H. G. (1966), *Angew. Chem.* **78**, 600.

Billen, D. (1968), *J. Mol. Biol.* **31**, 477.

Bissot, T. C., Parry, R. W., and Campbell, D. H. (1957), *J. Am. Chem. Soc.* **79**, 796.

Biswas, B. B., and Myers, J. (1960), *Nature* **186**, 238.

Bjork, G. R., and Svensson, I. (1967), *Biochim. Biophys. Acta* **138**, 430.

Bloch, K. (1965), *Science* **150**, 19.

Bloch, K., and Goodwin, T. W. (1959), cited in *Ciba Foundation Symposium on Biosynthesis of Terpenes and Sterols*, Boston, Little, Brown, p. 45.

Bock, R. M. (1967), in *Methods in Enzymology*, Vol. XII; *Nucleic Acids*, Part A, L. Grossman and K. Moldave, Eds., Academic Press, New York, p. 224.

Bogdanov, A. A., Prokof'ev, M. A., Antonovich, E. G., Terganova, G. V., and Anisimova, V. M. (1962), *Biokhimiya* **27**, 266.

Bonney, R. J., Mittelman, A., and Hall, R. H. (1971), paper submitted for publication.

Borek, E., Ryan, A., and Rockenbach, J. (1955), *J. of Bacteriol.* **69**, 460.

Borek, E., and Srinivasan, P. R. (1966), *Ann. Rev. Biochem.* **35**, 275.

Brammer, K. W. (1963), *Biochim. Biophys. Acta* **72**, 217.

Brawerman, G., and Chargaff, E. (1951), *J. Am. Chem. Soc.* **73**, 4052.

Brawerman, G., Hufnagel, D. A., and Chargaff, E. (1962), *Biochim. Biophys. Acta* **61**, 340.

Bredereck, H., Haas, H., and Martini, A. (1948), *Chem. Ber.* **81**, 307.

Breshears, S. R., Wang, S. S., Bechtolt, S. G., and Christensen, B. E. (1959), *J. Am. Chem. Soc.* **81**, 3789.

Brimacombe, R. L. C., Griffin, B. E., Haines, J. A., Haslam, W. J., and Reese, C. B. (1965), *Biochemistry* **4**, 2452.

Brookes, P., and Lawley, P. D. (1960), *J. Chem. Soc.* 539.

Brookes, P., and Lawley, P. D. (1962), *J. Chem. Soc.* 1348.

Broom, A. D., and Robins, R. K. (1965), *J. Am. Chem. Soc.* **87**, 1145.

Broom, A. D., Townsend, L. B., Jones, J. W., and Robins, R. K. (1964), *Biochemistry* **3**, 494.

Brossmer, R., and Röhm, E. (1963), *Angew. Chem. Intern. Ed.* **2**, 742.

Brossmer, R., and Röhm, E. (1964), *Angew. Chem. Intern. Ed.* **3**, 66.

Brostoff, S. W., and Ingram, V. M. (1967), *Science* **158**, 666.

Broude, N. E., Budowsky, E. I., and Kochetkov, N. K. (1967), *Mol. Biol. (USSR)* **1**, 214.

Brown, D. H., and Novelli, G. D. (1968), *Biochem. Biophys. Res. Commun.* **31**, 262.

Brown, D. J., Hoerger, E., and Mason, S. F. (1955), *J. Chem. Soc.* 211.

Brown, D. M., Burdon, M. G., and Slatcher, R. P. (1965), *Chem. Commun.* 77.

Brown, D. M., Burdon, M. G., and Slatcher, R. P. (1968), *J. Chem. Soc.* 1051.

Brown, D. M., and Phillips, J. H. (1965), *J. Mol. Biol.* **11**, 663.

Brown, D. M., and Schell, P. (1961), *J. Mol. Biol.* **3**, 709.

Brown, D. M., and Schell, P. (1965), *J. Chem. Soc.* 208.

Brown, D. M., Todd, A. R., and Varadarajan, S. (1956), *J. Chem. Soc.* 2384.

Brown, E. B., and Johnson, T. B. (1923), *J. Am. Chem. Soc.* **45**, 2702.

Brown, F. B., Cain, J. C., Gant, D. E., Parker, L. F. J., and Smith, E. L. (1955), *Biochem. J.* **59**, 82.

Brown, G. M., and Attardi, G. (1965), *Biochem. Biophys. Res. Commun.* **20**, 298.

Brown, J. (1955), *J. Appl. Chem.* **5**, 358.

Brownlee, G. G., Sanger, F., and Barrell, B. G. (1968), *J. Mol. Biol.* **34**, 379.

Buckley, W. B., Witkus, E. R., and Berger, C. A. (1962), *Nature* **194**, 1200.

Budowsky, E. I., Shibayev, V. N., and Eliseeva, G. I. (1968a), in *Synthetic Procedures in Nucleic Acid Chemistry*, Vol. 1, W. W. Zorbach and R. S. Tipson, Eds., Interscience–John Wiley and Sons, New York, p. 436.

Budowsky, E. I., Shibayeva, R. P., Sverdlov, E. D., and Monastyrskaya, G. S. (1968b), *Mol. Biol. (USSR)* **2**, 321.

Budowsky, E. I., Simukova, N. A., and Gus'kova, L. I. (1968c), *Biochim. Biophys. Acta* **166**, 755.

BIBLIOGRAPHY

Budowsky, E. I., Simukova, N. A., Shibayeva, R. P., and Kochetkov, N. K. (1965), *Biokhimiya* **30**, 902.

Budowsky, E. I., Sverdlov, E. D., Shibayeva, R. P., Monastyrskaya, G. S., and Kochetkov, N. K. (1968d), *Mol. Biol. (USSR)* **2**, 329.

Burdon, R. H. (1966), *Nature* **210**, 797.

Burrows, W. J., Armstrong, D. J., Skoog, F., Hecht, S. M., Boyle, J.T.A., Leonard, N. J., and Occolowitz, J. (1968), *Science* **161**, 691.

Burton, K. (1967), *Biochem. J.* **104**, 686.

Burton, K., and Riley, W. T. (1966), *Biochem. J.* **98**, 70.

Byrne, R., Levin, J. G., Bladen, H. A., and Nirenberg, M. W. (1964), *Proc. Natl. Acad. Sci. (U.S.)* **52**, 140.

Capra, J. D., and Peterkofsky, A. (1966), *J. Mol. Biol.* **21**, 455.

Carbon, J. A. (1960), *J. Org. Chem.* **25**, 1731.

Carbon, J. A. (1965), *Biochim. Biophys. Acta* **95**, 550.

Carbon, J. A., and Curry, J. B. (1968), *Proc. Natl. Acad. Sci. (U.S.)* **59**, 467.

Carbon, J. A., and David, H. (1968), *Biochemistry* **7**, 3851.

Carbon, J. A., David, H., and Studier, M. H. (1968), *Science* **161**, 1146.

Carbon, J. A., Hung, L., and Jones, D. (1965), *Proc. Natl. Acad. Sci. (U.S.)* **53**, 979.

Cavalieri, L. F., Tinker, J. F., and Bendich, A. (1949), *J. Am. Chem. Soc.* **71**, 533.

Ceccarini, C., Maggio, R., and Barbata, G. (1967), *Proc. Natl. Acad. Sci. (U.S.)* **58**, 2235.

Cerna, J., Rychlik, I., and Sorm, F. (1964), *Collection Czech. Chem. Commun.* **29**, 2832.

Cerutti, P. (1968), *Biochem. Biophys. Res. Commun.* **30**, 434.

Cerutti, P., Holt, W., and Miller, N. (1968a), *J. Mol. Biol.* **34**, 505.

Cerutti, P., Ikeda, K., and Witkop, B. (1965), *J. Am. Chem. Soc.* **87**, 2505.

Cerutti, P., Kondo, Y., Landis, W. R., and Witkop, B. (1968b), *J. Am. Chem. Soc.* **90**, 771.

Cerutti, P., and Miller, N. J. (1967), *J. Mol. Biol.* **26**, 55.

Chambers, R. W. (1965), *Biochemistry* **4**, 219.

Chambers, R. W. (1966), in *Progress in Nucleic Acid Research and Molecular Biology*, Vol. 5, J. N. Davidson and W. E. Cohn, Eds., Academic Press, New York, p. 349.

Chambers, R. W., Kurkov, V., and Shapiro, R. (1963), *Biochemistry* **2**, 1192.

Chambon, P., Weill, J. D., Doly, J., Strosser, M. T., and Mandel, P. (1966), *Biochem. Biophys. Res. Commun.* **25**, 638.

Champe, S. P., and Benzer, S. (1962), *Proc. Natl. Acad. Sci. (U.S.)* **48**, 532.

Chargaff, E., Lipshitz, R., and Green, C. (1952), *J. Biol. Chem.* **195**, 155.

Chen, C-M., and Hall, R. H. (1969), *Phytochemistry* **8**, 1687.

Chen, C-M., Logan, D. M., McLennan, B.D., and Hall, R. H. (1968), *Plant Physiol.* **43**, S-18.

Chheda, G. B. (1969), *Life Sci.*, **8**, 979.

Chheda, G. B., and Hall, R. H. (1966), *Biochemistry* **5**, 2082.

Chheda, G. B., and Hall, R. H. (1969), *J. Org. Chem.* **34**, 3492.

Chheda, G. B., Hall, R. H., Magrath, D. I., Mozejko, J., Schweizer, M. P., Stasiuk, L., and Taylor, P. R. (1969a), *Biochemistry* **8**, 3278.

Chheda, G. B., Hall, R. H., and Tanna, P. M. (1969b), *J. Org. Chem.* **34**, 3498.

Chheda, G. B., and Mittelman, A. (1967a), Abstracts, 154th Meeting, American Chemical Society.

Chheda, G. B., and Mittelman, A. (1967b), *Fed. Proc.* **26**, 730.

Chheda, G. B., Mittelman, A., and Grace, J. T. (1969c), *J. Pharm. Sci.* **58**, 75

Chheda, G. B., Mittelman, A., and Grace, J. T. (1968), Abstracts, 156th Meeting, American Chemical Society, Atlantic City, N.J.

Chrispeels, M. J., Boyd, R. F., Williams, L. S., and Neidhardt, F. C. (1968), *J. Mol. Biol.* **31**, 463.

Clark, B. F. C., Dube, S. K., and Marcker, K. A. (1968), *Nature* **219**, 484.

Cline, R. E., Fink, R. M., and Fink, K. (1959), *J. Am. Chem. Soc.* **81**, 2521:

Codington, J. F., Fecher, R., Maguire, M. H., Thomson, R. Y., and Brown, G. B. (1958), *J. Am. Chem. Soc.* **80**, 5164.

Cohn, W. E. (1949), *Science* **109**, 377.

Cohn, W. E. (1955), in *The Nucleic Acids*, Vol. 1, E. Chargaff and J. N. Davidson, Eds., Academic Press, New York, p. 211.

Cohn, W. E. (1960), *J. Biol. Chem.* **235**, 1488.

Cohn, W. E. (1967), in *Methods in Enzymology*, Vol. XII; *Nucleic Acids*, Part A, L. Grossman and K. Moldave, Eds., Academic Press, New York, p. 101.

Cohn, W. E., Kurkov, V., and Chambers, R. W. (1963), *Biochem. Prep.* **10**, 135.

Cohn, W. E., and Uziel, M. (1961), *Biochem. Prep.* **8**, 116.

Cohn, W. E., and Volkin, E. (1951), *Nature* **167**, 483.

Cornish, H. H., and Christman, A. A. (1957), *J. Biol. Chem.* **228**, 315.

Cory, S., Marcker, K. A., Dube, S. K., and Clark, B. F. C. (1968), *Nature* **220**, 1039.

Craddock, V. M., and Magee, P. N. (1963), *Biochem. J.* **89**, 32.

Craddock, V. M., Villa-Trevino, S., and Magee, P. N. (1968), *Biochem. J.* **107**, 179.

Cramer, F., Randerath, K., and Schäfer, E. A. (1963), *Biochim. Biophys. Acta* **72**, 150.

Cramer, F., and Seidel, H. (1963), *Biochim. Biophys. Acta* **72**, 157.

Cramer, F., and Seidel, H. (1964), *Biochim. Biophys. Acta* **91**, 14.

Crestfield, A. M., and Allen, F. W. (1957); *Chromatographic Methods*, **2**, 9.

Das, H. K., Goldstein, A., and Lowney, L. I. (1967), *J. Mol. Biol.* **24**, 231.

Davidson, D., and Baudisch, O. (1926), *J. Am. Chem. Soc.* **48**, 2379.

Davis, F. F., and Allen, F. W. (1957), *J. Biol. Chem.* **227**, 907.

Davis, F. F., Carlucci, A. F., and Roubein, I. F. (1959), *J. Biol. Chem.* **234**, 1525.

Davoll, J., and Lowy, B. A. (1951), *J. Am. Chem. Soc.* **73**, 1650.

Davoll, J., and Lowy, B. A. (1952), *J. Am. Chem. Soc.* **74**, 1563.

Davoll, J., Lythgoe, B., and Todd, A. R. (1948a), *J. Chem. Soc.* 1685.

Davoll, J., Lythgoe, B., and Todd, A. R. (1948b), *J. Chem. Soc.* 967.

Dawid, I. B. (1965), *J. Mol. Biol.* **12**, 581.

Dekker, C. A. (1965), Symposium on Chemistry of Nucleosides and Nucleotides, 150th Meeting, American Chemical Society, Atlantic City, N.J.

Dekker, C. A., and Todd, A. R. (1950), *Nature* **166**, 557.

diCarlo, F. J., Schultz, A. S., and Kent, A. M. (1952), *J. Biol. Chem.* **199**, 333.

Dion, H. W., Calkins, D. G., and Pfiffner, J. J. (1954), *J. Am. Chem. Soc.* **76**, 948.

Dlugajczyk, A., and Eiler, J. J. (1966a), *Biochim. Biophys. Acta* **119**, 11.

BIBLIOGRAPHY

Dlugajczyk, A., and Eiler, J. J. (1966b), *Proc. Soc. Exptl. Biol. Med.* **123**, 453.

Doctor, B. P., Loebel, J. E., Sodd, M. A., and Winter, D. B. (1969), *Science* **163**, 693.

Doi, R. H., Kaneko, I., and Igarashi, R. T. (1968), *J. Biol. Chem.* **243**, 945.

Dornow, A., and Petsch, G. (1954), *Ann*, **588**, 45.

Dorsey, J. K., and Porter, J. W. (1968), *J. Biol. Chem.* **243**, 4667.

Doskočil, J., and Šormová, Z. (1965), *Biochim. Biophys. Acta* **95**, 513.

Dube, S. K., Marcker, K. A., Clark, B. F. C., and Cory, S. (1968), *Nature* **218**, 232.

Dubin, D. T., and Gunalp, A. (1967), *Biochim. Biophys. Acta* **134**, 106.

Dunn, D. B. (1959), *Biochim. Biophys. Acta* **34**, 286.

Dunn, D. B. (1960), *Biochim. Biophys. Acta* **38**, 176.

Dunn, D. B. (1961a), *Biochim. Biophys. Acta* **46**, 198.

Dunn, D. B. (1961b), *Fifth Intern. Congr. Biochim.*, Moscow **10**, 68 (also unpublished data).

Dunn, D. B. (1963), *Biochem. J.* **86**, 14P.

Dunn, D. B., and Flack, I. H. (1967a), Abstracts 4th Federation Europe Biochemical Society, p. 82.

Dunn, D. B., and Flack, I. H. (1967b), *Phytochem.* **6**, 459.

Dunn, D. B., and Hall, R. H. (1968), in *Handbook of Biochemistry*, H. A. Sober, Ed., Chemical Rubber Co., Cleveland, G-3 to G-90.

Dunn, D. B., Hitchborn, J. H., and Trim. A. R. (1963), *Biochem. J.* **88**, 34P.

Dunn, D. B., and Smith, J. D. (1958), *Biochem. J.* **68**, 627.

Dunn, D. B., Smith, J. D., and Simpson, M. V. (1960b), *Biochem. J.* **76**, 24P.

Dunn, D. B., Smith, J. D., and Spahr, P. F. (1960a), *J. Mol. Biol.* **2**, 113.

Echelin, P., and Morris, I. (1965), *Biol. Rev.* **40**, 143.

Eisen, A. Z., Weissmann, S., and Karon, M. (1962), *J. Lab. Clin. Med.* **59**, 620.

Elion, G. B. (1962), *J. Org. Chem.* **27**, 2478.

Elion, G. B., Burgi, E., and Hitchings, G. H. (1952), *J. Am. Chem. Soc.* **74**, 411.

Elion, G. B., Ide, W. S., and Hitchings, G. H. (1946), *J. Am. Chem. Soc.* **68**, 2137.

Elion, G. B., Lange, W. H., and Hitchings, G. H. (1956), *J. Am. Chem. Soc.* **78**, 217.

Elmore, D. T. (1950), *J. Chem. Soc.* 2084.

Engelbrecht, L. (1967), *Wiss. Z. Univ. Rostock* (*Math.-Naturwiss. Reihe*) **16**, 647.

Ergle, D. R., and Katterman, F. R. H. (1961), *Plant Physiol.* **36**, 811.

Ergle, D. R., Katterman, F. R. H., and Richmond, T. R. (1964), *Plant Physiol.* **39**, 145.

Erickson, H., and Beer, M. (1967), *Biochemistry* **6**, 2694.

Falaschi, A., and Kornberg, A. (1965), *Proc. Natl. Acad. Sci.* (*U.S.*) **54**, 1713.

Farkas, J., Kaplan, L., and Fox, J. J. (1964), *J. Org. Chem.* **29**, 1469.

Fasman, G. D., Lindblow, C., and Seaman, E. (1965), *J. Mol. Biol.* **12**, 630.

Feldmann, H., Dütting, D., and Zachau, H. G. (1966), *Z. Physiol. Chem.* **347**, 236.

Felix, K., Jilke, I., and Zahn, R. K. (1956), *Hoppe-Seylers Z. Physiol. Chem.* **303**, 140.

Fellner, P., and Sanger, F. (1968), *Nature* **219**, 236.

Fenrych, W., Falerych, W., Huang, H., and Johnson, B. C. (1968), *Fed. Proc.* **27**, 795.

Ferris, J. P., and Orgel, L. E. (1966), *J. Am. Chem. Soc.* **88**, 3829.

Fikus, M., Wierzchowski, K. L., and Shugar, D. (1962), *Photochem. Photobiol.* **1**, 325.

Fink, K., and Adams, W. S. (1968), *Arch. Biochem. Biophys.* **126**, 27.

Fink, K., Adams, W. S., Davis, F. W., and Nakatani, M. (1963), *Cancer Res.* **23**, 1824.

Fink, K., Adams, W. S., and Pfleiderer, W. (1964), *J. Biol. Chem.* **239**, 4250.

Fink, R. M., Cline, R. E., McGaughey, C., and Fink, K. (1956), *Anal. Chem.* **28**, 4.

Fischer, E. (1897), *Chem. Ber.* **30**, 2400.

Fischer, E., and Roeder, G. (1901), *Chem. Ber.* **34**, 3751.

Fiskin, A., and Beer, M. (1965), *Biochim. Biophys. Acta* **108**, 159.

Fissekis, J. D., Myles, A., and Brown, G. B. (1964), *J. Org. Chem.* **29**, 2670.

Fittler, F., and Hall, R. H. (1966), *Biochem. Biophys. Res. Commun.* **25**, 441.

Fittler, F., Kline, L. K., and Hall, R. H. (1968a), *Biochemistry* **7**, 940.

Fittler, F., Kline, L. K., and Hall, R. H. (1968b), *Biochem. Biophys. Res. Commun.* **31**, 571.

Flaks, J. G., and Cohen, S. S. (1957), *Biochim. Biophys. Acta* **25**, 667.

Fleissner, E. (1967), *Biochemistry* **6**, 621

Fleissner, E., and Borek, E. (1963), *Biochemistry* **2**, 1093.

Foft, J. W., Hsu, W-T., and Weiss, S. B. (1968), *Fed. Proc.* **27**, 341.

Forget, B. G., and Weissman, S. M. (1967), *Science* **158**, 1695.

Forget, B. G., and Weissman, S. M. (1968), *J. Biol. Chem.* **243**, 5709.

Fox, J. J., and Shugar, D. (1952), *Biochim. Biophys. Acta* **9**, 369.

Fox, J. J., and Van Praag, D. (1960), *J. Am. Chem. Soc.* **82**, 486.

Fox, J. J., Van Praag, D., Wempen, I., Doerr, I. L., Cheong, L., Knoll, J. E., Eidinoff, M. L., Bendich, A., and Brown, G. B. (1959), *J. Am. Chem. Soc.* **81**, 178.

Fox, J. J., Watanabe, K. A., and Bloch, A. (1966), in *Progress in Nucleic Acid Research and Molecular Biology*, Vol. 5, J. N. Davidson and W. E. Cohn, Eds., Academic Press, New York, p. 251.

Fox, J. J., Wempen, I., Hampton, A., and Doerr, I. L. (1958), *J. Am. Chem. Soc.* **80**, 1669.

Fox, J. J., Yung, N., Davoll, J., and Brown, G. B. (1956), *J. Am. Chem. Soc.* **78**, 2117.

Fox, J. J., Yung, N., Wempen, I., and Doerr, I. L. (1957), *J. Am. Chem. Soc.* **79**, 5060.

Fox, J. J., Yung, N., Wempen, I., and Hoffer, M. (1961), *J. Am. Chem. Soc.* **83**, 4066.

Freese, E., Bautz, E., and Bautz-Freese, E. (1961), *Proc. Natl. Acad. Sci. (U.S.)* **47**, 845.

Freese, E., and Bautz-Freese, E. (1965), *Biochemistry* **4**, 2419.

Fresco, J. R. (1963), in *Informational Macromolecules*, H. J. Vogel, V. Bryson, and J. O. Lampen, Eds., Academic Press, New York, p. 121.

Friedman, O. M., Mahapatra, G. N., and Stevenson, R. (1965), *Biochim. Biophys. Acta* **103**, 286.

Fujimoto, D., Srinivasan, P. R., and Borek, E. (1965), *Biochemistry* **4**, 2849.

Fuller, W., and Hodgson, A. (1967), *Nature* **215**, 817.

Furukawa, Y., and Honjo, M. (1968), *Chem. Pharm. Bull.* (Tokyo) **16**, 1076.

Furukawa, Y., Kobayashi, K., Kanai, Y., and Honjo, M. (1965), *Chem. Pharm. Bull.* (Tokyo) **13**, 1273.

Gal-Or, L., Mellema, J. E., Moudrianakis, E. N., and Beer, M. (1967), *Biochemistry* **6**, 1909.

Gassen, H. G., and Witzel, H. (1965), *Biochim. Biophys. Acta* **95**, 244.

Gefter, M., Hausmann, R., Gold, M., and Hurwitz, J. (1966), *J. Biol. Chem.* **241**, 1995.

Gefter, M., and Russell, R. L. (1969), *J. Mol. Biol.* **39**, 145.

Geiduscheck, E. P. (1961), *Proc. Natl. Acad. Sci. (U.S.)* **47**, 950.

BIBLIOGRAPHY

Gerster, J. F., Jones, J. W., and Robins, R. K. (1963), *J. Org. Chem.* **28**, 945.

Gerster, J. F., and Robins, R. K. (1965), *J. Am. Chem. Soc.* **87**, 3752.

Gerster, J. F., and Robins, R. K. (1966), *J. Org. Chem.* **31**, 3258.

Gilham, P. T. (1962), *J. Am. Chem. Soc.* **84**, 687.

Gillam, I., Millward, S., Blew, D., von Tigerstrom, M., Wimmer, E., and Tener, G. M. (1966), *Biochemistry* **6**, 3043.

Gin, J. B., and Dekker, C. A. (1968), *Biochemistry* **7**, 1413.

Giner-Sorolla, A., and Bendich, A. (1958), *J. Am. Chem. Soc.* **80**, 3932.

Ginsberg, T., and Davis, F. F. (1968), *J. Biol. Chem.* **243**, 6300.

Girshovich, A. S., Grachev, M. A., and Obukhova, L. V. (1968), *Mol. Biol. (USSR)* **2**, 351.

Girshovich, A. S., Knorre, D. G., Nelidova, O. D., and Ovander, M. N. (1966), *Biochim. Biophys. Acta* **119**, 216.

Glebov, R. N., Zaitseva, G. N., and Belozerskii, A. N. (1965), *Biokhimiya* **30**, 586.

Goehler, B., and Doi, R. H. (1966), *Proc. Natl. Acad. Sci. (U.S.)* **56**, 1047.

Goehler, B., and Doi, R. H. (1968), *J. Bacteriol.* **95**, 793.

Goehler, B., Kaneko, I., and Doi, R. (1966), *Biochem. Biophys. Res. Commun.* **24**, 446.

Gold, M., Gefter, M., Hausmann, R., and Hurwitz, J. (1966), *J. Gen. Physiol.* **49**, 5.

Gold, M., Hausmann, R., Maitra, U., and Hurwitz, J. (1964), *Proc. Natl. Acad. Sci. (U.S.)* **52**, 292.

Gold, M., and Hurwitz, J. (1963), *Cold Spring Harbor Symp. Quant. Biol.* **28**, 149.

Gold, M., and Hurwitz, J. (1964a), *J. Biol. Chem.* **239**, 3858.

Gold, M., and Hurwitz, J. (1964b), *J. Biol. Chem.* **239**, 3866.

Gold, M., Hurwitz, J., and Anders, M. (1963), *Proc. Natl. Acad. Sci. (U.S.)* **50**, 164.

Goldstein, G. (1967), *Anal. Biochem.* **20**, 477.

Goldstein, J., Bennett, T. P., and Craig, L. C. (1964), *Proc. Natl. Acad. Sci. (U.S.)* **51**, 119.

Goldwasser, E., and Heinrikson, R. L. (1966), in *Progress in Nucleic Acid Research and Molecular Biology*, Vol. 5, J. N. Davidson and W. E. Cohn, Eds., Academic Press, New York, p. 399.

Gonano, F. (1967), *Biochemistry* **6**, 977.

Goodman, H. M., Abelson, J., Landy, A., Brenner, S., and Smith, J. D. (1968), *Nature* **217**, 1019.

Gordon, J., Boman, H. G., and Isaksson, L. A. (1964), *J. Mol. Biol.* **9**, 831.

Gordon, M. P., Intrieri, O. M., and Brown, G. B. (1958), *J. Am. Chem. Soc.* **80**, 5161.

Gough, H. M., and Lederberg, S. (1966), *J. Bacteriol.* **91**, 1460.

Grace, J. T., Hakala, M. T., Hall, R. H., and Blakeslee, J. (1967), *Proc. Am. Assc. Cancer Res.* **8**, 23.

Gray, M. W., and Lane, B. G. (1967), *Biochim. Biophys. Acta* **134**, 243.

Gray, M. W., and Lane, B. G. (1968), *Biochemistry* **7**, 3441.

Green, M., and Cohen, S. S. (1957), *J. Biochem.* **225**, 397.

Greenberg, H., and Penman, S. (1966), *J. Mol. Biol.* **21**, 527.

Griffin, B. E., Haines, J. A., and Reese, C. B. (1967), *Biochim. Biophys. Acta* **142**, 536.

Griffin, B. E., Todd, A. R., and Rich, A. (1958), *Proc. Natl. Acad. Sci. (U.S.)*, **44**, 1123.

Grillo, R. S., and Polsky, R. (1966), *Exptl. Cell. Res.* **44**, 375.

Grimm, W. A. H., Fujii, T., and Leonard, N. J. (1968), in *Synthetic Procedures in Nucleic Acid Chemistry*, W. W. Zorbach and R. S. Tipson, Eds., Interscience–J. Wiley and Sons, New York, p. 212.

Grimm, W. A. H., and Leonard, N. J. (1967), *Biochemistry* **6**, 3625.

Grünberger, D., Holy, A., Smrt, J., and Sörm, F. (1968), *Coll. Czech. Chem. Comm.* **33**, 3858.

Gulland, J. M., and Holiday, E. R. (1936), *J. Chem. Soc.* 765.

Guyot, M. M., and Mentzer, C. (1958), *Compt. Rend.* **246**, 436.

Haines, J. A., Reese, C. B., and Todd, A. R. (1962), *J. Chem. Soc.* 5281.

Haines, J. A., Reese, C. B., and Todd, A. R. (1964), *J. Chem. Soc.* 1406.

Hall, J. B., and Allen, F. W. (1960), *Biochim. Biophys. Acta* **45**, 163.

Hall, R. H. (1962), *J. Biol. Chem.* **237**, 2283.

Hall, R. H. (1963a), *Biochem. Biophys. Res. Commun.* **12**, 361.

Hall, R. H. (1963b), *Biochem. Biophys. Res. Commun.* **12**, 429.

Hall, R. H. (1963c), *Biochem. Biophys. Res. Commun.* **13**, 394.

Hall, R. H. (1963d), *Biochim. Biophys. Acta* **68**, 278 (and unpublished).

Hall, R. H. (1964a), *Biochemistry* **3**, 769.

Hall, R. H. (1964b), *Biochemistry* **3**, 876.

Hall, R. H. (1965), *Biochemistry* **4**, 661.

Hall, R. H. (1967), in *Methods in Enzymology*, Vol. XII; *Nucleic Acids* Part A, L. Grossman and K. Moldave, Eds., Academic Press, New York, p. 305.

Hall, R. H. (1968), *Proceedings of the Sixth International Conference on Plant Growth Substances*, F. Wightman, and G. Setterfield, Eds., Runge Press, Ottawa, p. 47.

Hall, R. H., and Chheda, G. B. (1965), *J. Biol. Chem.* **240**, PC 2754.

Hall, R. H., Csonka, L., David, H., and McLennan, B. D. (1967a), *Science* **156**, 69.

Hall, R. H., and Fleysher, M. H. (1968), in *Synthetic Procedures in Nucleic Acid Chemistry*, W. W. Zorbach and R. S. Tipson, Eds., Interscience–John Wiley and Sons, New York, p. 517.

Hall, R. H., and Gale, G. O. (1960), *Proc. Soc. Exptl. Biol. Med.* **103**, 234.

Hall, R. H., Mittelman, A., Horoszewicz, J. S., and Grace, J. T. (1967b), *Ann. N.Y.Acad. Sci.* **143**, 799.

Hall, R. H., and Robins, M. J. (1968a), in *Synthetic Procedures in Nucleic Acid Chemistry*, W. W. Zorbach, and R. S. Tipson, Eds., Interscience–J. Wiley and Sons, New York, p. 11.

Hall, R. H., and Robins, M. J. (1968b), in *Synthetic Procedures in Nucleic Acid Chemistry*, W. W. Zorbach and R. S. Tipson, Eds., Interscience–J. Wiley and Sons, New York, p. 210.

Hall, R. H., Robins, M. J., Stasiuk, L., and Thedford, R. (1966), *J. Am. Chem. Soc.* **88**, 2614.

Hall, R. H., and Srivastava, B. I. S. (1968), *Life Sci.* **7**, 7.

Hampton, A., Biesele, J. J., Moore, A. E., and Brown, G. B. (1956), *J. Am. Chem. Soc.* **78**, 5695.

Hancock, R. L. (1967), *Can. J. Biochem.* **45**, 1513.

Hancock, R. L., McFarland, P., and Fox, R. R. (1967), *Experientia* **23**, 806.

BIBLIOGRAPHY

Hanes, C. S., and Isherwood, F. A. (1949), *Nature* **164**, 1107.

Hanze, A. R. (1967), *J. Am. Chem. Soc.* **89**, 6720.

Harada, F., Gross, H. J., Kimura, F., Chang, S. H., Nishimura, S., and RajBhandary, U. L. (1968), *Biochem. Biophys. Res. Commun.* **33**, 299.

Harkins, T. R., and Freiser, H. (1958), *J. Am Chem. Soc.* **80**, 1132.

Harris, G., and Wiseman, A. (1962), *Biochim. Biophys. Acta* **55**, 374.

Hasegawa, S., Fujimura, S., Shimizu, Y., and Sugimura, T. (1967), *Biochim. Biophys. Acta* **149**, 369.

Hayashi, H., Fisher, H., and Söll, D. (1969), *Biochemistry* **8**, 3680.

Hayashi, Y., Osawa, S., and Miura, K. (1966), *Biochim. Biophys. Acta* **129**, 519.

Hayatsu, H., Takeishi, K-I., and Ukita, T. (1966), *Biochim. Biophys. Acta* **123**, 445.

Hayatsu, H., and Ukita, T. (1966), *Biochim. Biophys. Acta* **123**, 458.

Hayatsu, H., and Ukita, T. (1967), *Biochem. Biophys. Res. Commun.* **29**, 556.

Hayward, R. S., and Weiss, S. B. (1966), *Proc. Natl. Acad. Sci. (U.S.)* **55**, 1161.

Hecht, S. M., Gupta, A. S., and Leonard, N. J. (1969), *Anal. Biochem.* **30**, 249.

Hecht, S. M., Helgeson, J. P., and Fujii, T. (1968), in *Synthetic Procedures in Nucleic Acid Chemistry*, W. W. Zorbach and R. S. Tipson, Eds., Interscience–J. Wiley and Sons, New York, p. 8.

Hedgcoth, C., and Jacobsen, M. (1968), *Anal. Biochem.* **25**, 55.

Heinrikson, R. L., and Goldwasser, E. (1963), *J. Biol. Chem.* **238** PC 485.

Heinrikson, R. L., and Goldwasser, E. (1964), *J. Biol. Chem.* **239**, 1177.

Heirwegh, K. P. M., Ramboer, C., and DeGroote, J. (1967), *Am. J. Med.* **42**, 913.

Helgeson, J. P., and Leonard, N. J. (1966), *Proc. Natl. Acad. Sci. (U.S.)* **56**, 60.

Hemmens, W. F. (1963), *Biochim. Biophys. Acta* **68**, 284.

Hemmens, W. F. (1964), *Biochim. Biophys. Acta* **91**, 332.

Heppel, L. (1967), in *Methods in Enzymology*, Vol. XII; *Nucleic Acids*, Part A, L. Grossman and K. Moldave, Eds., Academic Press, New York, p. 316.

Hilbert, G. E., Jansen, E. F., and Hendricks, S. B. (1935), *J. Am. Chem. Soc.* **57**, 552.

Hilbert, G. E., and Johnson, T. B. (1930), *J. Am. Chem. Soc.* **52**, 1152.

Hitchings, G. H., Elion, G. B., Falco, E. A., and Russell, P. B. (1949), *J. Biol. Chem.* **177**, 357.

Hitchings, G. H., and Falco, E. A. (1944), *Proc. Natl. Acad. Sci. (U.S.)* **30**, 294.

Ho, N. W., Y., and Gilham, P. T. (1967), *Biochemistry* **6**, 3632.

Holland, J. J., Taylor, M. W., and Buck, C. A. (1967), *Proc. Natl. Acad. Sci. (U.S.)* **58**, 2437.

Holley, R. W., Apgar, J., Everett, G. A., Madison, J. T., Marquisee, M., Merrill, S. H., Penswick, J. R., and Zamir, A. (1965), *Science* **147**, 1462.

Holy, A., and Scheit, K-H. (1967), *Biochim. Biophys. Acta* **138**, 230.

Honjo, M., Kanai, Y., Furukawa, Y., Mizuno, Y., and Sanno, Y. (1964), *Biochim. Biophys. Acta.* **87**, 696.

Hori, M. (1967), in *Methods in Enzymology*, Vol. XII; *Nucleic Acids*, Part A, L. Grossman and K. Moldave, Eds., Academic Press, New York, p. 381.

Horn, E. E., and Herriott, R. M. (1962), *Proc. Natl. Acad. Sci. (U.S.)* **48**, 1409.

Hoskinson, R. M., and Khorana, H. G. (1965), *J. Biol. Chem.* **240**, 2129.

Hotchkiss, R. D. (1948), *J. Biol. Chem.* **175**, 315.

Howard, G. A., Lythgoe, B., and Todd, A. R. (1947), *J. Chem. Soc.* 1052.

Howgate, P., Jones, A. S., and Tittensor, J. R. (1968), *J. Chem. Soc.* (C), 275.

Hsu, W-T., Foft, J. W., and Weiss, S. B. (1967), *Proc. Natl. Acad. Sci.* (*U.S.*) **58**, 2028.

Hsu, Yu-C. (1964), *Nature* **203**, 152.

Huang, R. C. (1967), *Fed. Proc.* **26**, 603.

Huang, R. C., and Bonner, J. (1965), *Proc. Natl. Acad. Sci* (*U.S.*) **54**, 960.

Hudson, L., Gray, M. W., and Lane, B. G. (1965), *Biochemistry* **4**, 2009.

Hurst, R. O., Marko, A. M., and Butler, G. C. (1953), *J. Biol. Chem.* **204**, 847.

Hurwitz, J., Anders, M., Gold, M., and Smith, I. (1965), *J. Biol. Chem.* **240**, 1256.

Hurwitz, J., Gold, M., and Anders, M. (1964), *J. Biol. Chem.* **239**, 3462.

Ichikawa, T., Kato, T., and Takenishi, T. (1965), *J. Het. Chem.* **2**, 253.

Ikehara, M., and Tada, H. (1968), in *Synthetic Procedures in Nucleic Acid Chemistry*, Vol. 1, W. W. Zorbach and R. S. Tipson, Eds., Interscience–John Wiley and Sons, New York, p. 188.

Ingram, V. M., and Sullivan, E. (1962), *Biochim. Biophys. Acta* **61**, 583.

Ishikura, H., Yamada, Y., Murao, K., Saneyoshi, M., and Nishimura, S. (1969), *Biochem. Biophys. Res. Commun.* **37**, 990.

Ivanova, O. I., Knorre, D. G., and Malygin, E. G. (1967), *Mol. Biol.* (*USSR*) **1**, 335.

Iwakata, S., and Grace, J. T. (1964), *N. Y. J. Med.* **64**, 2279.

Jacob, F., and Monod, J. (1961), *J. Mol. Biol.* **3**, 318.

Jacobson, R. A., and Bonner, J. (1968), *Biochem. Biophys. Res. Commun.* **33**, 716.

Janion, C., and Shugar, D. (1965), *Acta Biochim. Polon.* **12**, 337.

Janot, M., Cavé, A., and Goutaul, R. (1959), *Bull. Soc. Chem.* France, 896.

Johns, J. (1911), *J. Biol. Chem.* **9**, 161.

Johnson, J. A., Thomas, H. J., and Schaeffer, H. J. (1958), *J. Am. Chem. Soc.* **80**, 699.

Johnson, T. B., and Coghill, R. D. (1925), *J. Am. Chem. Soc.* **47**, 2838.

Johnson, T. B., and Heyl, F. W. (1907), *Am. Chem. J.* **37**, 628.

Johnson, T. B., and Litzinger, A. (1936), *J. Am. Chem. Soc.* **58**, 1940.

Johnson, T. B., and McCollum, E. Y. (1906), *J. Biol. Chem.* **1**, 437.

Johnson, T. B., and Speh, C. F. (1907), *Am. Chem. J.* **38**, 602.

Jones, A. S., Tittensor, J. R., and Walker, R. P. (1966), *J. Chem. Soc.* 1635.

Jones, J. W., and Robins, R. K. (1963), *J. Am. Chem. Soc.* **85**, 193.

Josse, J., and Kornberg, A. (1962), *J. Biol. Chem.* **237**, 1968.

Jukes, T. H. (1966), *Biochem. Biophys. Res. Commun.* **24**, 744.

Kallen, R. G., Simon, M., and Marmur, J. (1962), *J. Mol. Biol.* **5**, 248.

Kalousek, F., and Morris, N. R. (1968), *J. Biol. Chem.* **243**, 2440.

Kano-Sueoka, T., Nirenberg, M., and Sueoka, N. (1968), *J. Mol. Biol.* **35**, 1.

Kaye, A. M., Fridlender, B., Salomon, R., and Bar-Meir, S. (1967), *Biochim. Biophys. Acta* **142**, 331.

Kaye, A. M., and Leboy, P. S. (1968), *Biochim. Biophys. Acta* **157**, 289.

Kaye, A. M., Salomon, R., and Fridlender, B. (1967), *J. Mol. Biol.* **24**, 479.

Kaye, A. M., and Winocour, E. (1967), *J. Mol. Biol.* **24**, 475.

Kelmers, A. D., Novelli, G. D., and Stulberg, M. P. (1965), *J. Biol. Chem.* **240**, 3979.

BIBLIOGRAPHY

Kenner, G. W., Taylor, C. W., and Todd, A. R. (1949), *J. Chem. Soc.* 1620.

Kevin, S. P., Witkus, E. R., and Berger, C. A. (1966), *Exptl. Cell Res.* **41**, 259.

Khwaja, T. A., and Robins, R. K. (1966), *J. Am. Chem. Soc.* **88**, 3640.

Kikugawa, K., Hayatsu, H., and Ukita, T. (1967a), *Biochim. Biophys. Acta* **134**, 221.

Kikugawa, K., Muto, A., Hayatsu, H., Miura, K-I., and Ukita, T. (1967b), *Biochim. Biophys. Acta* **134**, 232.

Kirkegaard, L. H., and Bock, R. M. (1968), private communication.

Kissman, H. M., Pidacks, C., and Baker, B. R. (1955), *J. Am. Chem. Soc.* **77**, 18.

Kissman, H. M., and Weiss, M. J. (1956), *J. Org. Chem.* **21**, 1053.

Kjellin-Straby, K., and Boman, H. G. (1965), *Proc. Natl. Acad. Sci. (U.S.)* **53**, 1346.

Klämbt, D., Thies, G., and Skoog, F. (1966), *Proc. Natl. Acad. Sci. (U.S.)* **56**, 52.

Klein, A., and Sauerbier, W. (1965), *Biochem. Biophys. Res. Commun.* **18**, 440.

Klein, W. (1945), in *Methods der Ferment-Forschung*, Bamann and Myrback, Eds., Academic Press, New York, p. 125.

Kline, L., Fittler, F., and Hall, R. H. (1969), *Biochemistry* **8**, 4361.

Knorre, D. G., Malygin, E. G., Mushinskaya, G. S., and Favorov, V. V. (1966), *Biokhimiya* **31**, 334.

Ko, S-H. D., Rabinowitz, M., and Goldberg, I. H. (1964), *Biochemistry* **3**, 1840.

Kochetkov, N. K., Budowsky, E. I., Broude, N. E., and Klebanova, L. M. (1967a), *Biochim. Biophys. Acta* **134**, 492.

Kochetkov, N. K., Budowsky, E. I., and Demushkin, V. P. (1967b), *Mol. Biol. (U.S.S.R.)* **1**, 583.

Kochetkov, N. K., Budowsky, E. I., Demushkin, V. P., Turchinsky, M. F., Simukova, N. A., and Sverdlov, E. D. (1967c), *Biochim. Biophys. Acta* **142**, 35.

Kochetkov, N. K., Budowsky, E. I., and Domkin, V. D. (1967d), *Mol. Biol. (USSR)* **1**, 558.

Kochetkov, N. K., Budowsky, E. I., and Shibayeva, R. P. (1963a), *Biochem. Biophys. Acta* **68**, 493.

Kochetkov, N. K., Budowsky, E. I., and Shibayeva, R. P. (1964), *Biochim. Biophys. Acta* **87**, 515.

Kochetkov, N. K., Budowsky, E. I., and Simukova, N. A. (1962a), *Biokhimiya* **27**, 519.

Kochetkov, N. K., Budowsky, E. I., and Simukova, N. A. (1962b), *Biochim. Biophys. Acta* **55**, 257.

Kochetkov, N. K., Budowsky, E. I., and Simukova, N. A. (1963b), *Dokl. Akad. Nauk SSSR* **153**, 597.

Kochetkov, N. K., Budowsky, E. I., Turchinsky, M. F., and Demushkin, V. P. (1963c), *Dokl. Akad. Nauk SSSR* **152**, 1005.

Kondo, Y., and Witkop, B. (1968), *J. Am Chem. Soc.* **90**, 764.

Kornberg, A. (1962), *Enzymatic Synthesis of DNA*, John Wiley and Sons, New York.

Kornberg, A., Zimmerman, S. B., Kornberg, S. R., and Josse, J. (1959), *Proc. Natl. Acad. Sci. (U.S.)* **45**, 772.

Koshimizu, K., Kusaki, T., Mitsui, T., and Matsubara, S. (1967), *Tetrahedron Letters* **14**, 1317.

Kriek, E. (1965), *Biochem. Biophys. Res. Commun.* **20**, 793.

Kriek, E., and Emmelot, P. (1963), *Biochemistry* **2**, 733.

Kriek, E., and Emmelot, P. (1964), *Biochim. Biophys. Acta* **91**, 59.

Kunieda, T., and Witkop, B. (1967), *J. Am. Chem. Soc.* **89**, 4232.

Kuno, S., and Lehman, I. R. (1962), *J. Biol. Chem.* **237**, 1266.

Kuo, T-T., Huang, T-C., and Teng, M-H. (1968), *J. Mol. Biol.* **34**, 373.

Kuriki, Y. (1964), *Biochim. Biophys. Acta* **80**, 361.

Kusama, K., Prescott, D. M., Froholm, L. O., and Cohn, W. E. (1966), *J. Biol. Chem.* **241**, 4086.

Lake, J. A., and Beeman, W. W. (1967), *Science* **156**, 1371.

Laland, S. G., Overend, W. G., and Webb. M. (1952), *J. Chem. Soc.* 3224.

Lane, B. G. (1963), *Biochim. Biophys. Acta* **72**, 110.

Lane, B. G. (1965), *Biochemistry* **4**, 212.

Lark, C. (1968a), *J. Mol. Biol.* **31**, 389.

Lark, C. (1968b), *J. Mol. Biol.* **31**, 401.

Lawley, P. D. (1961), *J. Chim. Phys.* 1011.

Lawley, P. D. (1966), in *Progress in Nucleic Acid Research and Molecular Biology*, Vol. 5, J. N. Davidson and W. E. Cohn, Eds., Academic Press, New York, p. 89.

Lawley, P. D. (1967), *J. Mol. Biol.* **24**, 75.

Lawley, P. D., and Brookes, P. (1962), *J. Mol. Biol.* **4**, 216.

Lawley, P. D., and Brookes, P. (1963), *Biochem. J.* **89**, 127.

Lawley, P. D., Brookes, P., Magee, P. N., Craddock, V. M., and Swann, P. F. (1968), *Biochim. Biophys. Acta* **157**, 646.

Lederberg, J. (1947), *Genetics* **32**, 505.

Lederberg, S. (1966), *J. Mol. Biol.* **17**, 293.

Ledinko, N. (1964), *J. Mol. Biol.* **9**, 834.

Lee, H-J., and Wigler, P. W. (1968), *Biochemistry* **7**, 1427.

Lee, J. C., Ho, N. W. Y., and Gilham, P. T. (1965), *Biochim. Biophys. Acta* **95**, 503.

Lee, J. C., and Ingram, V. M. (1967), *Science* **158**, 1330.

Lee, J. C., and Ingram, V. M. (1968), *Fed. Proc.* **27**, 800.

Lee, S., Brown, G. L., and Kosinski, Z. (1967), *Biochem. J.* **103**, 25C.

Lehman, I. R., and Pratt, E. A. (1960), *J. Biol. Chem.* **235**, 3254.

Lengfeld, F., and Stieglitz, J. (1893), *Am. Chem. J.* **15**, 504.

Leonard, N. J., and Deyrup, J. A. (1962), *J. Am. Chem. Soc.* **84**, 2148.

Leonard, N. J., Achmatowicz, S., Loeppky, R. N., Carraway, K. L., Grimm, W. A. H., Szweykowska, A., Hamzi, H. Q., and Skoog, F. (1966), *Proc. Natl. Acad. Sci. (U.S.)* **56**, 709.

Leonard, N. J., and Fujii, T. (1964), *Proc. Natl. Acad. Sci. (U.S.)* **51**, 73.

Leonard, N. J., Hecht, S. M., Skoog, F., and Schmitz, R. Y. (1968), *Proc. Natl. Acad. Sci. (U.S.)* **59**, 15.

Leppla, S. H., Bjoraker, B., and Bock, R. M. (1968), in *Methods in Enzymology*, Vol. XII; *Nucleic Acids*, Part B, L. Grossman and K. Moldave, Eds., Academic Press, New York, p. 236.

Letham, D. S. (1966a), *Life Sci.* **5**, 551.

Letham, D. S. (1966b), *Life Sci.* **5**, 1999.

426

BIBLIOGRAPHY

Letham, D. S. (1968) in *Biochemistry and Physiology of Plant Growth Substances*, F. Wightman and G. Setterfield, Eds., Runge Press, Ottawa, p. 19.

Letham, D. S., Shannon, J. S., and McDonald, I. R. C. (1967), *Tetrahedron* **23**, 479.

Levene, P. A., and Bass, L. W. (1931), *Nucleic Acids*, Chemical Catalog Co., New York.

Levene, P. A., Bass., L. W., and Simms, H. S. (1926a), *J. Biol. Chem.* **70**, 229.

Levene, P. A., and La Forge, F. B. (1912), *Chem. Ber.* **45**, 608.

Levene, P. A., Simms, H. S., and Bass, L. W. (1926b), *J. Biol. Chem.* **70**, 243.

Levene, P. A., and Tipson, R. S. (1934), *J. Biol. Chem.* **104**, 385.

Levene, P. A., and Tipson, R. S. (1935), *J. Biol. Chem.* **111**, 313.

Lichtenstein, J., and Cohen, S. S. (1960), *J. Biol. Chem.* **235**, 1134.

Linsmaier, E., and Skoog, F. (1965), *Physiol. Plantarum* **18**, 100.

Lipsett, M. N. (1965), *J. Biol. Chem.* **240**, 3975.

Lipsett, M. N. (1967), *J. Biol. Chem.* **242**, 4067.

Lipsett, M. N., and Doctor, B. P. (1967), *J. Biol. Chem.* **242**, 4072.

Lipsett, M. N., Norton, J. S., and Peterkofsky, A. (1967), *Biochemistry* **6**, 855.

Lipsett, M. N., and Peterkofsky, A. (1966), *Proc. Natl. Acad. Sci.* (*U.S.*) **55**, 1169.

Lis, A. W., and Lis, E. W. (1962), *Biochim. Biophys. Acta* **61**, 799.

Lis, A. W., and Passarge, W. E. (1966), *Arch. Biochem. Biophys.* **114**, 593.

Lis, A. W., and Passarge, W. E. (1969), *Physiol. Chem. and Phys.* **1**, 68.

Litt, M., and Hancock, V. (1967), *Biochemistry* **6**, 1848.

Littlefield, J. W., and Dunn, D. B. (1958), *Biochem. J.* **70**, 642.

Littlefield, J. W., and Gould, E. A. (1960), *J. Biol. Chem.* **235**, 1129.

Litwack, M. D., and Weissmann, B. (1966), *Biochemistry* **5**, 3007.

Loeb, M. R., and Cohen, S. S. (1959), *J. Biol. Chem.* **234**, 364.

Loveless, A. (1969), *Nature* **223**, 206.

Ludlum, D. B., Warner, R. C., and Wahba, A. J. (1964), *Science* **145**, 397.

Luzzati, D. (1962), *Biochem. Biophys. Res. Commun.* **9**, 508.

Lynch, B. M., Robins, R. K., and Cheng, C. C. (1958), *J. Chem. Soc.* 2973.

Lyttleton, J. W., and Petersen, G. B. (1964), *Biochim. Biophys. Acta* **80**, 391.

Macon, J. B., and Wolfenden, R. (1968), *Biochemistry* **7**, 3453.

Madison, J. T. (1967), in *Methods in Enzymology*, Vol. XII; *Nucleic Acids*, Part A, L. Grossman and K. Moldave, Eds., Academic Press, New York, p. 137.

Madison, J. T., Everett, G. A., and Kung, H. (1966), *Science* **153**, 531.

Madison, J. T., and Holley, R. W. (1965), *Biochem. Biophys Res. Commun.* **18**, 153.

Madison, J. T., and Kung, H. (1967), *J. Biol. Chem.* **242**, 1324.

Magasanik, B., Vischer, E., Doniger, R., Elson, D., and Chargaff, E. (1950), *J. Biol. Chem.* **186**, 37.

Magee, P. N., and Farber, E. (1962), *Biochem. J.* **83**, 114.

Magee, P. N., and Lee, K. Y. (1964), *Biochem. J.* **91**, 35.

Magrath, D. I., and Shaw, D. C. (1967), *Biochem. Biophys. Res. Commun.* **26**, 32.

Mandel, L. R., and Borek, E. (1963a), *Biochemistry* **2**, 555.

Mandel, L. R., and Borek, E. (1963b), *Biochemistry* **2**, 560.

Mandel, L. R., Srinivasan, P. R., and Borek, E. (1966), *Nature* **209**, 586.

Markham, R., and Smith, J. D. (1949), *Biochem. J.* **45**, 294.

Markham, R., and Smith, J. D. (1951), *Biochem. J.* **49**, 401.

Markham, R., and Smith, J. D. (1952a), *Biochem. J.* **52**, 552.

Markham, R., and Smith, J. D. (1952b), *Biochem. J.* **52**, 558.

Marmur, J., and Doty, P. (1962), *J. Mol. Biol.* **5**, 109.

Marshak, A., and Vogel, H. J. (1950), *Fed. Proc.* **9**, 85.

Martin, D. M. G., and Reese, C. B. (1968), *J. Chem. Soc.* (C), 1731.

Martin, D. M. G., Reese, C. B., and Stephenson, G. F. (1968), *Biochemistry* **7**, 1406.

Matsubara, S., Armstrong, D. J., and Skoog, F. (1968), *Plant Physiol.* **43**, 451.

McCloskey, J. A., Lawson, A. M., Tsuboyama, K., Krueger, P. M., and Stillwell, R. N. (1968), *J. Am. Chem. Soc.* **90**, 4182.

McLaren, A. D., and Shugar, D. (1964), *Photochemistry of Proteins and Nucleic Acids*, The Macmillan Co., New York.

McLennan, B. D., Logan, D. M., and Hall, R. H. (1968), *Proc. Am. Assoc. Cancer Res* **9**, 47.

Merril, C. R. (1968), *Biopolymers* **6**, 1727.

Michelson, A. M., and Cohn, W. E. (1962), *Biochemistry* **1**, 490.

Michelson, A. M., and Pochon, F. (1966), *Biochim. Biophys. Acta* **114**, 469.

Midgley, J. E. M. (1965), *Biochim. Biophys. Acta* **108**, 340.

Miles, H. T. (1956), *Biochim. Biophys. Acta* **22**, 247.

Miles, H. T. (1961), *J. Org. Chem.* **26**, 4761.

Miles, H. T., Howard, F. B., and Frazier, J. (1963), *Science* **142**, 1458.

Miller, C. O. (1965), *Proc. Natl. Acad. Sci.* (*U.S.*) **54**, 1052.

Miller, C. O. (1967), *Science* **157**, 1055.

Miller, C. S. (1955), *J. Am. Chem. Soc.* **77**, 752.

Miller, E. C., Juhl, U., and Miller, J. A. (1966), *Science* **153**, 1125.

Miller, N., and Cerutti, P. (1967), *J. Am. Chem. Soc.* **89**, 2767.

Mitchel, R. E. J. (1968), Ph.D. Thesis, University of British Columbia, Vancouver, Canada.

Mittelman, A., Bonney, R., Hall, R. H., and Grace, J. T. (1969), *Biochim. Biophys. Acta* **179**, 242.

Mittelman, A., Grace, J. T., and Hall, R. H. (1967a), *Proc. Am. Assoc. Cancer Res.* **8**, 47.

Mittelman, A., Hall, R. H., Yohn, D. S., and Grace, J. T. (1967b), *Cancer Res.* **27**, 1409.

Mizoguchi, T., Levin, G., Woolley, D. W., and Stewart, J. M. (1968), *J. Org. Chem.* **33**, 903.

Mizuno, Y., Itoh, T., and Tagawa, H. (1965), *Chem. Ind.* 1498.

Mizutani, T., Miyazaki, M., and Takemura, S. (1968), *J. Biochem.* (Japan) **64**, 839.

Molinaro, M., Sheiner, L. B., Neelon, F. A., and Cantoni, G. L. (1968), *J. Biol. Chem.* **243**, 1277.

Monier, R., Stephenson, M. L., and Zamecnik, P. C. (1960), *Biochim. Biophys. Acta* **43**, 1.

Monseur, X. G., and Adriaens, E. L. (1960), *J. Pharm. Belg.* 279.

Montgomery, J. A., and Hewson, K. (1960), *J. Am. Chem. Soc.* **82**, 463.

Moore, A. M., and Thomson, C. H. (1955), *Science* **122**, 594.

Moore, P. B. (1966), *J. Mol. Biol.* **18**, 38.

Morisawa, S., and Chargaff, E. (1963), *Biochim. Biophys. Acta* **68**, 147.

Mothes, K. (1967), *Wiss. Z. Univ. Rostock* (Math.-Naturwiss. Reihe) **16**, 619.

Moudrianakis, E. N., and Beer, M. (1965a), *Proc. Natl. Acad. Sci.* (*U.S.*) **53**, 564.

428

BIBLIOGRAPHY

Moudrianakis, E. N., and Beer, M. (1965b), *Biochim. Biophys. Acta* **95**, 23.

Muench, K. H., and Berg, P. (1966a), *Biochemistry* **5**, 970.

Muench, K. H., and Berg, P. (1966b), *Biochemistry* **5**, 982.

Muench, K. H., and Safille, P. A. (1968), *Biochemistry* **7**, 2799.

Murashige, T., and Skoog, F. (1962), *Physiol Plantarum* **15**, 473.

Muto, A., Miura, K., Hayatsu, H., and Ukita, T. (1965), *Biochim. Biophys. Acta* **95**, 669.

Nagasawa, N., Kumashiro, I., and Takenishi, T. (1967), *J. Org. Chem.* **32**, 251.

Nagata, C., Imamura, A., Saito, H., and Fukui, K. (1963), *Gamn.* **54**, 109.

Naito, T., Kawakami, T., Sano, M., and Hirata, M. (1961), *Chem. Pharm. Bull.* (Tokyo) **9**, 249.

Nakaya, K., Takenaka, O., Horinishi, H., and Shibata, K. (1968), *Biochim. Biophys. Acta* **161**, 23.

Naylor, R., Ho, N. W. Y., and Gilham, P. T. (1965), *J. Am. Chem. Soc.* **87**, 4209.

Neidhardt, F. C., and Earhart, C. F. (1966), *Cold Spring Harbor Symp. Quant. Biol.* **31**, 557.

Nelson, J. A., Ristow, S. C., and Holley, R. W. (1967), *Biochim. Biophys. Acta* **149**, 590.

Ness, R. K., and Fletcher, H. G. (1960), *J. Am. Chem. Soc.* **82**, 3434.

Nichols, J. L., and Lane, B. G. (1966a), *Can. J. Biochem.* **44**, 1633.

Nichols, J. L., and Lane, B. G. (1966b), *Biochim. Biophys. Acta* **119**, 649.

Nichols, J. L., and Lane, B. G. (1967), *J. Mol. Biol.* **30**, 477.

Nichols, J. L., and Lane, B. G. (1968a), *Can. J. Biochem.* **46**, 109.

Nichols, J. L., and Lane, B. G. (1968b), *Biochim. Biophys. Acta* **166**, 605.

Nichols, J. L., and Lane, B. G. (1968c), *Can. J. Biochem.* **46**, 1487.

Nikolskaya, I. I., Tkatcheva, Z. G., Vanyushin, B. F., and Tikchonenko, T. I. (1968), *Biochim. Biophys. Acta* **155**, 626.

Nirenberg, M. W., and Leder, P. (1964), *Science* **145**, 1399.

Nishihara, M., Chrambach, A., and Aposhian, H. V. (1967), *Biochemistry* **6**, 1877.

Nishimura, S., Yamada, Y., and Ishikura, H. (1969), *Biochim. Biophys. Acta* **179**, 517.

Nishimura, T., and Iwai, I. (1964), *Chem. Pharm. Bull.* (Tokyo) **12**, 352.

Nishimura, T., Shimizu, B., and Iwai, I. (1964), *Chem. Pharm. Bull.* (Tokyo) **12**, 1471.

Norton, J., and Roth, J. S. (1967), *J. Biol. Chem.* **242**, 2029.

Novelli, G. D. (1967), *Ann Rev. Biochem.* **36**, 449.

Ochiai, M., Marumoto, R., Kobayashi, S., Shimazu, H., and Morita, K. (1968), *Tetrahedron* **24**, 5731.

Ofengand, J. (1965), *Biochem. Biophys. Res. Commun.* **18**, 192.

Olenick, J. G., and Hahn, F. E. (1964), *Biochim. Biophys. Acta* **87**, 535.

Olomucki, M., Desvages, G., Thoai, N-V., and Roche, J. (1965), *Compt. Rend.* **260**, 4519.

Osawa, S. (1960), *Biochim. Biophys. Acta* **43**, 110.

Osborne, D. J. (1965), *J. Sci. Food. Agr.* **16**, 1.

Osborne, D. J., and McCalla, D. R. (1961), *Plant Physiol.* **36**, 219.

Park, R. W., Holland, J. F., and Jenkins, A. (1962), *Cancer Res.* **22**, 469.

Partridge, S. M. (1948), *Biochem. J.* **42**, 238.

Pataki, G. (1967), *J. Chromatog.* **29**, 126.

Payot, P., and Grob, C. A. (1954), *Helv. Chim. Acta* **37**, 1266.

Pedersen, C., and Fletcher, H. G. (1960), *J. Am. Chem. Soc.* **82**, 5210.

Penswick, J. R., and Holley, R. W. (1965), *Proc. Natl. Acad. Sci. (U.S.)* **53**, 543.

Peterkofsky, A. (1968), *Biochemistry* **7**, 472.

Peterson, E. A., and Sober, H. A. (1959), *Anal. Chem.* **31**, 857.

Pfleiderer, W. (1961), *Ann.* **647**, 167.

Philippsen, P., Thiebe, R., Wintermeyer, W., and Zachau, H. G. (1968), *Biochem. Biophys. Res. Commun.* **33**, 922.

Phillips, J. H., and Brown, D. M. (1967), in *Progress In Nucleic Acid Research and Molecular Biology*, Vol. 7, J. N. Davidson and W. E. Cohn, Eds., Academic Press, New York, p. 349.

Phillips, J. H., Brown, D. M., Adam, R., and Grossman, L. (1965), *J. Mol. Biol.* **12**, 816.

Phillips, J. H., Brown, D. M., and Grossman, L. (1966), *J. Mol. Biol.* **21**, 405.

Phillips, J. H., and Kjellin-Straby, K. (1967), *J. Mol. Biol.* **26**, 509.

Pizer, L. I., and Cohen, S. S. (1962), *J. Biol. Chem.* **237**, 1251.

Pleiss, M., Ochiai, H., and Cerutti, P. A. (1969), *Biochem. Biophys. Res. Commun.* **34**, 70.

Pochon, F., and Michelson, A. M. (1965), *Proc. Natl. Acad. Sci. (U.S.)* **53**, 1425.

Pochon, F., and Michelson, A. M. (1967), *Biochim. Biophys. Acta* **145**, 321.

Pollak, J. K., and Arnstein, H. R. V. (1962), *Biochim. Biophys. Acta* **55**, 798.

Price, T. D., Hinds, H. A., and Brown, R. S. (1963), *J. Biol. Chem.* **238**, 311.

Prystaš, M., and Šorm, F. (1968), in *Synthetic Procedures in Nucleic Acid Chemistry*, Vol. 1, W. W. Zorbach and R. S. Tipson, Eds., Interscience–John Wiley and Sons, New York, p. 403.

RajBhandary, U. L., and Chang, S. H. (1968), *J. Biol. Chem.* **243**, 598.

RajBhandary, U. L., Chang, S. H., Stuart, A., Faulkner, R. D., Hoskinson, R. M., and Khorana, H. G. (1967), *Proc. Natl. Acad. Sci. (U.S.)* **57**, 751.

Rake, A. V., and Tener, G. M. (1966), *Biochemistry* **5**, 3992.

Ray, D. S., and Hanawalt, P. C. (1964), *J. Mol. Biol.* **9**, 812.

Reichard, P. (1955), *Acta Chem. Scand.* **9**, 1275.

Revel, M., Herzberg, M., Becarevic, A., and Gros, F. (1968), *J. Mol. Biol.* **33**, 231.

Revel, M., and Littauer, U. Z. (1966), *J. Mol. Biol.* **15**, 389.

Richter, E., Loeffler, J. E., and Taylor, E. C. (1960), *J. Am. Chem. Soc.* **82**, 3144.

Robbins, P. W., and Hammond, J. B. (1962), *J. Biol. Chem.* **237**, PC 1379.

Robbins, P. W., and Kinsey, B. M. (1963), *Fed. Proc.* **22**, 229.

Roberts, M., and Visser, D. W. (1952), *J. Am. Chem. Soc.* **74**, 668.

Robins, M. J., Bowles, W. A., and Robins, R. K. (1964), *J. Am. Chem. Soc.* **86**, 1251.

Robins, M. J., Hall, R. H., and Thedford, R. (1967), *Biochemistry* **6**, 1837.

Robins, M. J., and Robins, R. K. (1965), *J. Am. Chem. Soc.* **87**, 4934.

Robins, R. K., Dille, K. J., Willits, C. H., and Christensen, B. E. (1953), *J. Am. Chem. Soc.* **75**, 263.

Rodeh, R., Feldman, M., and Littauer, U. Z. (1967), *Biochemistry* **6**, 451.

Rogozinska, J. H. (1967), *Bull. Acad. Polon. Sci.* **15**, 313.

Rogozinska, J. H., Helgeson, J. P., and Skoog, F. (1964), *Physiol. Plantarum* **17**, 165.

Romanko, E. G., Klein, Kh. Ya., and Kulayeva, O. N. (1968), *Biokhimiya* **33**, 547.

Roscoe, D. H., and Tucker, R. G. (1964), *Biochem. Biophys. Res. Commun.* **16**, 106.

Roscoe, D. H., and Tucker, R. G. (1966), *Virology*, **29**, 157.

Rosenberg, E. (1965), *Proc. Natl. Acad. Sci. (U.S.)* **53**, 836.

Rosset, R., Monier, R., and Julien, J. (1964), *Bull. Soc. Chim. Biol.* **46**, 87.

Rotherham, J., and Schneider, W. C. (1960), *Biochim. Biophys. Acta* **41**, 344.

Rottman, F., and Cerutti, P. (1966), *Proc. Natl. Acad. Sci. (U.S.)* **55**, 960.

Rushizky, G. W., and Sober, H. A. (1962), *Biochim. Biophys. Acta* **55**, 217.

Salganik, R. I., Dashkevich, V. S., and Dymshits, G. M. (1967), *Biochim. Biophys. Acta* **149**, 603.

Salser, J. S., and Balis, M. E. (1967), *Biochim. Biophys. Acta* **149**, 220.

Saneyoshi, M., and Nishimura, S. (1967), *Biochim. Biophys. Acta* **145**, 208.

Sanger, F., Brownlee, G. G., and Barrell, B. G. (1965), *J. Mol. Biol.* **13**, 373.

Saski, T., and Mizuno, Y. (1967), *Chem. Pharm. Bull.* (Tokyo) **15**, 894.

Scannell, J. P., Crestfield, A. M., and Allen, F. W. (1959), *Biochim. Biophys. Acta* **32**, 406.

Scheit, K-H., and Holy, A. (1967), *Biochim. Biophys. Acta* **149**, 344.

Scherp, H. W. (1946), *J. Am. Chem. Soc.* **68**, 912.

Schildkraut, C. L., Marmur, J., and Doty, P. (1962), *J. Mol. Biol.* **4**, 430.

Schindler, P. (1949), *Helv. Chim. Acta* **32**, 979.

Schneider, I. R., Diener, T. O., and Safferman, R. S. (1964), *Science* **144**, 1127.

Schramm, G., Grötsch, H., and Pollmann, W. (1961), *Angew. Chem.* **73**, 619.

Schramm, G., Lunzmann, G., and Bechmann, F. (1967), *Biochim. Biophys. Acta* **145**, 221.

Schulman, L. H., and Chambers, R. W. (1968), *Proc. Natl. Acad. Sci. (U.S.)* **61**, 308.

Schuster, H. (1960), *Biochem. Biophys. Res. Commun.* **2**, 320.

Schuster, H. (1961), *J. Mol. Biol.* **3**, 447.

Schuster, H., and Vielmetter, W. (1961), *J. Chim. Phys.* **58**, 1005.

Schweizer, M. P., Chheda, G. B., Baczynskyj, L., and Hall, R. H. (1969), *Biochemistry* **8**, 3283.

Schweizer, M. P., Chheda, G. B., Hall, R. H., Baczynskyj, L., and Biemann, K. (1968), Abstracts, 156th American Chemical Society Meeting, Atlantic City, N.J.

Seidel, H. (1967), *Biochim. Biophys. Acta* **138**, 98.

Sekiya, T., Yoshida, M., and Ukita, T. (1967), *Biochim. Biophys. Acta* **149**, 610.

Shannon, J. S., and Letham, D. S. (1966), *New Zealand J. Sci.* **9**, 833.

Shapiro, H. S., and Chargaff, E. (1960), *Biochim. Biophys. Acta* **39**, 68.

Shapiro, R., and Agarwal, S. C. (1966), *Biochem. Biophys. Res. Commun.* **24**, 401.

Shapiro, R., and Agarwal, S. C. (1968), *J. Am. Chem. Soc.* **90**, 474.

Shapiro, R., Agarwal, S. C., and Cohen, B. I. (1967), Abstract N-7, XXIst IUPAC Congress, Sept. 4–10.

Shapiro, R., and Chambers, R. W. (1961), *J. Am. Chem. Soc.* **83**, 3920.

Shapiro, R., Cohen, B. I., Shiuey, S-J., and Mauer, H. (1969), *Biochemistry* **8**, 238.

Shapiro, R., and Gordon, C. N. (1964), *Biochem. Biophys. Res. Commun.* **17**, 160.

Shapiro, R., and Hachmann, J. (1966), *Biochemistry* **5**, 2799.

Shapiro, R., and Pohl, S. H. (1968), *Biochemistry* **7**, 448.

Shaw, E. (1950), *J. Biol. Chem.* **185**, 439.

Shaw, E., and Woolley, D. W. (1949), *J. Biol. Chem.* **181**, 89.

Shaw, G., Smallwood, B. M., and Wilson, D. V. (1966), *J. Chem. Soc.* 921.

Shaw, G., and Warrener, R. N. (1958), *J. Chem. Soc.* 157.

Shaw, G., Warrener, R. N., Maguire, M. H., and Ralph, R. K. (1958), *J. Chem. Soc.* 2294.

Sheid, B., Srinivasan, P. R., and Borek, E. (1968), *Biochemistry* **7**, 280.

Shimadate, T., Yoshimura, J., and Sato, T. (1960), *Nippon Kagaku Zasshi* **78**, 208.

Shimizu, B., Asai, M., and Nishimura, T. (1967), *Chem. Pharm. Bull.* (Tokyo) **15**, 1847.

Shiro, T., Yamanoi, A., Konishi, S., Okumura, S., and Takahashi, M. (1962), *Agr. Biol. Chem.* (Tokyo) **26**, 785.

Shugar, D., and Fox, J. J. (1952), *Biochim. Biophys. Acta* **9**, 199.

Shugar, D., and Szer, W. (1962), *J. Mol. Biol.* **5**, 580.

Shugart, L., Chastain, B. H., Novelli, G. D., and Stulberg, M. P. (1968a), *Biochem. Biophys. Res. Commun.* **31**, 404.

Shugart, L., Novelli, G. D., and Stulberg, M. P. (1968b), *Biochim. Biophys. Acta* **157**, 83.

Simmonds, H. A. (1969), *Clin. Chim. Acta* **23**, 319.

Simon, L. N., Glasky, A. J., and Rejal, T. H. (1967), *Biochim. Biophys. Acta* **142**, 99.

Simpson, R. B. (1964), *J. Am. Chem. Soc.* **86**, 2059.

Singh, H., and Lane, B. G. (1964a), *Can. J. Biochem.* **42**, 87.

Singh, H., and Lane, B. G. (1964b), *Can. J. Biochem.* **42**, 1011.

Sinsheimer, R. L. (1954), *Science* **120**, 551.

Sinsheimer, R. L. (1960), *Ann. Rev. Biochem.* **29**, 503.

Sinsheimer, R. L., and Koerner, J. F. (1951), *Science* **114**, 42.

Sinsheimer, R. L., and Koerner, J. F. (1952), *J. Biol. Chem.* **198**, 293.

Siperstein, M. D., and Fagan, V. M. (1966), *J. Biol. Chem.* **241**, 602.

Skinner, C. G., Claybrook, J. R., Talbert, F. D., and Shive, W. (1956), *Arch. Biochem. Biophys.* **65**, 567.

Skoog, F., Hamzi, H. Q., Szweykowska, A. M., Leonard, N. J., Carraway, K. L., Fujii, T., Helgeson, J. P., and Loeppky, R. N. (1967), *Phytochemistry* **6**, 1169.

Sluyser, M., and Bosch, L. (1962), *Biochim. Biophys. Acta* **55**, 479.

Small, G. D., and Gordon, M. P. (1968), *J. Mol. Biol.* **34**, 281.

Smith, J. D. (1967), in *Methods in Enzymology*, Vol. XII, *Nucleic Acids*, Part A, L. Grossman and K. Moldave, Eds., Academic Press, New York, p. 350.

Smith, J. D., and Dunn, D. B. (1959a), *Biochim. Biophys. Acta* **31**, 573.

Smith, J. D., and Dunn, D. B. (1959b), *Biochem. J.* **72**, 294.

Smith, J. D., and Stoker, M. G. P. (1951), *Brit. J. Exptl. Pathol.* **32**, 433.

Smrt, J., Skoda, J., Lisy, V., and Šorm, F. (1966), *Biochim. Biophys. Acta* **129**, 210.

Söll, D., Cherayil, J. D., and Bock, R. M. (1967), *J. Mol. Biol.* **29**, 97.

Söll, D., Cherayil, J., Jones, D. S., Faulkner, R.D., Hampel, A., Bock, R. M., and Khorana, H. G. (1966), *Cold Spring Harbor Symp. Quant. Biol.* **31**, 51.

Srinivasan, P. R., and Borek, E. (1963), *Proc. Natl. Acad. Sci. (U.S.)* **49**, 529.

Srinivasan, P. R., and Borek, E. (1964a), *Biochemistry* **3**, 616.

Srinivasan, P. R., and Borek, E. (1964b), *Science* **145**, 548.

Srinivasan, P. R., and Borek, E. (1966), in *Progress in Nucleic Acid Research*, Vol. 5, J. N. Davidson and W. E. Cohn, Eds., Academic Press, New York, p. 157.

Srivastava, B. I. S., and Ware, G. (1965), *Plant Physiol.* **40**, 62.

BIBLIOGRAPHY

Staehelin, M. (1959), *Biochim. Biophys. Acta* **31**, 448.

Staehelin, M. (1964), *J. Mol. Biol.* **8**, 470.

Staehelin, M. Rogg, H., Baguley, B. C., Ginsberg, T., and Wehrli, W. (1968), *Nature* **219**, 1363.

Starr, J. L., and Fefferman, R. (1964), *J. Biol. Chem.* **239**, 3457.

Stent, G. S., and Brenner, S. (1961), *Proc. Natl. Acad. Sci. (U.S.)* **47**, 2005.

Strauss, B. S., and Robbins, M. (1968), *Biochim. Biophys. Acta* **161**, 68.

Strehler, B. L., Hendley, D. D., and Hirsch, G. P. (1967), *Proc. Natl. Acad. Sci. (U.S.)* **57**, 1751.

Stulberg, M. P., and Isham, R. R. (1967), *Proc. Natl. Acad. Sci. (U.S.)* **57**, 1310.

Subak-Sharpe, H., Shepard, W. M., and Hay, J. (1966), *Symp. Quant. Biol.* **33**, 583.

Sueoka, N., Kano-Sueoka, T., and Gartland, W. J. (1966), *Symp. Quant. Biol.* **31**, 571.

Sugimura, T., Fujimura, S., Hasegawa, S., and Kawamura, Y. (1967), *Biochim. Biophys. Acta* **138**, 438.

Suzuki, T., and Hochster, R. M. (1964), *Can. J. Microbiol.* **10**, 867.

Suzuki, T., and Hochster, R. M. (1966), *Can. J. Biochem.* **44**, 259.

Svensson, I., Björek, G., Björek, W., Johansson, K. E., and Johansson, A. (1968), *Biochem. Biophys. Res. Commun.* **31**, 216.

Sypherd, P. S. (1968), *J. Bacteriol.* **95**, 1844.

Szer, W., and Shugar, D. (1961), *Acta Biochim. Polon.* **8**, 235.

Szer, W., and Shugar, D. (1966), *Acta Biochim. Polon.* **13**, 177.

Szer, W., and Shugar, D. (1968a), in *Synthetic Procedures in Nucleic Acid Chemistry*, Vol. 1, W. W. Zorbach and R. S. Tipson, Eds., Interscience–John Wiley and Sons, New York, p. 58.

Szer, W., and Shugar, D. (1968b), in *Synthetic Procedures in Nucleic Acid Chemistry*, Vol. 1, W. W. Zorbach and R. S. Tipson, Eds., Interscience–John Wiley and Sons, New York, p. 433.

Takahashi, I., and Marmur, J. (1963a), *Nature* **197**, 794.

Takahashi, I., and Marmur, J. (1963b), *Biochem. Biophys. Res. Commun.* **10**, 289.

Takemura, S., Mizutani, T., and Miyazaki, M. (1968), *J. Biochem.* (Tokyo) **64**, 827.

Takemura, S., Murakami, M., and Miyazaki, M. (1969), *J. Biochem.* (Tokyo) **65**, 553.

Tamaoki, T., and Lane, B. G. (1968), *Biochemistry* **7**, 3431.

Tatum, E. L. (1945), *Proc. Natl. Acad. Sci. (U.S.)* **31**, 215.

Taylor, E. C., and Cheng, C. C. (1959), *Tetrahedron Letters* **12**, 9.

Taylor, E. C., Cheng, C. C., and Vogl, O. (1959a), *J. Org. Chem.* **24**, 2019.

Taylor, E. C., Vogl, O., and Cheng, C. C. (1959b), *J. Am. Chem. Soc.* **81**, 2442.

Taylor, H. F. W. (1948), *J. Chem. Soc.* 765.

Taylor, M. W., Buck, C. A., Granger, G. A., and Holland, J. J. (1968), *J. Mol. Biol.* **33**, 809.

Tewari, K. K., and Wildman, S. G. (1966), *Science* **153**, 1269.

Thedford, R., Fleysher, M. H., and Hall, R. H. (1965), *J. Med. Chem.* **8**, 486.

Theil, E. C., and Zamenhof, S. (1963), *J. Biol. Chem.* **238**, 3058.

Thiebe, R., and Zachau, H. G. (1968), *European J. Biochem.* **5**, 546.

Thomas, A. J. (1959), *Arch. Biochem. Biophys.* **79**, 162.

Thomas, A. J., and Sherratt, H. S. A. (1956), *Biochem. J.* **62**, 1.

Thomas, H. J., Temple, C., and Montgomery, J. A. (1968), in *Synthetic Procedures in Nucleic Acid Chemistry*, Vol. 1, W. W. Zorbach and R. S. Tipson, Eds., Interscience–John Wiley and Sons, New York, p. 22.

Tomasz, M., and Chambers, R. W. (1964), *J. Am. Chem. Soc.* **86**, 4216.

Tomasz, M., and Chambers, R. W. (1966), *Biochemistry* **5**, 773.

Tomasz, M., Sanno, Y., and Chambers, R. W. (1965), *Biochemistry* **4**, 1710.

Tomlinson, R. V., and Tener, G. M. (1963), *Biochemistry* **2**, 697.

Townsend, L. B., and Robins, R. K. (1962), *J. Org. Chem.* **27**, 990.

Townsend, L. B., Robins, R. K., Leoppky, R. N., and Leonard, N. J. (1964), *J. Am. Chem. Soc.* **86**, 5320.

Traube, W. (1900), *Chem. Ber.* **33**, 1371.

Traube, W., and Dudley, H. W. (1910), *Chem. Ber.* **43**, 3839.

Tsuboi, M., Kyogoku, Y., and Shimanouchi, T. (1962), *Biochim. Biophys. Acta* **55**, 1.

Tsutsui, E., Srinivasan, P. R., and Borek, E. (1966), *Proc. Natl. Acad. Sci. (U.S.)* **56**, 1003.

Turchinsky, M. F., Sverdlov, E. D., and Budowsky, E. I. (1967), *Mol. Biol. (USSR)* **1**, 623.

Ueda, T. (1960), *Chem. Pharm. Bull.* (Tokyo) **8**, 455.

Ueda, T., and Fox, J. J. (1963), *J. Am. Chem. Soc.* **85**, 4024.

Ueda, T., and Fox, J. J. (1964), *J. Org. Chem.* **29**, 1770.

Ueda, T., Iida, Y., Ikeda, K., and Mizuno, Y. (1966), *Chem. Pharm. Bull.* (Tokyo) **14**, 666.

Ueda, T., and Nishino, H. (1968), *J. Am. Chem. Soc.* **90**, 1678.

Ulbricht, T. L. V. (1959), *Tetrahedron* **6**, 225.

Uryson, S. O., and Belozerskii, A. N. (1959), *Dokl. Akad. Nauk. (SSSR)* **125**, 1144.

Uziel, M., and Gassen, H. G. (1968), *Fed. Proc.* **27**, 342.

Uziel, M., Koh, C. K., and Cohn, W. E. (1968), *Anal. Biochem.* **25**, 77.

Van Montagu, M., and Stockx, J. (1965), *Arch. Intern. Physiol. Biochim.* **73**, 158.

Van Schaik, N., and Pitout, M. J. (1966), *S. Afr. J. Sci.* **62**, 53.

Vanyushin, B. F., and Belozerskii, A. N. (1959), *Dokl. Akad. Nauk. (SSSR)* **129**, 944.

Vanyushin, B. F., Belozerskii, A. N., Kokurina, N. A., and Kadirova, D. X. (1968), *Nature* **218**, 1066.

Vaughan, M. H., Soeiro, R., Warner, J. R., and Darnell, J. E. (1967), *Proc. Natl. Acad. Sci. (U.S.)* **58**, 1527.

Venkstern, T. V., Krutilina, A. I., Li, L., Axelrod, V. D., Mirzabekov, A. D., and Bayev, A. A. (1968), *Mol. Biol. (USSR)* **2**, 394.

Venner, H. (1960), *Chem. Ber.* **93**, 140.

Verwoerd, D., Kohlhage, H., and Zillig, W. (1961), *Nature* **192**, 1038.

Vielmetter, W., and Schuster, H. (1960), *Z. Naturforsch* **15b**, 298, 304.

Villa-Trevino, S., and Magee, P. N. (1966), *Biochem. J.* **100**, 36P.

Vischer, E., Zamenhof, S., and Chargaff, E. (1949), *J. Biol. Chem.* **177**, 429.

Visser, D. W. (1968), in *Synthetic Procedures in Nucleic Acid Chemistry*, Vol. 1, W. W. Zorbach and R. S. Tipson, Eds., Interscience–John Wiley and Sons, New York, p. 428.

Visser, D. W., Barron, G., and Beltz, R. (1953), *J. Am. Chem. Soc.* **75**, 2017.

Vold, B. S., and Sypherd, P. S. (1968), *Proc. Natl. Acad. (U.S.)* **59**, 453.

Volkin, E. (1954), *J. Am. Chem. Soc.* **76**, 5892.

434

BIBLIOGRAPHY

Volkin, E., Khym, J. X., and Cohn, W. E. (1951), *J. Am. Chem. Soc.* **73**, 1533.

Vorbrüggen, H., and Strehlke, P. (1969), *Angew. Chem.* (Int. Ed.) **8**, 977.

Vournakis, J. N., and Scheraga, H. A. (1966), *Biochemistry* **5**, 2997.

Wagner, E. K., Penman, S., and Ingram, V. M. (1967), *J. Mol. Biol.* **29**, 371.

Wainfan, E., Srinivasan, P. R., and Borek, E. (1965), *Biochemistry* **4**, 2845.

Wakamatsu, H., Yamada, Y., Saito, T., Kumashiro, I., and Takenishi, T. (1966), *J. Org. Chem.* **31**, 2035.

Walerych, W. S., Venkataraman, S., and Johnson, B. C. (1966), *Biochem. Biophys. Res. Commun.* **23**, 368.

Waller, C. W., Fryth, P. W., Hutchings, B. L., and Williams, J. H. (1953), *J. Am. Chem. Soc.* **75**, 2025.

Walsh, B. T., and Wolfenden, R. (1967), *J. Am. Chem. Soc.* **89**, 6221.

Wang, S. Y. (1959a), *J. Am. Chem. Soc.* **81**, 3786.

Wang, S. Y. (1959b), *J. Org. Chem.* **24**, 11.

Watanabe, K. A., and Fox, J. J. (1966), *Angew. Chem.* **78**, 589.

Waters, L. C., and Novelli, G. D. (1967), *Proc. Natl. Acad.* (*U.S.*) **57**, 979.

Weil, J. H. (1965), *Bull. Soc. Chim. Biol.* **47**, 1303.

Weil, J. H., Befort, N., Rether, B., and Ebel, J. (1964), *Biochem. Biophys. Res. Commun.* **15**, 447.

Weisblum, B., Cherayil, J. D., Bock, R. M., and Söll, D. (1967), *J. Mol. Biol.* **28**, 275.

Weisblum, B., Gonano, F., von Ehrenstein, G., and Benzer, S. (1965), *Proc. Natl. Acad. Sci.* (*U.S.*) **53**, 328.

Weiss, M. J., Joseph, J. P., Kissman, H. M., Small, A. M., Schaub, R. E., and McEvoy, F. J. (1959), *J. Am. Chem. Soc.* **81**, 4050.

Weiss, S. B., Hsu, W-T., Foft, J. W., and Scherberg, N. H. (1968), *Proc. Natl. Acad. Sci.* (*U.S.*) **61**, 114.

Weiss, S. B., and Legault-Demare, J. (1965), *Science* **149**, 429.

Weissmann, B., Bromberg, P. A., and Gutman, A. B. (1957a), *J. Biol. Chem.* **244**, 407.

Weissmann, B., Bromberg, P. A., and Gutman, A. B. (1957b), *J. Biol. Chem.* **224**, 423.

Weissmann, B., and Gutman, A. B. (1957), *J. Biol. Chem.* **229**, 239.

Weissman, S., Eisen, A. Z., and Karon, M. (1962), *J. Lab. Clin. Med.* **59**, 852.

Wempen, I., Brown, G. B., Ueda, T., and Fox, J. J. (1965), *Biochemistry* **4**, 54.

Wempen, I., Duschinsky, R., Kaplan, L., and Fox, J. J. (1961), *J. Am. Chem. Soc.* **83**, 4755.

Wempen, I., and Fox, J. J. (1967a), in *Methods in Enzymology*, Vol. XII; *Nuelcic Acids*, Part A, L. Grossman and K. Moldave, Eds., Academic Press, New York, p. 59.

Wempen, I., and Fox, J. J. (1967b), in *Methods in Enzymology*, Vol. XII; *Nucleic Acids*, Part A, L. Grossman and K. Moldave, Eds., Academic Press, New York, p. 76.

Wheeler, H. L., and Johnson, T. B. (1903), *Am. Chem. J.* **29**, 492.

Wheeler, H. L., and Johnson, T. B. (1904), *Am. Chem. J.* **31**, 591.

Wheeler, H. L., and Liddle, L. M. (1908), *Am. Chem. J.* **40**, 547.

Wheeler, H. L., and Merriam, H. F. (1903), *Am. Chem. J.* **29**, 478.

Whitehead, C. W. (1952), *J. Am. Chem. Soc.* **74**, 4267.

Whitfeld, P. R., and Markham, R. (1953), *Nature* **171**, 1151.

Whitfeld, P. R., and Witzel, H. (1963), *Biochim. Biophys. Acta* **72**, 338.

Wiberg, J. S. (1967), *J. Biol. Chem.* **242**, 5824.

Wilhelm, R. C., and Ludlum, D. B. (1966), *Science* **153**, 1403.

Wilson, R. G., and Caicuts, M. J. (1966), *J. Biol. Chem.* **241**, 1725.

Winocour, E. (1967), *Virology* **31**, 15.

Winocour, E., Kaye, A. M., and Stollar, V. (1965), *Virology* **27**, 156.

Wright, R. S., Tener, G. M., and Khorana, H. G. (1958), *J. Am. Chem. Soc.* **80**, 2004.

Wyatt, G. R. (1951a), *Biochem. J.* **48**, 581.

Wyatt, G. R. (1951b), *Biochem. J.* **48**, 584.

Wyatt, G. R. (1955), in *The Nucleic Acids*, Vol. 1, E. Chargaff and J. N. Davidson, Eds., Academic Press, New York, p. 243.

Wyatt, G. R., and Cohen, S. S. (1953), *Biochem. J.* **55**, 774.

Yamazaki, A., Kumashiro, I., and Takenishi, T. (1967a), *J. Org. Chem.* **32**, 3032.

Yamazaki, A., Kumashiro, I., and Takenishi, T. (1967b), *J. Org. Chem.* **32**, 3258.

Yamazaki, A., Kumashiro, I., and Takenishi, T. (1967c), *J. Org. Chem.* **32**, 1825.

Yang, S. S., and Comb, D. G. (1968), *J. Mol. Biol.* **31**, 139.

Yang, W-K., and Novelli, G. D. (1968), *Proc. Natl. Acad. Sci. (U.S.)* **59**, 208.

Yaniv, M., and Barrell, B. G. (1969), *Nature* **222**, 278.

Yoshida, M., Furuichi, Y., Kaziro, Y., and Ukita, T. (1968a), *Biochim. Biophys. Acta* **166**, 636.

Yoshida, M., Kaziro, Y., and Ukita, T. (1968b), *Biochim. Biophys. Acta* **166**, 646.

Yoshida, M., and Ukita, T. (1965a), *J. Biochem.* (Tokyo) **57**, 818.

Yoshida, M., and Ukita, T. (1965b), *J. Biochem.* (Tokyo) **58**, 191.

Yoshida, M., and Ukita, T. (1966), *Biochim. Biophys. Acta* **123**, 214.

Yoshida, M., and Ukita, T. (1968a), *Biochim. Biophys. Acta* **157**, 455.

Yoshida, M., and Ukita, T. (1968b), *Biochim. Biophys. Acta* **157**, 466.

Yu, C-T., and Allen, F. W. (1959), *Biochim. Biophys. Acta* **32**, 393.

Yu, C-T., and Zamecnik, P. C. (1963a), *Biochem. Biophys. Res. Commun.* **12**, 457.

Yu, C-T., and Zamecnik, P. C. (1963b), *Biochim. Biophys. Acta* **76**, 209.

Zachau, H. G., Dütting, D., and Feldmann, H. (1966a), *Angew. Chem.* **78**, 392.

Zachau, H. G., Dütting, D., and Feldmann, H. (1966b), *Z. Physiol. Chem.* **347**, 212.

Zachau, H. G., Dütting, D., Feldmann, H., Melchers, F., and Karau, W. (1966c), *Symp. Quant. Biol.* **31**, 417.

Zahn, R. K. (1958), Hoppe-Seylers *Z. Physiol. Chem.* **313**, 77.

Zaitseva, G. N., Dmitriva, T. M., Sui, C-F., and Belozerskii, A. N. (1962), *Dokl. Akad. Nauk. (SSSR)* **147**, 1211.

Zakharyan, R. A., Venkstern, T. V., and Bayev, A. A. (1967), *Biokhimiya* **32**, 1068.

Zamir, A., Holley, R. W., and Marquisee, M. (1965), *J. Biol. Chem.* **240**, 1267.

Ziff, E. B., and Fresco, J. R. (1968), *J. Am. Chem. Soc.* **90**, 7338.

Zimmerman, E. F., and Holler, B. W. (1967), *J. Mol. Biol.* **23**, 149.

AUTHOR INDEX

438

444

446

AUTHOR INDEX

Ukita, T., 271, 356, 357, 362, 366, 367, 368, 369, 370, 383
Ulbricht, T. L. V., 73, 203
Uryson, S. O., 283
Uziel, M., 15, 247, 255, 257, 303

Van Montagu, M., 127, 128
Van Praag, D., 48, 75, 81, 116, 120, 122, 123, 130, 176, 194, 195, 202, 203, 353
Van Schaik, N., 283
Vanyushin, B. F., 281, 283, 284, 285, 287
Varadarajan, S., 53
Vaughan, M. H., 4, 5, 100, 301
Venkataraman, S., 296
Venkstern, T. V., 4, 264, 373, 374
Venner, H., 187, 198, 199
Veres, K., 335
Verwoerd, D., 349, 351
Vielmetter, W., 353, 383
Villa-Trevino, S., 147, 243, 307
Vischer, E., 215, 286
Visser, D. W., 160, 178, 179
Vogel, H. J., 213
Vogl, O., 32, 65
Vold, B. S., 9
Volkin, E., 172, 289, 293
von Ehrenstein, G., 7
Vorbrüggen, H., 174
Vournakis, J. N., 259

Wacker, A., 399
Wagner, E. K., 100, 301
Wahba, A. J., 3
Wainfan, E., 284, 285
Wakamatsu, H., 27
Walerych, W. S., 296
Walker, R. T., 379
Waller, C. W., 97

Walsh, B. T., 86, 94, 96
Wang, S. S., 37
Wang, S. Y., 76, 377
Ware, G., 335
Warner, J. R., 4, 5, 100, 301
Warner, R. C., 3
Warrener, R. N., 72, 158
Watanabe, K. A., 127, 338
Waters, L. C., 8
Webb, M., 282, 284
Wehrli, W., 104, 119, 128, 173, 269
Weil, J. H., 377
Weill, J. D., 111
Weisblum, B., 7
Weiss, M. J., 86, 96
Weiss, S. B., 8, 309, 310
Weissman, S., 315
Weissman, S. M., 5
Weissmann, B., 89, 90, 137, 140, 251, 313, 314, 315
Wempen, I., 46, 48, 69, 85, 93, 94, 96, 99, 103, 113, 116, 120, 122, 123, 130, 157, 163, 176, 192, 193, 194, 195, 202, 203, 353
Wheeler, H. L., 45, 50, 53, 68, 73, 75
Whitehead, C. W., 70
Whitfeld, P. R., 366, 371
Wiberg, J. S., 293
Wierzchowski, K. L., 121
Wigler, P. W., 129, 131
Wildman, S. G., 283
Wilhelm, R. C., 3
Williams, J. H., 97
Williams, L. S., 10
Willits, C. H., 27, 32, 33, 56, 64, 65
Wilson, D. V., 320
Wilson, R. G., 353
Winocour, E., 287

Winter, D. B., 311
Wintermeyer, W., 279, 371
Wiseman, A., 340
Witkop, B., 182, 183, 384, 385, 387, 388
Witkus, E. R., 337
Witzel, H., 18, 366
Wolfenden, R., 86, 89, 94, 96, 386
Woolley, D. W., 64, 340
Wright, R. S., 84
Wyatt, G. R., 213, 215, 239, 240, 282, 284, 285, 286

Yamada, Y., 27, 276, 277
Yamanoi, A., 133, 151
Yamazaki, A., 55, 132, 133, 139, 143, 151
Yang, S. S., 9
Yang, W-K., 277
Yaniv, M., 268
Yohn, D. S., 305
Yoshida, M., 271, 362, 366, 367, 368, 369, 370
Yoshimura, J., 84
Yu, C-T., 170, 377
Yung, N., 113, 162, 163, 192, 193

Zachau, H. G., 104, 128, 262, 276, 278, 279, 323, 371
Zahn, R. K., 282
Zaitseva, G. N., 15, 90, 92, 97, 137, 166, 172
Zakharyan, R. A., 373, 374
Zamecnik, P. C., 22, 172, 377
Zamenhof, S., 286, 291
Zamir, A., 155, 259, 267, 398
Ziff, E. B., 121, 393, 394
Zillig, W., 349, 351
Zimmerman, E. F., 4, 301
Zimmerman, S. B., 292

SUBJECT INDEX

447